TECHNICAL COMMUNICATION for COLLEGE and CAREER

TECHNICAL COMMUNICATION for COLLEGE and CAREER

How to Write in Academia, Business, Engineering, Research, Science, and Technology

Jack Fishstrom

Apex Persuasion
www.apexpersuasion.com
jack@apexpersuasion.com

Published 2019 by Apex Persuasion
Printed in the United States of America

21 20 19 1 2 3 4

ISBN 978-1-7330404-0-2
Library of Congress Control Number: 2019940193

Table of Contents

APPENDICES

Introduction and Organization

This book is intended for writers of non-fiction, technical documents. If you conduct research, interpret data, create designs, analyze concepts, solve problems, present proposals, and write about such tasks, this book is for you. Its objective is to demonstrate and justify the best approach available for presenting clear, concise, and well-organized documents on complex material, so that the widest possible audience can comprehend the information quickly and easily. The lessons in this book are applicable to writing done both during and after college. The principles are the same, only the payments differ. For the former, you pay tuition for the privilege of receiving a grade on your writing. For the latter, you are paid a salary for the effort extended to produce a valuable document. Perhaps a rarity, this is one book that you can use in both life stages. In short, this is a guide to writing in college courses and occupational settings.

Can the same book be applicable to both college courses and one's professional career? Of course it can. It should be. After all, college is your training ground for success in your occupation. The vast majority of college students are using college as a springboard to a paying job. There's no shame in that. It's natural. College is a way station; college students await the day when they will be paid to come to the office or laboratory, do cool stuff, and write about that stuff. Unless your career path takes you to an unusual place (chef? gardener? pilot?), you will write at your professional job, and probably make presentations as well. Writing some type of document, such as letters, memos, emails, proposals, white papers, and so on, is part of making your living. This book prepares you to do that writing across a wide spectrum of job types situated in academia, business, engineering, research, science, and technology (ABERST). ABERST represents the universe of professional occupations to benefit from learning the lessons of this book.

Indeed, this book is for people who want to excel at technical writing as applied to ABERST. In particular, the following types of people will benefit from reading this book:

Anyone in college now—

This is a helpful, comprehensive reference book to guide you to become a better writer and achieve stellar grades on high-profile assignments

Anyone who has finished college and finds that writing is part of her current or anticipated job duties—

This book provides a succinct synthesis of the salient teaching points I teach my university students, organized into five potent parts, so you can efficiently write reports that please your managers and bosses

Anyone who knows one of my past students and is impressed by that person's communication skills—

This book dispenses the same "magic juice" that your friend or acquaintance demonstrates every time she makes a great impression with a report or speech

Anyone who is supervising, coordinating, or managing employees who must write reports in support of your collaborative work—

This book should be given to those employees to assist them in writing reports for you that are concise, effective, and professional in structure and format

Anyone who has been arguing with someone they work with over the proper way to organize reports, compose sentences, or create visual aids—

This book can resolve the dispute by indicating best practices and illuminating the way forward

Put another way, this is the book for you if you want to take personal control of learning the material associated with studying technical writing at a 4-year university. Whether you are enrolled in college or not, and whether you are attending my courses or not, you can learn the basics from this book via self study, with a simple, 5-day plan for reading all of it. Specifically, this guide to ABERST technical writing is organized into five parts, each one associated with 1 day of the week, as follows:

Part 1 Delivering Just the Right Document
Part 2 Writing Elegant Prose
Part 3 Fulfilling Your Communication Purpose
Part 4 Persuading Your Reader
Part 5 Mastering Important Genres

As an extra bonus, I have included detailed and valuable guidance on five related and important topics: common composition mistakes; resumes, job letters, and thank-you notes; formatting essentials; conference posters; and oral presentations and projection slides. You will find these topics covered in Appendices A, B, C, D, and E, respectively. They are not part of the 5-day plan, so you can read them at your convenience, when the material will be most helpful. Perhaps the following week.

Yes, you can read this book in 5 days (only 1 work week) and get the material under your belt, so you can apply it to all your writing assignments for the remaining semester or to your communication tasks at your job in ABERST. Devoting 1 day to each part is a feasible strategy to quick mastery and immersion. And even if you need 1 week to read each part, that still only adds up to 5 weeks.

Feel free, nonetheless, to approach the book differently. How you move through the book is up to you. You can follow my "crash course" suggestion, or you can stretch out the reading. The timing is up to you, depending on who you are as a writer and learner and what you need at this stage of life.

Another strategy is to use this book solely as a reference, skipping around to sections that provide just-in-time information, as you need it for a communication task you face immediately. And, it should remain a useful reference guide after you have read it through once. Keep it handy for quick referral as you wish to review and double-check particular topics, including those in the five appendices.

Just before you jump in, you can choose to read the following introduction to the five parts of the book. This brief overview is a map of the journey ahead. It can be perused so you can anticipate the topics to be covered and have a sense of the overall contents of the book. This type of overview is something you will want to provide in your own documents, as you will learn in Parts 1 and 3. Thus, this book models most of the principles it promotes explicitly, so it is self-reinforcing, which is beneficial to readers who wish to absorb the material. The learning starts right here.

Part 1. Delivering Just the Right Document

Much of ABERST technical writing is assignment-based, that is, some instructions have been given to the writer for preparing the particular document. In a college class, the instructor often provides detailed requirements. In most job settings, the contents and constraints of documents (such as information to be covered and formatting expectations) are provided to employees or contractors. And in scientific and engineering settings, in particular, clients, customers, sponsors, funding agencies, and publishers quite often dictate the detailed specifications of the expected technical reports. Because documents are regularly prescribed to the author, a large component of writing excellence can be attributed to simply following instructions for the document. Yes, you read that correctly: A great ABERST technical writer scores huge points with her audience by simply, yet carefully, following the given directions.

Accordingly, Part 1 of this book addresses the essential best practices for understanding and adhering to the instructions for a document. If I had to put a number on it (in terms of course grade, employee value, or scientific and technological competence), I would say that excellent technical writing might be approximately 20% driven by this adherence aspect, so I spend the first 20% of the book on this.

Part 2. Writing Elegant Prose

Millions of readers will admit they are not gifted writers, but all of them can spot a "typo" (typographical error) in an instant. Everyone can see the mistakes. They stand out. They obstruct the reading process. Sentence fragments seem sophomoric and ruin an author's credibility. Meandering sentences and bulky paragraphs exasperate even the most committed readers: teacher, boss, colleague, ally, fan, or friend.

Therefore, few would deny that another component of excellent writing is the mechanics of your prose. The ability to craft a clear sentence, layer sentences into paragraphs, and link paragraphs into sections can serve you well when trying to impress your readers and convey your message (and get the writing job done). And by *impress*, I combine the following writing objectives: catch the reader's attention, maintain that attention, and deliver an unequivocal message without confusion, misinterpretation, and difficulty (or worse, agony). If we could just write with polish and panache, consistent with all the important written-language rules, we would sound erudite, organized, and interesting. Elegant prose will help you win over your readers, as they give you their attention and respect...but only for a while. It takes you only so far. It is necessary (to not scare anyone away), but it is not sufficient (to keep them reading). It buys you time with a reader, so you can then be efficient with that reader's time and convey your message cogently and concisely, both of which hinge on other necessary document components.

As important as the purity of your writing is to your task, this aspect of competent ABERST technical writing accounts for only 20% of good writing, similar to the percentage attributed to delivering the document as instructed. You may be shocked once again to learn that *this* is not the magical key to unlocking the writing vault; it is just 20%. I cover it in Part 2 so you can start practicing and honing it, even as you work through the remaining three parts of the book.

Let me reiterate an important point: My placement of Writing Elegant Prose as Part 2 and not elsewhere in the book is significant. Often student writers are eager to get down to the writing and put some sentences into the document (even without understanding more fundamental issues of good writing). Many students and new writers understandably believe (often with concomitant fear and dread) that the key building blocks to high-quality documents are the fundamentals of grammar, sentence mechanics, and style. "If only my words would dance and my sentences would sing," thinks the aspiring technical writer. This "artistry," however, is not enough. No document will be masterful unless it is consistent with the principles set forth in the other four parts of this book, beyond Part 2. Why did I not place Writing Elegant Prose at the end, then? As said above, you have no reason to delay; get started right away with elegant writing.

Because elegant prose does not typically, magically appear in all its glory upon the first attempt to splatter the page with words, an integral part to writing elegant prose is also knowing how to read, revise, and proofread your early drafts. Thus, writing elegant prose incorporates *both* techniques and rules of fine writing *and* skills and habits of correcting and improving your writing (the "draft and revise" writing cycle). A great writer must know how to edit and proofread. Thus, Part 2 of this book helps you become a skillful writer and a proficient proofreader and self-editor. Importantly, editing starts with removing all material irrelevant to your communication purpose. While this strategy is introduced in Part 2 as foundational to writing elegant prose, it is expounded upon in Part 3 as integral to successful communication in its entirety.

Part 3. Fulfilling Your Communication Purpose

The important ABERST technical writing to be done in your near and distant future is going to be purposeful writing. Each document will have a purpose of communication, or, to put it differently, you will have a purpose to fulfill by writing and transmitting the document to a reader. This certainty should simplify your work, but that is not always the case. Typical inexperienced writers have a hard time identifying their communication purpose and an even harder time expressing that purpose succinctly. Part 3 of this book guides you through the process of identifying your communication purpose, announcing it in your document, and fulfilling it via the substance of your document.

Importantly, the communication purpose that matters in the context of practical, non-fiction writing is organizational. That is, the organizational context of the document matters. This allows me to exclude documents you are writing outside of any organizational context, such as scribbling notes to yourself, writing a personal to-do list, or entering information into a daily log or journal. These are intimately personal writing efforts, without a reader other than the author.

This book, in contrast, covers documents that are written to other readers, within specific organizational contexts where some information must be conveyed from one person to another. This is not personal writing but, rather, social writing. Such socially purposeful writing is *the* writing that may matter most to your academic and professional careers. Making sure your readers immediately recognize the social aspect of your writing, the organizational context, the purpose of communication, and the approach you will take to fulfilling that purpose is vital to a successful, high-quality document. Part 3 has the secrets to this success. Moreover, success with purpose is itself another 20% of masterful technical writing.

To be clear, notes to self and journal entries are outside the scope of this book. When it comes to putting carefully developed ideas into a structured document, you will be writing with a specific purpose to a specific reader or set of readers. You will use documents to inform or persuade (also actuate) those readers. Different types, or genres, of documents are used to cover the myriad purposes and organizational contexts that may arise, usually with one genre serving as more suitable for a particular purpose than another genre. These differences are covered in Part 5. But in Part 3 of the book, I walk you through the winning strategy that uses the writing purpose as a foundation upon which to build the whole document. Fictional writing, in contrast, entertains or excites; poetry amuses or soothes; and autobiographies and memoirs illuminate or inspire. Those are worthy writing purposes, and your technical writing might, indirectly, achieve some of those objectives, but they are not the typical primary purposes of writing in ABERST. Instead, non-fiction technical writing fulfills an organizational purpose. Without such a purpose, a document cannot be written. With such a purpose, a document materializes into a successful structure built on a solid foundation. With that purpose-based structure, you will inform or persuade a reader, or readers, in relation to a subject pertaining to an academic analysis, a business endeavor, an engineering project, a research effort, a science study, or a technology task.

Part 4. Persuading Your Reader

Writing that is purposeful in an organizational context, furthermore, is intended to help others in some way via a document's message, such as justify a decision, direct a plan of action, initiate a project, clarify a request, solve a problem, or present an optimal design. In order to inform, aid, and actuate readers, thereby facilitating some action and progress on a project, you must persuade them. Persuasion is at the heart of many academic, business, and technical (engineering, research, science, and technology) documents. I address each of these three occupational areas briefly below.

In academic writing, you may present an interpretation of a book, or an analysis of culture, or a thesis about improving voting procedures in your state, and you will need to argue for your point of view, which is typically labeled a "thesis" by your instructors.

In business writing, you may write a report analyzing the last quarter's sales numbers with a hypothesis that links the sudden decline to a simultaneous change in personnel, or you might suggest to your boss that your company implement an innovative logistics and inventory control software program, or you may set forth a dramatic vision that entails your organization's branching out into a related, yet different, business area.

In technical settings (engineering, research, science, and technology), you may present a design of a chemical processing factory, or you may analyze an aircraft accident and present your theory of the cause, or you may offer a new configuration for a micro gas chromatograph (MGC), smaller than any such previously made MGC.

In all the aforementioned writing situations, your success in communication hinges upon your ability to persuade the readers to see things your way, to concur with your conclusions, and to approve and implement your proposal. Skill in argumentative writing is crucial to achieving a successfully persuasive outcome.

Ultimately you must persuade your readers or your effort is futile, so this part could easily be all there is to non-fiction writing, or 100% of technical writing itself; but you cannot persuade unless readers commit their time and attention to your document, which requires a strong start. After a positive beginning that hooks a reader, moreover, the reading must continue to be worthwhile. Thus, in order to persuade, you must keep the audience reading by offering a document that is expertly composed, organized, illustrated, and formatted. Persuasion depends on these other strengths working cooperatively, and these are the skill areas covered in the other four parts of this book. Even persuasion, as important as it is, is not itself all it takes to be a skillful and masterful writer. Therefore, once again, persuasion is *only* 20% of good technical writing, and the remaining 80% is covered elsewhere in this book.

Moreover, some documents (a small slice of the paperwork pie) are not inherently persuasive, but merely informative. This leads us to the next part of the book: distinguishing between persuasive and informative documents, and identifying the ideal particular document type, or genre, for the purpose before you.

Part 5. Mastering Important Genres

Technical writing, unfortunately, is not a "one size fits all" proposition. You cannot hope to use one document template for the duration of your working life. On the contrary, you will need to be comfortable with a handful of document types, namely, the genres within non-fiction writing. The handful includes emails, business letters, short memoranda, long memoranda, status updates, progress reports, procedure manuals, articles, published notes, proposals, white papers, posters, slides for oral presentations, and formal reports. Depending on your organizational purpose and intended reader(s), one genre will be most suitable. You match the genre, that is, document type, to the writing purpose and audience. It is as simple as that.

Mastering the genres, therefore, involves knowing how to link your purpose and audience with an ideal document type, and that choice, in turn, will further dictate structure, format, and prose style. An excellently written report in the wrong format may be thoroughly ineffective for the intended audience. Sorry to tell you. It might not fulfill its intended purpose because the appropriate readers may not receive it, or may be misled while they read, or may not read it at all, choosing instead to shuffle it away somewhere to gather dust. Mix-ups between purpose, audience, and genre are not uncommon; history is full of business and engineering mishaps based on communication breakdowns often linked to such mix-ups. Correctly matching genre to purpose and audience, therefore, is an important element of masterful technical writing for ABERST.

In addition, an important sub-genre of technical writing is the category collectively referred to as "visual aids." These visual elements within a document are often essential and always beneficial to presenting one's message and communicating complex information. Visual aids complement nearly every well-written ABERST report, and one or two visual aids within a report sometimes capture the critical information that must be conveyed to readers. Furthermore, a visual aid must be designed to be clear and complete on its own, as if the surrounding document were not there to help readers understand the visual aid. For these reasons, visual aids require their own set of precepts and guidelines for both design and use, and they can be treated as a type of genre. Although they are integrated into the other genres, a dedicated effort to master this sub-genre will yield a huge return in terms of improved technical communication. All together, matching the genre to your purpose and audience while enhancing with visual aids can mean the difference between an average and a superior ABERST document. This final part of the book represents the remaining 20% of excellent technical writing.

Now that you have an overall sense of the fivefold organization of this book, you can start to read it. Dive in. And do not forget that the five appendices cover equally important topics that should help you communicate during college, acquire the perfect professional position, and excel in that position:

A. Common Composition Mistakes
B. Resumes, Job Cover Letters, and Thank-you Notes
C. Formatting Essentials
D. Conference Posters
E. Oral Presentations and Projection Slides

I operate under the assumption that you have limited time; you need a comprehensive yet conversational guidebook to the essentials of professional, non-fiction writing, and you need neither an extensive bibliography nor numerous cross-references of facts, data, and research findings. (All of which do exist to support the principles and rules covered herein, in case you are interested.) So, that's it: no footnotes, no references, and no citations. This is just a how-to book with straightforward and substantive guidance on mastering effective technical communication within ABERST, for college and career.

Part 1

Delivering Just the Right Document

Readers are not likely to quote your technical document in greeting cards or wall calendars, nor are they likely to turn your document into a film or piece of performance art. They are not that concerned with the words per se; they want the information inside the document to enable or assist them in doing their jobs. They need the report to help them do their work, not learn about you or find a new appreciation of language. This means you are being judged less for eloquence and virtuosity, and more for competence and efficiency. Technical writing in academia, business, engineering, research, science, and technology (ABERST) is about getting the message across quickly, clearly, and dutifully, that is, to the right readers. And it is about conforming to those readers' expectations, to the extent they have some. You must give readers the document they want and need. Technical writing is reader-centered writing, not self- or author-centered writing. Therefore, as a technical writer, your first obligation is to follow—to the letter—all directions and guidelines given to you for a particular document. This is the first component, with four to follow, of delivering just the right document.

Chapter 1.1

Fulfill the Letter of the Assignment

The silliest yet most-avoidable mistake made in college-level writing, as well as professional writing done in ABERST, is turning in a document that does not conform to the basic requirements of the assignment or request. This mistake can cost dearly, unlike a few mistakes in punctuation or spelling. At the extreme, a document will be given a failing grade or dismissed from further consideration when it does not resemble the expected submission. Each term, I give a few failing grades, and the problems can all be traced back to this single oversight: the submitted reports did not fulfill the instructions for the assignment. I have never given a failing grade to someone for any other reason. This is the big one, and it's simple and obvious to most students. Working professionals have similarly experienced dire results when submitting documents outside the bounds of specified requirements. To assist you in this writing element, I need not spend much time providing helpful tips or elaboration. The point is simple: follow the instructions of the assignment to the letter.

Let's look at a sample assignment and a proper response:

> Write a post-laboratory report in memorandum format, maximum 4 pages, with 1-inch margins and 12-point type.

Do exactly that. Do not write a journal-style article that is 7 pages long, with ½-inch margins and 11-point type. How can you tell the difference? First, count the pages and check the formatting. Second, write a post-laboratory memo, not an essay, not a letter, not a research paper, not an article, and not a poem.

Let me emphasize the close connection between college and career introduced above: Instructions are crucial in both realms. Students who miss page-limit restrictions really disappoint teachers . . . because we are mega-maniacal egoistic authoritarians with power issues. Oh, and also, page limits are imposed for rational reasons. Teachers actually think rationally about the assignment requirements ahead of time. Some reasons for page limits are the following: the document genre has inherent length limits, the expected workload for the course demands that exact amount of pages, and successful teaching outcomes necessitate the amount of writing requested. To explain further, page limits in college often reflect the time instructors have available to read student work and the amount of pages necessary for an instructor to adequately assess a student's strengths and weaknesses. They also challenge a writer's ability to identify the essential components and remove the inessential, that is, to write succinctly. In professional arenas, page limits often reflect physical constraints and available space in the medium, competing priorities among various material that needs to be aggregated or published, standard conventions for types of writing, and readers' or editors' expectations for detail and elaboration. No writer likes to cut words that have been produced through pains of labor and creativity, but every author must. In sum, page limits are imposed upon authors from outside for myriad rational reasons, and authors must adhere to these limits if they wish their writing effort to succeed. The same applies to formatting requirements. Margins allow for annotation of the document, and type-size requirements ensure that the printed document is legible for expected readers.

If you are curious about the origins of the page limits and other formatting requirements, just ask for details. Most teachers, supervisors, and clients will be happy to explain them to you. In fact, they are happy you noticed the aspects that make up the letter of the assignment. That is the point of this first guideline to you: teachers and other important readers want you to follow the assignment requirements, so reading them, following them, and asking about them as necessary before you turn in your work is always appreciated.

Importantly, the key message just stated above deserves reinforcement: imposing document requirements on students prepares them for professional obligations and situations. Let me give some poignant real-world examples to underscore the importance of following directions.

Large government contracts that are put out to bid always start with a request for proposal (RFP). The same is done with foundation grants and other private-funded projects. Those RFPs will specify a maximum number of pages, for example, 20, that cannot be exceeded by the proposal, which is the document that the authors submit to bid (apply) for the contract, funding, or project. Any document exceeding the page limit will be rejected without further review. Why would an organization want you to work on a $20 million project when you did not bother to count to 20?

A significant mistake is often made by start-up companies and people just embarking in business: they fail to realize the full importance of the document requirements. Even when presented as "guidelines" or "advice," they are rarely optional. Prudence demands that all requirements, guidelines, and advice provided to authors by the requesting organization or agency be followed to the letter. They are not suggestions; they are commands. And, authors are responsible for tracking them down when they have not been personally delivered. People experienced in ABERST know the places to look and the questions to ask to ensure that they are apprised of every single document requirement.

Furthermore, when submitting articles for publication to journals, authors are referred to a detailed list of basic document requirements and limitations, distinct from demands of high-quality substantive content, which is assumed. Once again, the list of submission "guidelines" (read: rules) will likely include maximum page limits or word counts. Indeed, the same importance is at play with word counts, sometimes 500, that must be adhered to. Your word-processing software will count the words for you, so be sure to make this happen. And the requirements might include font, type size, margin width, spacing, and inclusion (or, omission) of highlighting (bold, italics, and underline). Everyone reading this is capable of counting to 500 or measuring a 1-inch margin, but you do not want your document to suggest you did not bother to do so, or worse, cannot. Bluntly speaking, in a competitive world, if you neglect to count to 500 or measure a 1-inch margin, you are likely to be deemed unqualified to, for example, analyze 10-year census data or design a municipal light-rail system.

In almost all instances when a document has been imposed upon an author, particular guidelines and requirements have been integrated into the assignment or request. The very type of document is usually proscribed. In such instances, you should follow the conventions for such a document. This may include both style conventions (format) and substantive conventions, that is, the appropriate content. So, both form and content matter. In Part 5 of this book,

I cover the conventional form and content of many common document types, which you can consult as needs arise.

Thus, even when given a wide berth or open-ended requirements for a document, a wise author should endeavor to follow conventions associated with standard, well-known document types. For example, if you have been asked to send in a cover letter with your resume when applying for employment, you need to prepare an individualized cover letter in a conventional form. You should not provide a life story, pictorial display, or confession. As another example, if you have been asked to provide a policy recommendation to a local government agency using your expertise, do not provide a literature review of the recent published articles on this topic. Rather, submit a policy white paper. If a business client says to you they might be interested in expanding the scope of services your company provides to them, and they want you to put together a formal letter with a detailed cost estimate, do not send an email. Write and submit a letter. Similarly, upon finishing the first year of providing this new service to the client, you are asked to send a report detailing the first year's results, do not send a presentation slide deck. Send a report.

All teachers of college courses are happy I wrote this small chapter. They wanted someone to say it. You may think it is too basic to be worthy of mention, but you will thank me one day when one or both of these two little pieces of fundamental advice (fulfill the letter of the assignment and follow the conventions of the requested document) serves you in earning a high course grade or successfully submitting a professional document that advances your career. I had the temerity to cover these basic precepts, simple as they are, to help you. In this same spirit, we can go a little deeper and turn to our next topic.

Before we do so, I address one final question that might arise in your mind: What if I have not been given an "assignment" per se? That is, you might be thinking, what if I am at a job, and I find myself in a situation where I feel strongly that a report needs to be sent to someone within or outside my organization? In such case, you are the sole person who has realized the need to generate a report for a person who has neither requested nor delineated the specifics of the document. This is a normal situation, when events prompt the need to write a document, (event-initiated), rather than getting a prompt by a particular request or assignment from a specific person or organization (recipient-initiated). In this case, when you have not received specific report guidelines and the advice of this first brief chapter is irrelevant, you have the rest of the book as your companion. Starting in the very next chapter, I offer useful advice for event-initiated writing. From here forward, therefore, the instructions are perfectly suited for writing tasks that are undefined and open-ended, along with those tasks that have specific requirements set forth at the outset. In fact, beginning in the very next chapter, you will be guided through a technique especially important for open-ended writing tasks that require event-initiated documents, namely, start the document strongly.

This small chapter you have just completed gives you some essential advice for preparing a successful document in ABERST, but it has not shown you how to write a single worthwhile word. The writing starts in the next chapter.

Chapter 1.2

Start the Document Strongly

Readers must never be in doubt as to the use of their time in reading your document; they must be absolutely certain the document has something of value to them, per their interests in the topic or their professional responsibilities to acquire the information within the document and, usually, take some responsive action.

An effective start is critical. Your opening establishes your relationship with your reader. You need to win your reader's trust and respect. At the start, you tell the reader you have something important to say, and you will say it with skill. Here are the two objectives of a document's start: catch the reader's attention and reassure the reader that this document is meant for her and will help her in some way; it will be worthwhile to read it. You fulfill these two objectives through a combination of (1) document identification data and (2) substantive statements of your topic and writing purpose. Importantly, to start a document successfully, you must quickly reveal both your topic and your purpose in writing about that topic.

In entertainment, fiction, and other types of creative writing (film, stage plays, mystery novels, science fiction, and so on), a writer may choose to not reveal the true topic initially. This can be strategic. Readers (or viewers) will accept this when their intention is simply to be entertained, and possibly thrilled. A twist can be spellbinding. Consider the film *Moon*, directed by Duncan Jones. Viewers may initially think the film is about some strange mysterious problems causing distress at a mining operation on the Moon, but as the story develops, viewers realize the film is about a different topic altogether. (I won't spoil it.) Or, consider, *Oedipus Rex* by Sophocles. Initially, Oedipus thinks his woes are of one kind, but as the tale unfolds, he learns that much bigger difficulties confront him.

But in non-fiction writing for college or professional careers in ABERST, readers do not have the patience to be misled, confused, teased, and surprised. Thus, in academic and professional non-fiction writing, you must reveal your topic and purpose from the outset.

Different genres have different requirements for openings. I will not cover them all here in this chapter because Part 5 is devoted to elucidating the different useful technical report genres. Nonetheless, document starts for most genres have some common fundamental items, so those I address in this chapter. Moreover, as I introduce these fundamental items for document starts, I will mention some of the more typical genres as I present examples. Also, knowing what to avoid in a document's start is helpful, so I begin with the common pitfall known as "fluff."

Typically, a first draft will have lots of "fluff" at the beginning, which can be excised so that the crucial objectives of a document' start can be achieved. Fluff, in the common particular forms of platitudes, definitions, hackneyed expressions, common knowledge, self-evident statements, clichés, and the like, is usually boring. It doesn't grab a reader's attention as much as it belittles her intelligence and makes her yearn for some real substance from the author. In addition, fluff may not indicate a writer's true topic. Fluff is to be avoided, as it lacks informative value.

Fluff examples:

- Commercial airplanes are integral to our modern-day life.
- Artificial intelligence is the ability of machines, typically computers, to make decisions using reasoning-like processes via man-made computer algorithms.
- We all hope freedom can be shared by more and more people on Earth in the coming years.

Fluff is a delay tactic. It is avoidance. You must be vigilant to identify and eradicate fluff. Readers need real, relevant, and specific information. This is especially true of readers who are busy, important, incredulous, skeptical, hostile, or unsympathetic to your position. Even if they are already convinced of your document's importance, they must be further reassured that it is ably researched and written and that it contains the information they seek in a clear, well-organized, error-free, and cogent manner. Out of respect for your effort in writing a report, most readers—whether supportive or skeptical—will give your report some attention, but not an infinite amount.

Indeed, because they are inherently impatient due to natural and multiple pressures on their time, readers (or listeners) must be given the information they need without having to wait until they turn to the final page (or hear the end of your presentation). Early on in a document, therefore, readers must be (1) introduced to the true topic, as stated above. In addition, further narrowing from the broad topic to the specifics of the report must be accomplished by (2) focusing on a sub-topic; (3) providing context, meaning, and importance of that sub-topic; (4) conveying a report purpose and desired reader response; and (5) highlighting the main points found in the report's body.

To this effect, most authorities in the field recommend beginning a report with a long-standing, tried-and-true component known variously as a problem statement, purpose statement, foreword, or introduction, referred to hereafter as "Foreword." Furthermore, the conventional heuristic for report beginnings comprises a companion component to immediately follow the Foreword, namely, a Summary. These two components in tandem, as representing the first two sections of text of any report, can accomplish the five crucial objectives of a report beginning as stated above.

The Foreword and Summary can take on various shapes and guises depending on the report genre, but in all instances, despite their names, these two components serve the same functions and fulfill the same objectives of communication. In formal reports, they are combined into an Executive Summary. In business letters, they are the first and second paragraphs. In memoranda, they may be alternatively labeled as Background, Purpose, Preface, Context, Introduction, Highlights, In Brief, or Overview. In journal papers or articles they are combined into an Abstract. In user manuals or operating procedures, they may be collected into a single section titled "Objective," or something similar. No single heading is universally used or critical. The important issue is to have some type of introductory section(s) and use those sections to convey the ideal information for a report's start.

I give the heuristic for a Foreword and a Summary below. But, more importantly, I wish to convince you of their necessity in all technical communication. To do this, let's assume a phone call, from Justin. His friend, Raj, is the call's recipient. You can imagine that you are Raj and you are receiving the call, which represents a communication (document).

Justin starts this telephone conversation:

Hey, Raj, what's up?

Nothing. Is this Justin?

Yeah.

How are you, my friend?

Pretty good. I'm at the airport.

Really? What are you doing there?

I just got back from Cancun.

No kidding. Awesome. Did you bring me some Cuervo?

Well, yeah, sort of. I'll tell you about that later. It's a funny story. Anyway, you know my car?

The Jeep?

That's right. It's been having that trouble with the transmission, remember, the grinding?

I think so, what about it?

Well, I was going to drive to the airport last week, but I couldn't shift it into reverse.

Where was that?

At my apartment, in the parking lot.

Hey, Justin, hold on a sec, I got another call.

But Raj, wait . . .

Long silence while Raj talks with his girlfriend. Justin is left "hanging." After two minutes, Raj returns.

Hey, Justin, sorry about that. I got to go; that's Wendy on the other line, and we have to work out some things. I'll catch you later. Thanks for calling.

To say the least, Justin never stated the purpose of his call. Raj took another incoming call and subsequently hung up on him before he could explain the reason he called, namely, to see if Raj could come pick him up at the airport. This was a friendly chit-chat in which Raj participated, and he had no obligation to do more than he did. The communication did not achieve its objective, but Raj is not to blame. Justin is. Raj didn't place the call, Justin did. Raj was kind enough, but his girlfriend's call took priority. In the brief time that Raj committed to this communication effort, Justin failed to convey his sense of urgency, specify the favor needed, and explain his intended display of gratitude by offering Raj a meal of pizza accompanied with beer and Cuervo-brand tequila.

Let me summarize the problems in the communication: Justin failed to indicate the topic he wanted to discuss (a favor to be asked); the sub-topic, namely, his problem that prompted the call (needing a ride from the airport); the context (his car is not operational); the purpose of his communication (asking his friend to come get him at the airport); and highlights of the remaining communication (Raj will receive pizza, beer, and tequila for doing this favor, and Justin has an urgent need for a prompt response from Raj).

This can happen with a technical report, leading to an ineffective effort of communication. Instead of getting to the point and beginning at the proper beginning, untrained technical writers delay, obfuscate, and circle around with warm-up information—fluff—that leaves the reader guessing or, worse, frustrated and annoyed. The overall effect is not a good impression.

Let's look at a typical example of a technical report that possesses the same failings as in the phone conversation. It is analogous to Justin sending a first draft of a report to Raj.

Sample Report Beginning *without* Strategic Initial Sections

Introduction

Automobile safety matters to us. In mid-2014, one of the major automobile manufacturers admitted to placing a defective part into millions of its vehicles and failing to detect this defect as the cause of engine failure that led to accidents, injuries, and fatalities over an 11-year period. Despite the company's efforts to investigate the vehicles involved in accidents and identify the problem, they were unsuccessful. Eventually, a plaintiff suing the company for injuries caused by a crash in its vehicle, upon its own investigation, determined the source of the engine failure to be a defective part. This was a gross embarrassment to the company, as it represented multiple engineering failures and a breach of good faith with its customers. Keep in mind that the single defect in the automobiles with the common defective part would, in some circumstances, create a moving stall; that is, the car's engine would be shut off suddenly while the car was in operation, with the accelerator engaged. Moreover, the airbags would be deactivated when the engine was switched to off, rendering them useless at the very moment when the occupants needed them most. The recall effort would ultimately encompass nearly 2.6 million vehicles, at an approximate cost of $106 million, along with a $35 million fine from the federal government. The company set aside an additional $4 billion to pay additional fines, settlements, and judgments stemming from the deaths and injuries due to this defect. Our team sat down to a meeting last week to discuss this case and tease out teaching points that we could use in our own management oversight, staff training, and development of updated operating procedures and protocols, in order to prevent something like this happening to us.

The meeting was attended by 4 people from engineering, 2 from training and development, 6 from manufacturing, 1 from quality assurance, 3 from legal, and 4 from finance. These are the meeting minutes for distribution.

> The meeting was actually split up over two separate days due to travel and other obligations of some of the participants. The dates were June 7, 2016, and January 10, 2017. The first meeting took place in the 4th floor conference room, and the second one was held in the legal strategy room. Tape recordings were made of the meetings, but this report provides a synopsis of the key points as the recordings are lengthy and not the best sound fidelity.

A quick analysis of this first page from an example technical report shows that a topic is certainly identified, that is, car defect repercussions, as well as the purported report content, namely, meeting minutes. Still this report does not meet all five objectives of a document's start explained earlier in this chapter and reviewed here:

> (1) Introduce the true topic
> (2) Focus on a sub-topic
> (3) Provide context, meaning, and importance of that sub-topic
> (4) Convey a report purpose and desired reader response
> (5) Highlight the main points found in the report's body

Specifically, looking at the sample above, the first page does not announce for the reader a sub-topic of special interest within the broader topic of car defects, nor does it provide much context for devoting attention to that sub-topic. In other words, it does not provide the additional narrowing and focus that readers require, as follows: it does not indicate the situation within the company that motivates the attention devoted to a sub-topic, nor does it announce its importance or urgency, if any. It similarly fails to explicitly reveal the report's purpose, that is, the author's reason for communicating to the reader(s). It gives the contents (meeting minutes), but that is not the functional purpose of the report (more on this in Part 3). Lastly, it neither intimates actions desired of readers nor highlights the remaining information in the report.

Revised and reorganized with the necessary information sorted into the two highly recommended beginning sections, Foreword and Summary, the report becomes clear and focused. The necessary elements of these two strategic initial sections are presented first, in separate subsections below, before I show the revised report, to prepare you to understand the separate items that produce an improved start to the report. Before I say more, however, I must acknowledge that the approach presented here was developed by many others who preceded me in this field, in particular three professors at the University of Michigan: J.C. Mathes, Dwight Stevenson, and Leslie Olsen. My contribution is to explain the strategy succinctly for the busy student and professional.

1.2.1 Foreword—Three Mandatory Elements

To facilitate a successful document start, the Foreword should contain three elements. Moreover, it should have nothing besides those three elements, or it will become cluttered and no longer serve its purpose of helping the author start strong, quickly, and concisely. These three elements are as follows, with explanation directly below:

- Problem statement
- Work statement
- Purpose statement

Problem Statement

The problem statement explains the imperfect situation that motivated the work done by the author. The imperfect situation—problem—serves as the basis for writing a report, which, in turn, conveys a message that has some value to the identified recipients and other anticipated readers, relevant to the problem. The problem does not have to entail a dire situation or catastrophe, although such would certainly motivate an investigation into a solution, mitigation, or suitable response. On the contrary, a problem can be something exciting and inspiring; it is a problem because it is not yet realized. The need for work to be done before this new, exciting idea can materialize is the "problem." The imperfection is the critical characteristic. In the ABERST world, imperfections exist in nearly every endeavor, and they prompt work, study, research, and investigation. Something imperfect is behind all efforts and all technical reports. If the situation were perfect, no one would need to look further, study options, or do more research. No report would be needed. All involved persons could go out to celebrate and then home to their family and friends. Imagine the perfect situation, hypothetically: All our vehicles are problem free; they never wear out; their fuel efficiency is sky high, and customers have never complained about a single item. Or, we were looking for a cure for pancreatic cancer, but we found it; the cure has no troublesome side effects; it seems to be 100% effective for all patient types, and its cost is trivial.

These perfect situations rarely exit. Thus, persons working in ABERST spend their days struggling with imperfect situations: we try to improve something's efficacy or durability, reduce its environmental impact or cost, make it easier to build or repair, or develop something brand new entirely such as flat-screen televisions, smart phones, game consoles, driverless automobiles, drone delivery vehicles, and so on. All work is getting done to bring an imperfect situation, of one type or another, closer to perfection for the purpose of, among other things, improving human health, cleaning up pollution, making life easier, and increasing profits (for business ventures). The problem statement for any report is a brief description of the imperfect situation that motivated the author to do her work described in the report.

Thus, a problem is anything that is real but not ideal (negative problem), or ideal but not yet realized (positive problem). It is always a difference between a present state and an ideal state. A problem, on the one hand, may involve any of these negative situations: injury, breakdown, failure, defect, loss, disappointment, or misfortune. On the other hand, a problem may involve any of these positive situations: goal, dream, concept, or hope. The key element to the problem is that it necessitates the very work that the author did prior to writing the report. It motivates the work and, hence, the report. It explains the work. Without a problem, the work would be an empty, meaningless exercise. In organizations in ABERST, employees are working on some project, whether large or small, of value, and this means that the project is addressing a problem: something bad that must be improved, or something good that must be achieved.

Every well-written report I have ever read in my years in ABERST has started with a problem statement, such as the following:

- Trucks in our company are useful for hauling freight, but they are boxy and bulky and therefore not fuel-efficient. Better fuel efficiency is needed now, but a way to do this is yet unknown.

- Our vehicle's landing gear is lightweight yet fragile; it has broken on several recent landings. An improvement to the landing gear that remains light but is stronger needs to be identified.
- Generating electricity via nuclear power was discredited decades ago, but the design of nuclear power plants has improved greatly since the disasters in the last century. Our organization seeks a way to re-establish trust with the public in commercial nuclear energy.
- Federal and state programs for improving the outcomes at public high schools have been ineffective, but many excellent ideas have been developed and tested at separate locations around the country. A consolidated effort is desired, but states hesitate to collaborate.
- Fully autonomous automobiles offer quite a lot of advantages to municipalities and vehicle owners alike, but both roads and cars must be modified extensively before such automobiles can be implemented widely, beyond the testing phase.

To review, the problem always involves two items in conflict with each other, stemming from either a present "negative" or a present "positive" condition or ambition, respectively.

Here are some additional examples:

Negative problems:

- ▸ Our company-wide software systems failed last month for 5 hours, leading to thousands of flight cancellations; and the full cause of this shut-down is still unknown.
- ▸ The emissions from our vehicles remain higher than we have promised by 2025, so further reduction methods must be identified.

Positive problems:

- ▸ Perhaps we can send a manned mission to Mars one day, but the necessary technology is not yet developed.
- ▸ This new business direction could be quite profitable, but we have not yet found the capital resources to fund our exploratory efforts let alone our full roll out.

Work Statement

The work statement is a brief description, in general terms, of the effort taken to address the imperfect situation and produce a needed outcome, whether a solution, policy, interpretation, judgment, design, or recommendation. This statement does not convey each and every task completed by the author. Rather, it reveals the general approach to tackling the imperfect situation, with the understanding that details will be forthcoming later in the report.

The work statement accomplishes two important objectives early on. First, it establishes credibility. It tells readers that the author has a good reason to speak about the imperfect situation, for she has investigated, studied, analyzed, or evaluated the facts of the matter. Because of the work, the author is an authority (notice the common root term) and her expertise is based on immersion into the matter. Second, it provides some narrowing, yet again, from a larger universe of possible efforts related to the topic to a smaller, discrete approach taken by the author. Narrowing to a sub-topic by simultaneously providing a focused work statement represents particularly efficient and elegant writing. I explain further with the following examples.

If cancer is the problem in some shape or form, an author may be doing genetic research, or chemotherapy studies, or environmental toxicology analyses; any type of work is possible. The work statement specifies, without going into excessive detail. It tells readers the approach taken by the author that makes her a person qualified to write a report on the sub-topic at hand.

If closing unproductive retail locations is the problem, an author may be working on developing marketing programs for an on-line shopping alternative, or taking inventory of remaining stock, or meeting with downsized employees to assess their relocation options.

Work statements can be as brief as possible so long as they reveal the general nature of the work completed on which the report is based. Here are some acceptably short work statements:

- Our team simulated the pollution's migration using advanced software.
- We conducted a pull test on the mounting prototype to assess functionality.
- Accelerated testing was completed to estimate durability and identify fatigue points.

- Using a linear programming predictive scheme, the possible outcomes were generated given five scenarios.
- A physical apparatus was used to model the rotational motion and test the control system.
- Using known equations for aircraft flight, we estimated the airplane's expected performance from the physical characteristics of the current design.
- We used a radio antenna to scan a small section of the night sky for 1 year.
- The team interviewed the pilots and the flight crew to extract facts as they could recall them while the plane was experiencing gliding flight.

Purpose Statement

As the final element of the Foreword, the purpose statement conveys the principal reason for writing the report; it announces the principal objective of the report, thereby telling readers the author's point for writing and possibly suggesting to some readers the type of response expected of them. As discussed in detail elsewhere in this book, the Purpose statement is the foundation for the whole report. It is a cornerstone, to say the least. Here, this vital element is addressed as an element of strongly starting the document; later, it is revisited as a component of two other cornerstones, namely, writing elegant prose and having a point to your writing.

As should be clear from the discussion immediately above, an author might do any of a number of activities within a large topic. Indeed, the Work statement helps to focus on a sub-topic. Similarly, an author may have many possible reasons for writing. No reader could possibly anticipate a report's objective from the statements of Problem and Work. One needs to complete the triumvirate by adding the Purpose statement. Thus, an explicit Purpose statement is helpful to all readers, as it reveals the author's reason for writing and delivering the document.

Reasons for writing are limited to a finite set of outcomes that can be achieved simply through communication of words in a document. With words in a document, an author can ask for help (request), provide a design (present), explain a process (instruct), offer an opinion (recommend), rationalize a decision or policy (justify), suggest a purchase (recommend), and other similar communication tasks. The list of communication purposes is finite. Here is the set (without listing all synonyms), which is revisited in Chapter 3.1.

- Propose
- Recommend
- Request
- Present
- Convey
- Transmit
- Answer
- Assert
- Offer
- Justify
- Authorize
- Inform
- Instruct
- Inventory
- Update

In Chapter 3.1, I offer detailed help in finding one's communication purpose and building a report upon it, contrasting communication outcomes with project goals. For this chapter, I limit myself to simply explaining that a communication purpose must be formulated and articulated as the third and final element in the Foreword. It can be written in different ways, with various terms, but no matter how expressed, the statement must explicitly reveal the author's focused and narrow reason for writing the specific report. To reiterate, a report is written to achieve some focused and narrow communication outcome. The Purpose statement is a clear and concise announcement of that intended communication outcome.

A few sample Purpose statements should be illustrative:

- In this report, we *present* results from our 3-year study of the psychological effects of receiving large numbers of friends' photos using social media to encourage your organization to initiate educational programs intended for teenagers to promote wise uses of social media based on scientific findings.

- In this memo, I *request* funding to hire a graduate student to search for an optimal control algorithm for the proposed autonomous quad-rotor drone under development in my laboratory.
- The purpose of this letter is to *update* you on the progress of the project to evaluate the newly discovered drug that may prevent post-traumatic stress if administered in advance of the anticipated stress-inducing activity.

I have mentioned that the Purpose statement possibly suggests to the recipient(s) the type of response expected of them. Looking at the examples above, you can see that, while the author's communication purpose is explicit, the anticipated or necessary response from key readers is either explicitly or implicitly conveyed also:

- In the social media example, the author hopes the reader will use the findings in the report in developing education programs.
- In the control algorithm example, the author hopes the reader will authorize funding.
- In the drug example, the author hopes the reader will be reassured that progress has been made on the project.

On any project, you may progress through different phases and find a need to write different reports. Each report will have its own communication purpose. For the project itself, you may continue to work toward a single goal, but for the different phases and different reports, your communication purpose will change. The purposes for writing will change as your need to achieve different communication outcomes changes. Each report must contain within its Foreword a statement of the purpose the report has been written and delivered, in terms of a narrow and focused communication outcome.

1.2.2 Summary—All Report Highlights

In most non-fiction writing, a Foreword is followed by a Summary, in one shape or another. Even in a simple form of non-fiction, the essay, the thesis statement always comes at the end of the Introduction, after an author has introduced the topic, narrowed it to a focused sub-topic, and mentioned the gap in knowledge or difficulty to be overcome. This funneling of information sets up the coup d'etat—boom—here is the thesis. When you present your thesis in an essay, the thesis is your main point. You are highlighting the main point, which is one key objective of a Summary.

> The Summary is the place to reveal all the highlights of the report.

In purposeful ABERST documents, however, the Summary by necessity must highlight more than just an author's single thesis. It must be comprehensive enough to inform and persuade all readers of your point of view. Therefore, the Summary is the place to reveal all the highlights of the report. The highlights will be all the main points of your message, which are collectively an elaboration of your communication purpose announced in the Foreword. If your purpose is to achieve communication of a particular message with your document, the main points are the heart of that message.

For example, imagine your purpose is to propose using a vendor you believe is ideally suited to provide some needed supply. In your Summary, you explicitly state you have selected Vendor A, not Vendors B, C, and D, and you give the key reasons Vendor A is preferred. You do not have to discuss the rejected vendors in much detail yet, but your Summary should indicate that Vendor A is superior in several respects to the other vendors. If a second-best vendor is nearly as good as Vendor A, this could be included as a highlight. Lastly, if the reader should know of any other critical information, such as costs, deadlines, assignment of personal responsibility, or required actions, the Summary is an appropriate place to provide such.

As another example, imagine your purpose is to present an innovative public policy analysis inventorying the infrastructure hurdles to be overcome to produce a transportation system highly supportive of electric vehicles. In your Summary, you would begin with an overall generalization of your findings regarding the hurdles found, followed by highlights of the specific hurdles uncovered by your research. If you have the additional purpose of providing recommendations for overcoming each hurdle, you would present a brief summary of your suggestion for each hurdle. You would also need to provide the key findings from your research that prove the hurdles are, indeed, accurately depicted as hurdles. These findings could be anecdotal, mathematical, technological, societal, or something else, depending on the hurdle. You might also say something about any putative hurdle that you suspect readers would think should be included in your list but you have excluded. Trying to anticipate a reader's dissent or questions and answering the important ones are helpful elements in a Summary.

In sum, to facilitate a good Summary, specific report highlights are required. As a checklist, you can look to the following for the key items to be included in a Summary. In addition to the essential items for a Summary, the two-part list incorporates a few optional items that might be included depending on the specific issues at hand:

Essential:

- Recommendation (single, compound, or multiple)
- Conclusion(s) (as counterpart to recommendation[s])
- Analytical, Empirical, Computational, or Otherwise Meaningful Results to Support each Conclusion (actual quantitative or qualitative values obtained, in context with standards, benchmarks, or baselines)

Optional:

- Alternatives to the Recommendation
- Significant Characteristics of the Research Method that Enhance Credibility
- Deadlines
- Assignment of Responsibility
- Expected Actions or Next Steps
- Cost or Personnel Implications (if not part of essential Meaningful Results)

One very important objective of a Summary is to allow some readers, especially those with lots of responsibility and little time, to peruse just this short section and take away your main ideas with some confidence that you can defend those main ideas. That is, your Summary must be convincing, though brief, so it should provide the highlights of your argumentative support for your conclusions and recommendations.

To return to the first example used above, in support of Vendor A's selection, you must provide the highlights of Vendor A's superior qualities, whether that is delivery times, warranty periods, costs, replacement and repair promises, service commitments, or something else. In the second example, if you are concluding that Hurdles A, B, and C are the three chief hurdles to overcome, you must supply data substantiating the assertion that Hurdle A is indeed a hurdle, same with Hurdles B and C. This is important in the Summary, and the remaining details of your project can be saved for later in the report.

Combining the Foreword and Summary

You probably had no doubt that an author should introduce her report in the start, but it may come as a surprise to learn that an author can also end the report at the start. Sounds surreal? Yes, it is, like Lewis Black's *End of the Universe*. (If you have not heard this comedy spiel yet, find the will power to tear yourself away from this book and listen to this renowned comedian's take on aging, coffee, and life.) How can you possibly end just when you are starting? You will read throughout this book that ABERST reports should be as short as possible. Well, if you end in the beginning, that is about as short as a document can be, right? You just get started and you tell the reader, "You can go now. You have read it all." How nice. This is the advantage of the combination of Foreword and Summary: for the busiest of readers without first-line responsibility for double checking all of the report's details, these two sections let them stop reading and still take away the report's main points. In other words, if you start the document strongly, with a Foreword and a Summary, you successfully "end" simultaneously.

With an understanding of the elements of the two initial strategic sections for any report in ABERST, you are now in a position to appreciate the revised start to the sample report shown above involving the automobile defect. Fluff is absent, and the document starts by achieving all five objectives.

Revised Sample Report Beginning *with* Two Strategic Initial Sections

Foreword

One of our peer organizations recently suffered huge financial and reputational losses due to an undetected defect that affected nearly 2.6 million vehicles. Because we steadfastly implement programs to prevent defects in our products, we also regularly assess these programs for weaknesses and potential improvements. In response to the recent press coverage of the peer organization's defect troubles, our own small team of quality engineers meeting monthly suggested that a special multidisciplinary team be formed to study the details of the peer's defect-based travails and determine learning points that could be valuable for our organization. This report presents the findings of this study with recommendations for implementation in the near- and long-term.

Summary

The multidisciplinary study team suggests our company implement four new practices with respect to defect prevention. First, a Deep Engineering Review team will be created, with an initial staff of ten persons. Shane Andoah, Vice President of Manufacturing, will serve as the initial director of this team. He will select his staff over the next 3 months and submit his proposed list of team members to the Executive Board. Second, all reported accidents, breakdowns, and customer complaints about our products will be reported to the newly created Deep Engineering Review team, which will be charged with investigating and reporting, at least initially, in no more than 30 days any known or possible implications for manufacturing defects. Third, a company-wide information program will be developed, as a coordinated effort between the Deep Engineering Review team and Dane Coaster, VP of Internal Communication. The elements of this program will be delivered in draft form, for review, to the Executive Board within 6 months.

Fourth, as a critical component of the aforementioned information program, an employee "speak-up" program will be implemented and advertised wherein any employee with any information that could be relevant to identifying and mitigating a manufacturing defect will be rewarded, not penalized, for the effort, with details of the reward program to be developed at a later date by an *ad hoc* committee including Ms. Andoah, Mr. Coaster, Leland Traverse (VP of Personnel), and some members of the Executive Board.

These four practices were selected as optimal for our organization following a review of their advantages and disadvantages. The single strongest advantage of each recommended practice is presented here, with further explanation of our selection rationale, as well as comparison to other practices not selected, in the sections that follow this Summary. The strength of the Deep Engineering Team is that it leverages existing experience and authority within the company. The advantage of using the aforementioned team as a clearinghouse for all potential defects as intimated by reported accidents, breakdowns, and customer complaints is that it centralizes the investigation process so that patterns and trends can be seen more quickly than if this task were decentralized. The benefit of the company-wide information program is that it provides a critical complement to the other three methods without adding new infrastructure or capital cost. And, lastly, the advantage of the "speak-up" program is that we use our greatest asset—our employees—in a capacity that recognizes their knowledge, experience, and commitment to the success of this company at an additional cost that should be trivial when compared to the potential long-term savings.

Chapter 1.3

Satisfy the Details of the Assignment

Similar to the guidance provided with respect to fulfilling the letter of the assignment, the guidance for satisfying the details of the assignment is straightforward and possibly obvious. The details, however, go further than the foundational basics expressed by the letter of the assignment. The details, in contrast, encompass any and all of the myriad expectations for substantive information necessarily presented in the document. This applies only to recipient-initiated reports and not event-initiated ones. When recipients make the effort to lay out an assignment description, an author can be absolutely certain that every single item of requested information is required by the recipient. The recipient will scrutinize the report to see that every little detail she is expecting is in fact present. The recipient looks for these details once she has realized the report is the one expected. This realization occurs while reading the initial strategic sections. After that, recipient will study the remaining "body" of the document to ensure that critical details are included. Thus, authors must fulfill the recipient's expectations of details of the assignment.

The importance of fulfilling all the details of the assignment cannot be over-emphasized. The effort yields significant benefits. One, an author who submits a report conforming to all of the requirements associated with that document will be in a propitious position to earn a high grade if in college or high praise if working in ABERST. Two, the details of the assignment are useful as a guide to providing any required information. They help an author create an outline for the report contents and inspire development of associated or complementary report sections not directly specified by the assignment but still valuable to include.

For each document, you will need to determine whether the assignment details provide all the guidance you need as to content, or just a "minimum" expectation. For example, in an RFP, the details may be the necessary and sufficient components of the report because providing anything more or less would be unhelpful to the proposal reviewers. They know exactly the material they want from each submitting author. Aside from this type of document, however, the details required by the assignment are usually necessary but not sufficient for producing a complete and cogent document. For many documents, the assignment details are just a start. Authors must provide more than the required details of the assignment, but they compose a helpful base upon which to build the full report.

Here is the advice of this chapter in a nutshell: For any recipient-initiated document, read the instructions given to you and follow them. If instructions were presented across several documents, for example, one memo, two emails, and a post-it note, go through ALL of them and synthesize the myriad assignment ingredients into a comprehensive list. If you were orally briefed on the assignment, review your notes and be fastidious about responding to each and every orally conveyed expectation. Stay nimble and remember that the instructions may arrive in various forms, from email, to phone calls, to 10-page RFPs, and your document's strength depends, in no small way, on how closely it conforms to the details of the assignment.

In short, you must include all the contents of an assignment (or, requested report) that are presented. Go through the assignment or request and highlight every required element. Do not leave something out that has been explicitly introduced. If you are asked to provide a deceleration strategy, provide one. If you are asked to provide a recommendation for a velocity measurement system, provide it. If you are asked to calculate a confidence interval from your standard deviation and sample mean, using a particular statistical formula, do it. If you are asked to provide key empirical data used for calculating total impulse of a rocket motor, provide that data.

Typically, instructors in college provide the prescription for an excellent technical document. You have been given the formula, the required structure and format, and the expected critical contents. Deliver the goods as required. In the many offices and cubicles across ABERST, moreover, busy professionals are provided numerous expectations and requirements for the various documents that need to be written. The difficulty is keeping them all in mind, having them handy, collecting them from disparate sources, and using them as a final checklist before completing the first draft.

Let us look at two examples, one from college and one from a business situation in a fast-paced technology organization. In addition, taking a second look at the example from the previous chapter, manufacturing defect, from the perspective of this chapter should serve to reinforce the lessons so far.

College Example

A professor assigns you the task of researching and supporting one student-based campus improvement project. She gives you a partial list of suggested topics and mentions that you can think up a topic on your own also. You must research one of these ideas and look into the entire process of implementation, including people involved, schedule, costs, benefits, detriments, and so on. Her suggested list includes these: improve outdoors wifi, add outdoors charging stations, install yoga or meditation room in all school buildings, offer more bike racks around campus, and initiate an area dedicated to food trucks at lunch time. The assignment is delivered originally via a small paragraph on a handout and posted on the course website, outlining the information just summarized above. Later, she adds the following in an email to the class: Produce a paper between 10 and 20 pages, with at least one diagram showing a sample implementation of your planned idea. Subsequently, in class, in response to a student's question, she adds the following: each paper should have a proper bibliography with at least 10 sources cited.

Across all these disparate instructions delivered at different points in time, she has collectively given you the letter of the assignment: 10- to 20-page research paper on a topic of your choice that would improve students' experiences on campus, similar to approved ones given.

In addition, you have details of the assignment as well: find at least 10 sources of authority on the topic, prepare a proper bibliography, and include at least one diagram showing one sample implementation of your planned idea.

Business Example

Let us say you work in a laboratory, and you've made 1,000 specimens using an established process. The goal is for all the samples to be identical. Moreover, your supervisor has asked for a report that describes the process and the resultant samples and analyzes the consistency of the samples. Your supervisor has asked that you randomly sample and examine ten of the specimens. You were asked to find and use a statistical method appropriate to a small sample size to evaluate the predicted consistency of the 1,000 specimens from your sample pool of 10.

From these instructions, you have very little in terms of the "letter of the assignment." Not much has been proscribed. Still you can infer the "spirit" of the assignment using your experience and common sense: you should prepare an internal company memorandum. Other options, such as journal article, presentation slide deck, and formal report, are not ideal choices, as your supervisor said the report is for her and did not mention anything about trying to publish an article, present an oral briefing, or produce a bound, formal monograph. Similarly, an email would likely neither be the supervisor's intention with "report" nor a suitable vehicle to convey the details requested. As for page length, because none has been specified, you can rely on the default maxim: make the report as short as possible while remaining cogent, clear, and complete.

Fortunately, despite the lack of "letter" guidance, you did receive a good deal of "detail" guidance. One of the first items requested is a description of the process used for making the specimens. This is mandatory. You also must explain the method used for randomly selecting ten samples to prove you did, in fact, randomly sample the specimens as assigned. Similarly, you must explain your analysis of the statistical consistency across specimens, but you were not given a specific method. You need to identify the method you selected, perhaps Student's T distribution, and both explain its suitability to the required task and present the results of running this statistical analysis. Within this analysis, you must describe the characteristics of the ten sample specimens and relate the data from those samples to your statistical computation results, as well as walk the reader through the steps of the statistical method.

Manufacturing Defect Example

Imagine that your company's chief executive officer (CEO), advised by general counsel and a few vice presidents, realized that prudence required a renewed effort in quality control and mitigation of any potential manufacturing defects that could lead to large vehicle recalls. The CEO assigned some high- and mid-level employees, including you, to a newly formed team to study the possibilities and gave the team a charge, probably in an informal way, through an email or a face-to-face meeting with the team's leader. The CEO may have asked for something like this: review at least a dozen defect prevention and mitigation methods from literature and industry, see what our peers are doing, evaluate these methods, and narrow the options down to four or five that would be suitable for our organization. Put together a document for my review in 90 days with the advantages and disadvantages of all possibilities and a case for presenting the recommended ones, with an implementation plan.

Because we have already studied this scenario from the perspective of a document's start, we know that the authors of the report would have done poorly had they started the report with a list and description of the twelve possibilities reviewed by the team. Nonetheless, that list of the twelve initial possibilities is a required detail, as is a thorough analysis of the advantages and disadvantages of all possibilities reviewed. Ultimately, the authors must provide a case for their four recommended methods for defect prevention and mitigation, including proof that they explored at least a dozen options. A full explanation for rejecting eight of the twelve methods represents some important, additional details required by this assignment.

As before with page limits, this advice for fulfilling the details of the assignment does not apply if your document is event-initiated rather than recipient-initiated. So, this chapter is not germane for each and every document. Skip it when you can, and concern yourself with the other chapters of this book.

When you do know that the document is recipient-initiated, however, and you have a thorough checklist of details that must be included, you can proceed to include all those items in the report. Putting those items in the right place is the topic of the next section, where further explanation is provided as to the relationship between the document's start and its body.

Chapter 1.4

Shape the Document for All Readers

The three preceding chapters have approached the concept of "right document" by emphasizing the importance of (1) fulfilling the fundamental recipient-assigned requirements, (2) starting strongly, and (3) providing all recipient-directed details. To continue to build the "right document" with a fourth component, authors would do well to remember the notable characteristics of their readers: they are not waiting on every word, taking time to enjoy the special cadences and rhythms of an author's expression of magnificent ideas. On the contrary, they are busy people working in some specialty within ABERST with deadlines to meet and problems to solve. They need the document to provide useful information, in a way that makes the information easy to find. If they want to skim or skip portions, they should be able to do that also. Thus, putting the details that recipients need into the remainder of the report in a manner that enables quick reading, immediate comprehension, and subsequent referral is a technical writer's next priority.

Shaping a report for easy reading wins over your audience. Readers loosen up and look upon a report shaped skillfully as a manageable task. The opposite reaction is never desired: if readers glance at your report and think it looks dense, packed, intimidating, impassable, and insurmountable, you have lost their trust and support. You do not want to present an encyclopedia-thick tome that frightens readers; rather, they should welcome your easy-reading, helpful, pleasantly proportioned blue-ribbon report. The report is a link for a mutually beneficial relationship.

This is an opportune moment to reflect on the value of this book for both college and career. The commonality between college professors and other professionals working in ABERST is evident in several respects. One, professors represent typical readers: busy and demanding of high-quality writing. They want to see your best work, and just the right quantity of it, so they can assign you a fair grade without spending unnecessary time struggling to evaluate a confusing and unclear document. Two, professors want you to learn a method of writing that will be useful in your future, not just during your education, so their document requirements match those elsewhere in ABERST. Three, the method of assigning a grade by assessing the document's quality overlaps with the methods used by all ABERST readers to evaluate a document: check if it gets to the point quickly, stays on topic, and delivers a clear and complete message without causing confusion and head scratching. Therefore, assuming you are assigned non-fiction, technical writing by professors, delivering the right document is achieved via the same method for both college and career. And the fourth component of this method entails putting the document into a helpful and pleasant shape for all possible readers.

Shaping is a combination of just three steps: organization, subordination, and documentation. Skillfully shaping a document is one of the cornerstones of technical writing, so it belongs in the first part of this book because it is critical to delivering just the right document. The three steps of shaping are discussed separately in the sections below.

1.4.1 Organization

As said before, readers must know from the start that they possess a document that they either need to read or will benefit from reading. ("Possession" may refer to either holding in one's hands or reading on one's screen.) You must keep these readers on your side, as they read through the start and into the report's body. The body must reassure readers that the document is valuable to them and fulfills the promises made at the start. Thus, the body must present the necessary details in an organized, easy-to-find manner. Readers should not have to hunt around. They should not have to read all the way through in order to find items of special interest.

This organizational approach works particularly well when one considers the readers who are not immediately evident but may, at some point, take hold of the document. With an endless potential for readers, a "perfect document" may be hard to create. But if the document has been arranged in a logical manner with successive layers of information placed one after another, it becomes nearly universal in its clarity.

This approach is akin to the arrangement of food items at grocery stores. Have you noticed that nearly all stores, across various corporations, place the vegetables and fruit on the far right, and the dairy on the far left, with dry goods throughout the middle? This makes it easy for anyone, no matter age, origin, gender, education, and so on, to feel familiar with the store even if this is a first visit to that particular location. Stores are similarly arranged so shoppers can find products and buy them. So too, technical documents, if well written, are similarly arranged so readers can find information and use it.

This is the reason that before efforts are applied to sentence writing and word selection, the "right document" is built upon an effective organizational structure. Organization at the highest level encompasses the report's start, body, and base. The start has been discussed above, but now it can be seen in the context of the overall three-part organization. This approach is promoted by many authors, but I learned it from my mentors at the University of Michigan: J.C. Mathes and Dwight Stevenson. I am highly indebted to them for paving the way in this endeavor.

In short, all documents in ABERST are ideally organized into three overall parts: start, body, and base. In Mathes and Stevenson's model, these are formally labeled as Overview, Discussion, and Documentation (see Figure 1.4.1).

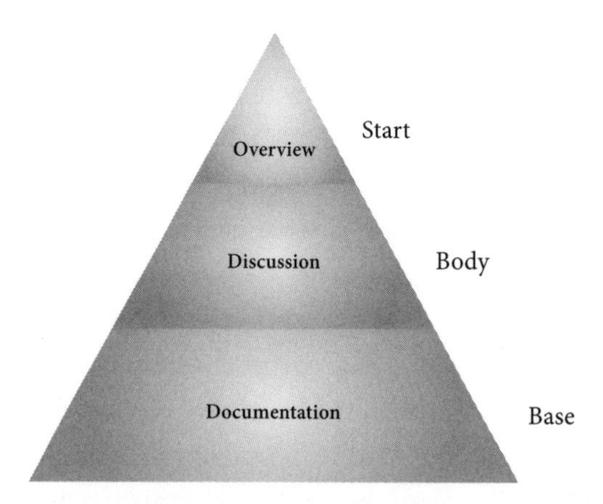

Figure 1.4.1 Overall Three-Part Structure of Technical Reports

You start strongly, as discussed in Chapter 1.2, by providing an overview of the document in a concise and properly organized manner. The Foreword and Summary are critical components of the Overview. The other component includes the appropriate document identification (ID) data such as title, author, recipient, and date. This information can be presented in various formats and styles, depending on the genre. In a memo, the ID data are presented in the "Date/To/From/Subject" Heading. In a formal report, the ID data are presented on the cover. On a business letter, these are provided at the top and bottom of the letter, in conventional locations. No matter the particular format, the ID data are collectively understood to serve as the document's Heading (see Figure 1.4.2).

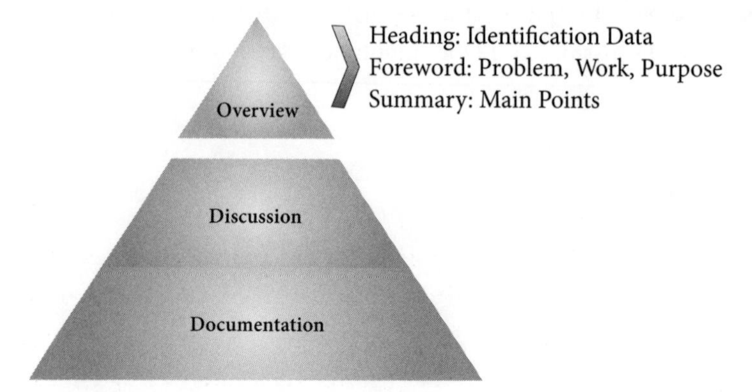

Figure 1.4.2 Overview's Three-Part Structure: Heading, Foreword, and Summary

The body of the document, often dubbed "Discussion," is the place to provide the elaborations, explanations, and justifications of all the main ideas highlighted in the Summary. This is the part where support details and step-by-step analyses are explained so that the full extent of the work is presented and assembled to succeed in delivering a cogent and comprehensive message. Any and all details relevant to supporting the main message of the report are placed here. To be clear, the Discussion is not a full log of every step of effort just for a record's sake. That would be tedious and probably encompass unnecessary material. Instead, it is a fully organized and strategic presentation of the necessary details, in order of importance. The need to subordinate within the Discussion is so crucial that the next section, Subordination, is devoted to that topic in itself.

In some sense, the Discussion is the full report. It must be written to stand alone, as if the Overview (and Documentation) did not exist. That is, should readers skip the Foreword and Summary, the Discussion itself is complete. The Overview merely abbreviates the Discussion, for those very busy readers who need only the main outline and not the details and for all readers who prefer to see a "map" of the route before they start on the full journey.

The Discussion sets forth each and every useful aspect of effort, thinking, calculating, measuring, testing, judging, double-checking, brainstorming, and so forth that the author and her team implemented prior to writing the report. In so much as an item remains relevant to the report after all is said and done, it is placed into the Discussion. If an item turns out to contribute little to the analysis presented in the report, it can be left out. Secondary and tertiary details that are still relevant are placed into the Documentation, which is covered in the final section of this chapter.

1.4.2 Subordination

Successful documents in ABERST depend on the presentation of meaningful details to support the author's purpose and produce a cogent message. As explained in the previous chapter, recipients may dictate some of those details, while others may be products of the author's planning and strategizing for the document. Whether recipient- or event-initiated, the details must be properly placed in the document, to follow effectively from the promise made to the readers during the document's two initial strategic sections. The promise is fully explicit in the highlights in the Summary, but elements of a promise appear in the Work statement while the Purpose statement is nothing but a promise, plain and simple. (In the Foreword, authors promise to elaborate on the work completed and fulfill the communication purpose.)

Importantly, the details promised do not belong in the start of the document. That would defeat the purpose of providing a concise Overview to start the report. Details would clog and elongate the short, strategic start. Thus, an author must be trained in organizing the necessary informative material into the remaining sections of the report. That skill is the focus of this section.

Subordination is the process of grouping all items into logical parts, and then dividing the information in a single part into further sub-parts. The prevalence of subordination in the Discussion is emphasized in Figure 1.4.3.

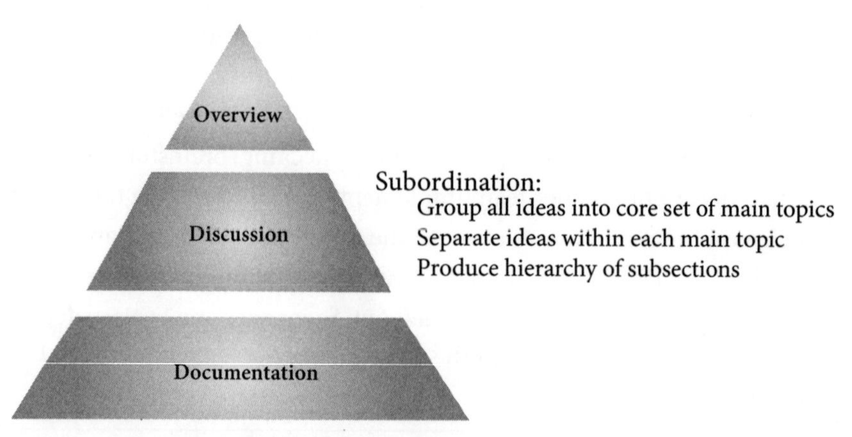

Figure 1.4.3 Subordination Within the Discussion

We can look at a subordination example quickly here, and the topic is reinforced in Part 3.

Imagine a report about modeling molecular reactions during combustion. This may be important to assist scientists and engineers in developing new, clean fuels for automobile engines. In this report, the author wants to present a novel approach to modeling, using computing power, a process that is very hard to study in real engines, due to the extremely short time intervals to be captured, not to mention the difficulty of the very hot environment. In this report, the author plans to discuss the three current modeling approaches before explaining her new (fourth) modeling approach. She also must explain the theory behind her model, the flaws with the other models, either individually or collectively, the conditions she used for her testing of her model, and her modeling results. When she presents the results, she must discuss the meaning of the results and acknowledge any problems or possible errors in her work. The above, of course, is just an abbreviated list of contents for the purpose of providing an example of subordination. A real report might have even more information. She begins by outlining the main topics, then creating subtopics under each of those:

Problems with Existing Modeling Methods

- Method A
- Method B
- Method C

Need for New, Better Modeling

- Reason A
- Reason B

Novel Modeling Method Presented Herein

- Theory Behind It
    - Input from Theory X
    - Input from Theory Y
- Potential Advantages
- Expected Results as Baseline (Predicted or Ideal, or Both)

Modeling Conditions Used for Test Modeling

- Condition 1
- Condition 2

Modeling Results

- Analysis
- Comparison to Expected Baseline (Predicted or Ideal)
- Errors and Explanation

Implication for Future Work

The details above are not as important at the overall approach. The subordination of subsections and groupings, or nesting, of components of the Discussion is the critical technique to be mastered so that documents written will be useful and manageable to all readers. Experts will appreciate the organization imposed on the complex topic, and novices will find their way to comprehension of a new subject thanks to your imposed hierarchical structure.

Subordination is a vital task performed by the author of a document in ABERST. If an author fails to do this task and puts the various details into a document in a higgledy-piggledy fashion, without rhyme or reason, the readers are forced to find the patterns of arrangement for themselves. They must forge the links and natural groupings between the disparate items of the Discussion. Instead, an author should do this for her readers, presenting an organized hierarchy of main ideas and sub-points.

The advice of this section boils down to this single idea: Outline your Discussion into a hierarchy of subordinated sections and subsections before worrying about even a single sentence, let alone any particular words. Develop the big picture for a report. See the skeleton or "map" of your document. Figure out the overall sections and their subsections. If subsections require further subsections, make them. This creates a multi-level report hierarchy within the Discussion. That is the essence of subordination. Further help in mastering this technique is provided in Part 3, in particular Chapters 3.4 and 3.5

Thus, for both organization at the highest level and subordination of all points and sub-points, structure is the guiding concept. Structure is key. Even here, in the first part of a book that covers technical writing, I am not talking about words, sentences, and punctuation. You do not know yet whether in this book I will discuss the difference between formidable and formative, or impressionable and influential. Those are just words. I begin with structure. You want to see the forest for the trees. Words are trees; structure is the forest.

Your experience reading this book reinforces the fundamental role of structure. You know, in fact, that this book is divided into five parts. That's structure. You know it, and you learned it right away. I helped make that clear. I emphasized it. And, it's important. As for other particulars in this book, you will get to them in time. But I didn't start with nitty-gritty details, as you would have had a difficult time processing them. Instead, the big picture (structure) was presented so you could comprehend and remember main principles. That's how powerful structure is. It sinks in, and quickly. So, as a writer, you worry about your structure (so readers will not have to). You take care of the forest, and the trees will sprout up and fill in the landscape in due time.

The concept of outlining reinforces the importance of structure. You have no doubt heard the precept to outline first, every time you have some writing assignment. That is based on the idea of structure. An outline is your structure in its basic form. It is the key beams and posts of a building, without any walls, windows, and decorating yet. Outlining enables you to shape your general design for the document before working on the details. When you are finally ready to consider details, you will find the last step of document shaping useful: Documentation.

1.4.3 Documentation

The base of any report is the Documentation. With the addition of this part, an ABERST report can attain its complete and proper shape. In this final part of a report, the author places various types of support material that would disrupt the flow and cohesion of the Discussion. While the document's Start contains only those few items required by the inherent nature of a brief Overview, and the Discussion contains all the promised details, still additional material remains that has an ancillary quality. This material is needed, but it can be referenced in the Discussion and placed at the back of the report, in the Documentation. Identifying the ancillary material and separating it from the necessary details of the Discussion takes practice and gets easier over time. As with learning to subordinate material, further help in understanding how to create a Documentation component is provided in Chapter 3.4. For now, the concept is introduced to illustrate the complete and proper shape of a document.

The Documentation part of the report is analogous to the basement of a house. It serves as an underlying support for the main living areas above it. Indeed, all the primary and secondary rooms of a house sit above the basement. Nonetheless, one can find some important items down there (for example, boilers, circuit boxes, and shut-off valves.) Likewise, the Documentation of a report contains helpful support material as well. Moreover, just as we do not usually bring guests directly into the basement when they come for a visit, we do not thrust our Documentation into the faces of our readers. Instead, we start the document with an Overview and deliver the details in a Discussion, and finally place the support material in the back, to be reviewed by individual readers as they find necessary. Similarly, guests come through the front door to *start* their visit, acquire a first impression, and obtain an *overview* of the house, then they linger socially in the kitchen, dining room, or living room (for *discussion*), and only journey to the basement if they are motivated by a specific reason or need.

Authors in ABERST must recognize that readers do not need to see all the details at once. Some details, in fact, are more valuable or meaningful than others. These key details belong in the Discussions. Authors must develop a skill in placing less-important information into the report's Documentation. Discussions should not be clogged with bulky or ancillary items that readers can choose to consult at a later time, but do not need to review during a first read through the document. Placed in the Documentation, those ancillary items are still available and present in the report, should a reader choose to consult them.

As an author provides a hierarchy of information in the Discussion, she must also evaluate each and every detail to assess whether or not the reader absolutely needs to see it here and now in the Discussion. If not, it can be placed in the Documentation. An author must make this decision as to the significance of this or that item (more advice in Chapter 3.4).

This means the author must anticipate readers' needs and expectations as to reviewing support and ancillary material that serves as the base of the report. Perhaps a reader wants to check a calculation. If so, that will be in the Documentation. Perhaps a reader wants to see a sketch of a rejected design. That will be in the Documentation. Perhaps a reader wants to see the raw data from a sensor, although the data were converted into meaningful units. The raw data will be in the Documentation.

Some items that would be suitable for Documentation are as follows:

- Raw data tables
- Supplemental data sets and graphic representations
- Calibrations
- Calculations
- Proofs
- Preliminary drawings
- Alternatives and variations
- Blank forms used (templates, matrices, and tables)
- Superseded information that may still be instructive (designs, devices, surveys, methods, and experimental setup)
- Procedures
- Schemata, diagrams, and input wire meshes
- Less-essential background (theory, history, and personal accounts)
- Photographs as record of events and activities
- Formulas
- Safety sheets or other verification
- Digressions of any kind

No such list like the one above could ever be complete. It serves as representative to inspire future authors. The thought process is going to be unique to each report. Similarly, although Documentation is the report segment that may be lengthy and bulky, it does not have to be. Sometimes a report requires very little material in the Documentation. It depends. But, unlike the other two segments, which must be as concise as possible, the Documentation is permitted to be extensive and packed with particulars of one kind or another (see list above) because that is its inherent nature. It is precisely the one segment that serves as a broad base upon which the other two segments are built. Nevertheless, when the Overview and Discussion require very little supplemental, ancillary, complementary, preliminary, or tertiary support material, the Documentation may be short or even neglected.

At least one concrete rule can be provided: Documentation is divided into two types of content, appendices and attachments (see Figure 1.4.4). Distinguishing between the two is simple: Appendices are items created by the author; attachments are created by others. Attachments, therefore, may be placed into the document just as they were "found" by the author. They should not be altered (aside from basic size adjustments or other improvements for clarity such as possible with a photocopier). The use of an attachment is to help the reader by providing some material they may never be able to access (one of a kind) or should not have to make the effort to access. An author is making the material accessible and at hand. Appendices are always original, so readers would not be able to access this material in the first place if it were not created and appended by the author.

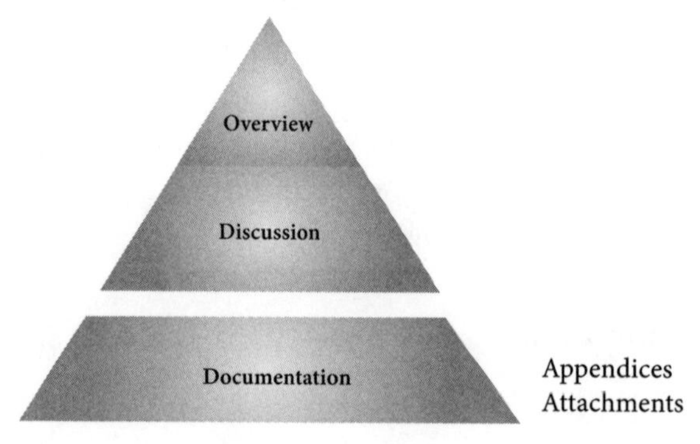

Figure 1.4.4 Documentation Includes Appendices and Attachments

Because appendices are an author's creation, they should be made to match the same high-quality characteristic of the rest of the report. Appendices reflect the effort and professionalism of the author. Although they are relegated to the back of the document, they will be reviewed by various readers and, therefore, should not diminish the good impression made by the previous two parts.

When you have a complete document, in the proper shape, you are nearly ready to deliver it. Before the final submission moment arises, you are strongly advised to proofread it one more time (see Chapter 2.3) and see if a friend or colleague can proofread it also. After that step, and when you are sure it is in the best shape you can make it, with all corrections and improvements implemented, you must dutifully deliver it to the correct readers, and that is the subject of the next chapter.

Chapter 1.5
Submit the Document Correctly

Technical documents in ABERST are usually written to someone specifically. That aspect is a major characteristic of the document, namely, it is a communication from the author to a specific reader or readers. An effective business letter is personally addressed to a recipient, typically someone at another organization. A memorandum begins with a heading for indicating "from" and "to." A formal report often has a cover with a line indicating "prepared for." Technical documents sent or addressed to no one in particular are often problematic.

Admittedly, some ABERST documents by their inherent nature have no specific readers, such as journal articles. Still, in the effort to have the article published, technical writers might be asked to send the draft article to a particular person at the journal who takes responsibility for reviewing and selecting the articles to be published.

Similarly, procedural documents, such as instruction manuals and standard operating procedures, once they are approved (by a particular manager in position to give approval) are written for no one specifically, but any and all people who might need to follow the instructions: buyers, owners, researchers, assemblers, technicians, assistants, mechanics, repair personnel, trouble shooters, and the like.

Acknowledging these few variations, this book focuses on the writing that is done for particular readers, and correctly submitting a document to those intended readers is the focus of this ending chapter to Part 1. A quick review of the four preceding chapters should help you see how the topic of submission aptly completes Part 1: Delivering Just the Right Document.

So far, you have become adept at writing a document that fulfills the letter of the assignment. It starts strongly, with a concise yet comprehensive overview of the whole document. Furthermore, you have satisfied all the details of the assignment, and your document is in great shape through organization, subordination, and documentation. All that remains is to submit it to the correct readers. If specific readers have initiated the document, you need not be puzzled as to whom it goes to. Similarly, if you were given explicit instructions to send the document to a different person other than the initiator of the assignment, do as you were requested. This is straightforward.

Still, you can make mistakes with recipient-initiated documents regarding submission: you can miss a deadline, or send the document to the wrong place, or fail to provide enough copies, or send it to an additional party who should not receive a copy at this stage of the project. Who knows? You must submit to only the correct persons in the correct manner. I can advise you only to be careful and thorough when submitting a recipient-initiated document. With event-initiated documents, however, I can offer the detailed instructions that follow.

In addition to the issues raised by recipient-initiated documents, a different set of concerns arises with event-initiated documents. In these instances, and these may be plentiful over the course of a career, the appropriate readers are not specified. Instead, you must decide for yourself the suitable recipients of your report. This means you need to both identify and get the attention of these persons so they know to read your document upon receipt. To do this, you follow three steps: identification, attention, and distribution.

1.5.1 Identification

Step 1, finding the correct recipient(s), involves several tasks. First, consider the point of the document and your purpose for writing (covered in detail in Chapter 3.1). In short, what do you need to achieve by writing this document? Second, ask yourself these questions: Which persons either inside or outside my organization must help me achieve this purpose? Which persons either inside or outside might benefit from reading this document? Who also needs to be apprised of the document, even if her involvement is indirect and secondary?

Third, another effective question to ask yourself is this: Which organizational unit or group is responsible for taking the action requested or demanded by the document? This may be funding, approving, training, building, repairing, auditing, investigating, testing, planning, coordinating, organizing, surveying, and so on. After identifying the unit or group, next find the person in that unit/group who is in charge, or who can make decisions.

I vehemently suggest you always deliver your document to the suitable decision-maker, otherwise the "wrong recipient," that is, any other employee who has no strong inclination to get involved or respond, is likely to ignore the report. Furthermore, your efficacy is usually better if you aim high, rather than low. Indeed, a person in a superior position (a "higher up" type, supervisor, manager, vice president, and the like) can easily and comfortably send a report to another person for handling, but someone without that sense of commitment and responsibility may easily ignore something that either (1) does not concern him, (2) is not his responsibility, or (3) is "above my pay grade," as they say. Many people, fortunately, will try to properly route or escalate a report that mistakenly ended up on their desk, but when this does not happen, the report is simply lost or buried, and you must prevent this interruption in progress to the best of your ability.

Importantly, managers and other responsible persons can be held liable for neglecting a matter or being ignorant of it, both at their level or below. This is not the case with all levels of employees, however. Although any ordinary employee who ignores your report because it really should have been sent to a different person in the organization will likely not be chastised, you as an author may be criticized for sending the document to the wrong person, and, at least, you will almost certainly regret the delay that ensues due to your poor choice of recipient (even if not truly your fault). You cannot truly win praise and appreciation for a well-written document that fails to produce a positive result by achieving its intended purpose if it flounders in a bureaucratic swamp because it was not submitted to the ideal recipient(s). Thus, you must find the right recipient rather than just anyone who may easily "shelve it."

It is OK to seek help in identification. While you absolutely must know your purpose to write a document, you may not know who should receive it, especially if you are new at an organization. If such is the case, you must ask your supervisor or peers for assistance.

Overall, identifying the correct recipient(s) is an author's responsibility, and it is a "technical writing" skill because it is fundamental to providing purposeful and useful documents within an organizational context.

1.5.2 Attention

In addition to identifying intended recipients, you must initiate personal contact with them, to alert them to the document's existence. In Step 2, recipients must be informed that a document you have written is on its way to them for reading and response. This is a crucial step and may be more complicated than it sounds. If these persons do not expect the report, and by definition they will NOT because it is event-initiated, you need to get their attention so they do not dismiss, discard, or otherwise misplace the report without reading it. Also, once you are in contact with the recipients, you must confirm a mutually workable method for submitting the document to them, and sometimes more than one method is required, tailored to preferences of different recipients.

Making contact with intended recipients may not be as simple as it sounds. In some instances, it may be harder than writing the report. This occurs because the report may need to go to someone who does not know you and may not be inclined to give your report any notice. It may be that the recipient is someone from a department with which you have no connection and know little about. Or, if the report needs to go to a high-level manager or executive, and you are a junior employee, this can be intimidating. In the most basic sense, when a report must be submitted to someone with whom you have no rapport and no prior dealings, it may be difficult to bridge the divide of unfamiliarity. Indeed, when a document must be submitted to a reader who is set apart from your "home base," this bridging of different backgrounds, positions, and specialties may not be so easy. Here are some examples:

- The author is in laboratory testing and she needs to send a document to the legal department.
- The author is from engine testing, and she needs to send a document to the vice president for public relations.
- The author is in government affairs, and she needs to send a document to the head of neurology.

- The author is from psychology, and she needs to send a document to finance and accounting.

These are just a few examples of how reports must often bridge wide divides of power, specialization, and priorities.

So, as difficult as it may be, you must make contact with the intended recipients and not just dump the report on them. Here are some attention-getting options to consider:

- If the intended recipients would be more responsive to the report if it were to arrive from someone higher in the organizational hierarchy than you, identify a superior who can deliver the report on your behalf. This person's name and position will garner the attention the document deserves.
- Make a phone call and speak directly to the intended recipients prior to sending the document to alert them to its imminent arrival. Confirm by conversation that they understand the document is important and should be read by them.
- Write an email, leave a voice message, or send a text or other intra-company message to the intended recipients prior to sending the document to alert them. Ask for an acknowledging reply so you know they received your email or other type of message. Until that reply is received, send a follow-up message every few days.

In an ideal world, this attention-getting step should not be necessary. In utopia, you should be able to submit your document to an unsuspecting recipient, and, given the strongly written start (per Chapter 1.2), this reader will be quickly on board with the importance and purpose of the document and her necessary follow-up action. But, this is not utopia. We live in a world of hard-working yet imperfect people and lots of chaos. You cannot count on your strong start with recipients who are not expecting the document. As perfectly written as the document may be, recipients with busy schedules, other priorities, and personal struggles might ignore it.

1.5.3 Distribution

After making contact and receiving acknowledgement from the intended recipients, you implement Step 3. In this step, you distribute the document to all recipients in whatever manner that was agreed upon in Step 2. The delivery method should be suitable to your needs, their needs, or both, ideally. If one recipient wants a hard copy and another wants an electronic version, give both their preferred choice. Alternatively, if you have placed the document in a shared drive, box, or cloud location, you must give the recipients the location and ensure they have all necessary information to enable access to that location. In short, you must be sure to submit the document in a manner that the recipients are aware of and comfortable with.

1.5.4 Conclusion

You have delivered just the right document by following the brief instructions of Part 1 of this book: your document fulfills both the letter and the details of the assignment, it starts strongly (and "ends" there as well), it is shaped for all readers, and its presence and importance is known to the necessary readers. It is now in readers' hands, and you can proceed to work on new tasks on the same project, or delve into a new project. Alternatively, this is an excellent time to begin studying Part 2 of this book.

Part 2
Writing Elegant Prose

Whether you enjoy writing or not, it is a necessary task in academia, business, engineering, research, science, and technology (ABERST). Many authors express the wish to be competent writers, while others simply say they want to get through the chore as fast as possible. I suspect, however, that even some of these less-enthusiastic authors desire deep down to be good writers, but they are discouraged for lack of good guidance or opportunity to practice and excel. Although not primarily writers by occupation, ABERST authors are still writers. Having been a technical writer myself for over 30 years, I have observed that technical writing is a valuable and compelling sub-field of writing. Documents within ABERST can be cogent, clever, compelling, and captivating. At its essence, technical writing can be elegant, involving polish and panache, style and substance. Thus, I operate within the assumption that ABERST students and professionals aspire to write elegantly, and they wish to emulate the great writers they have read and admired. No one aims low in ABERST. We all want our documents to impress readers, to flow, to charm, and to make sense.

One obstacle, however, is that ABERST students and professionals do not have time to embark on a second career in writing. You want the education and training you have had so far to be sufficient. You want your current level of proficiency to be up to the job. More than likely, however, you need some assistance. I offer that here, so your words, sentences, and paragraphs truly reflect both your intelligence and your excellent work. This second part of the book is directed at explaining the essentials of elegant writing, geared to ABERST. I have selected a limited number of topics that I believe to be the building blocks that are most critical to producing clear and clever writing. These are the tricks of the trade used by fiction writers and other published authors boiled down to a powerful package of principles as applied to ABERST, for the benefit of technical authors.

This part of the book has three chapters and an associated appendix. Chapter 2.1 explains a fundamental starting point for elegant writing, namely, drawing a boundary line. Specifically, you use purpose as the boundary line, and potential information for your document falls on one side or the other: it is either eliminated or retained by assessing it against your communication purpose. This allows you to focus your hard work of crafting clear prose on only those pieces of information that truly belong in the document. This streamlines your writing effort to make it all the more effective.

In Chapter 2.2, I present the composition guidelines and rules I have selected to be the most relevant to your needs as a technical writer. These are the secrets to writing elegant prose in a smart and snappy style. Because achieving elegance and style usually requires an iterative process, Chapter 2.3 explains the important role played by editing in producing powerful prose, and advice for effective proofreading is offered. Finally, Appendix A covers the common mistakes found in ABERST documents, for which you should be vigilant while proofreading. This appendix offers the final step in learning how to write elegant prose and provides opportunities for you to practice some of the techniques explained. It has a hands-on component. You can write directly on the pages where blank lines have been placed, so that you can practice and hone your skills right away. I am presenting some of the exercises that my university students are required to complete when they take my courses. I am including some of them in this book because I am convinced this interactive approach improves mastery and retention of the material. Your expertise will be more comprehensive with such practice than if you were merely reading. With that introduction of the strategy underlying this part, you are ready to begin with Chapter 2.1.

Chapter 2.1

Let Purpose Propel Prose

The most beautifully crafted sentences will be of no use if they are compiled into an inappropriate structure for your writing purpose. Thus, as an author, you must identify your purpose, followed by selecting the structure that best suits that purpose. Structure will give you a skeleton (shape) upon which to assemble all the document's details in an effective order (as covered in Chapter 1.4). The organization is the skeleton, and the details are the meat, so writing becomes a process of putting meat on the bones. If you prefer a vegetarian analogy, you can think of this as adding lots of vegetables into a multi-layered lasagna. The meat and the vegetables need some structure to hold them together and make them presentable.

The topic of document-specific organization is covered in depth in Part 5, where I address ten specific document genres and the special requirements and strategies suited to each. In this second part, however, I discuss the alignment of your prose with your purpose in a general way, before addressing the particular guidelines for elegant prose writing.

Once you know your purpose, you will understand whether you need to inform your readers or persuade them. You will also know the request you are making of them with regard to actuating their response to the writing. This self-reflective knowledge basically dictates the many important ideas to be conveyed in your document. Those can be written out in a straightforward manner.

Purpose is introduced in Chapter 1.2 and expounded upon in Chapter 3.1, but here I address the subject to reinforce the value of purpose as a boundary for elegant prose: any information within your boundary of purpose stays in your document; any material outside the boundary gets thrown out. Purpose helps you eliminate unnecessary material to make your document concise: tight and to the point.

Therefore, you streamline and simplify your burden of writing by identifying your writing purpose. This identification should not be a mystery. As intimated in various places in Part 1, the communication purpose has three components: the message (persuasion or information) you as author need to convey, the person (personalization) who needs to get this message, and the intended action (actuation) you hope that person takes by virtue of your document.

Below are some examples of purposes that present a main point, as either (1) a thesis for *persuasion* or (2) an original organization of *information*:

- The main point of a 19th-century novel (persuasion) as I see it, explained to my professor (personalization), so that she will give me a high grade for assignment execution (actuation).
- The optimal size of a heat exchanger (persuasion), explained to my supervisor (personalization), so she can initiate the manufacturing of that exact heat exchanger (actuation).
- My opinion on the best way to teach literacy (persuasion), shared with faculty members in primary education (personalization), to win their support for a pilot program (actuation) to implement literacy lessons in this ideal manner.
- A standard procedure for testing a model airplane in a wind tunnel (information), provided to company employees (personalization), to facilitate their implementing of the procedure to complete the testing (actuation).

The order of the three elements of purpose is not crucial because your report need not contain this statement per se. It is only a planning tool for you. It is intended to help you find the boundaries of your report, not something that has to be adhered to without deviation. If it helps you to think of personalization, then actuation, before information, go ahead. Look at this example where the order follows personalization, actuation, and information.

- If in college, you need to ask your instructor (personalization) to grant you an extension of a deadline (actuation), so you can have one more week to travel to another location to conduct additional research for your report (information). And, if you consider this purpose, you will probably agree that it is best suited to an email (genre choices covered further in Chapter 5.1).

As demonstrated by the examples above, this threefold analytical method can be applied to any communication task. On the job, you might need to announce a company picnic: picnic is next Friday at noon at Freedom Park (information), to all employees (personalization), so they can attend at right time and place (actuation). You might use two genres of documents for this announcement: group email and flyer either placed in mailboxes or hung on office bulletin boards. You might also use a group email invitation (or, e-vite) and a scrolling message on in-house television monitors. These latter two methods are multi-media variations of the email and the flyer, respectively.

After identifying your threefold writing purpose, you are in a position to select the perfectly matched structure and document genre. These go hand in hand.

Gifted writers know that they must understand the message they wish to convey, so they can choose the best structure for conveying that message. Are they going to dole out clues to a crime little by little, to keep readers engaged in trying to solve the crime along with the detective in the story? Alternatively, does the author wish to tell a story about two siblings, estranged for decades, who find reconciliation after some unusual event in one of their lives? This message requires a structure with a chronology and a choice of words and tone to contain the emotional shifts throughout the timeline of the story.

As with these gifted writers, you must be able to see the big picture before you worry about the individual "pixels" that make up the separate images that compose the big picture. In other words, your document may have many interrelated components, but only one big picture guiding the whole. In ABERST, the big picture is your writing purpose. That's it. You do not need to find anything more profound or mysterious or provocative. Just the writing purpose. If you can identify that purpose, at the outset of your effort, the big picture will be clear, and that clarity, in turn, will prompt and motivate all the necessary document elements to communicate your writing purpose.

In Part 3 of this book, I explain how to fulfill your writing purpose in the document. But, for this part, I implore you to identify your writing purpose (in its tripartite form) to help you write your prose. It helps you determine your style and tone, extent of details, organization and structure, required graphics and visuals, and the overall format. Do not spend 10 pages on flawless prose when the purpose demands something closer to a 2-page document. Similarly, you are unlikely to achieve your purpose if you write only 1 page, when 10 pages are required, with flow charts and data tables. Therefore, knowing your writing purpose will allow you to pre-determine the approximate extent you will need to organize your information into sections, paragraphs, and sentences. Before you edit every sentence and check for perfect punctuation, you need to make sure you are providing the proper and appropriate overall content, consistent with the big picture that underlies the whole document in the first place.

As I mentioned in Chapter 1.2, to impress readers that you are a competent writer who will deliver a message to them, you must catch their attention and maintain it. This, again, is catalyzed or promoted by identifying your purpose. The purpose catches the attention. They should understand early in the document (see Chapter 1.2) your purpose for writing, and they will be committed to reading further. That commitment from them is unnecessary if they learn the purpose from you and it is irrelevant to them. That is OK. You do not want disinterested or inappropriate persons reading your document. You want the right persons reading it, as covered above in Chapter 1.5. So, your prose is powerful when it conveys your purpose quickly. Then, you proceed to let the full array of information unfold in a clear and cogent way, per the guidelines in the other chapters of this second part.

As bad as spelling and punctuation errors are, nothing is worse than unnecessary information in an ABERST technical report. Sentences or paragraphs with information that distracts from fulfilling your writing purpose are taxing and harmful. So, the general maxim is this: know your purpose and include only material that supports that purpose. Everything else must be deleted. This rejected material will vary per purpose and genre, but, generally speaking, you will eliminate all commentary and personal information. Waste no time writing up this unnecessary background information or dramatic digressions. Instead, you will be fact-driven, straightforward, and business oriented. You will provide the necessary work-related information to convey your message to the immediate reader and plausible, future readers. Here are examples of commentary and personal information, in italics, that do not promote the writing purpose:

- For a report with a purpose of presenting a conceptual design of a manufacturing plant to make sorbitol from corn processing waste: *Plant flow diagrams are not included in this interim report. I sit in the fifth cubicle from the window, in the area with all the software engineers and business analytics, on the 7th floor. This allows me to share a printer and copier with two other business units, including a 36-inch-wide plotter. The plotter has been broken for the past 3 weeks.*
- For a report presenting an ideal control system for a satellite antenna's motor rotation function: *The night I tested the control algorithm, looking for the ideal gains, combining optical gains for the potential, integral, and derivative control components, my co-worker was unable to be present for the physical testing, so sweeping through a set of 30 trials took twice as long as expected.*
- For a report highlighting the identified critical activity delays in attaching a wing to the fuselage of an airplane: *I transferred from assembly line quality assurance, after asking my supervisor where he thought I might be able to learn new skills at the company and regain my enthusiasm for working on aircraft manufacturing. So, I approached this fact-finding mission as helping both the company and myself.*

In sum, understanding your communication purpose is crucially helpful because it allows you to eliminate any worry about producing a whole mess of unnecessary prose that has nothing to do with your purpose. You can simplify and get laser-focused. With only a select set of sentences to include, you can adhere to all the critical composition rules, to which I turn now.

Chapter 2.2

Review Composition Rules

In this chapter, I cover some fundamental building blocks of writing so that this book provides a single-source guide to technical writing as it needs to be done in college courses and on the job in ABERST. If you have hung with me this far, you are giving me your trust to explain the practical essentials, and that is how I handle the topics below, which fall roughly into the overarching category of composition. If you are using this book for first-year composition or English in college, it should serve well, unless your instructor has specific assignments in mind that are based on information in other writing guides and textbooks. Ideally, you have completed at least one prior English writing course where a more conventional grammar or English primer was provided. That would be an ideal foundation upon which to rest the lessons of this book, so you are ready to transition from high-school English and early-college essay writing to university writing, followed quickly by professional writing in ABERST. Indeed, if your goal is to write non-fiction technical documents at a university or on the job, this is not likely your first foray into writing and communicating in written English. I can fairly suppose that you have had some prior education and practice in this field. So, this chapter is a review of the key fundamentals to elegant prose writing geared to instructing a rising professional.

I admit this is not an exhaustive grammar book, and other books cover the topics in this single chapter in more detail and with worksheets and exercises. If you want to find another guide or textbook on writing or grammar as an additional resource, you can find some good ones. In fact, I can recommend my favorites. Send me an email, and I will reply with some suggestions. If you want some hands-on practice, moreover, you can find some in Appendix A. In the meantime, your writing should benefit from the instructions and explanations in the nine sections that follow:

2.2.1 Paragraphs

Good writing may seem to be built upon clever words and correct grammar, but as I have said elsewhere in this book the real recipe to great writing involves other, more crucial ingredients. Paragraphs are one of those other, often overlooked components of report writing that play an essential role in the success of your communication effort. Paragraphs complement other foundational writing components—purpose and structure—by representing the separate but related points that you wish to present to fulfill your purpose. Within the ideal structure, paragraphs allow you to convey each and every one of your important points—one point per paragraph, and vice versa. As you move through your message, adding point upon point, you are simply writing one paragraph after another. If you were to map out the high-level points of your report, you would not be counting words or even sentences but, rather, paragraphs. The main teaching lesson about paragraphs, therefore, is this: each high-level point is presented as a paragraph and nothing less or more.

This definition of paragraphs begs the question, what is a high-level point? Perhaps a high-level point is best defined by what it is not: a particular detail. The difference between a high-level point and a particular detail can be understood as follows: high-level points, on the one hand, are generalized explanations that readers would understand as relevant to the report and meaningful, while being devoid of particular details; particular details, on the other hand, are straightforward facts that on their own do not overtly advance your message. Examples:

- **High-level point:** To confirm this hypothesis, we conducted both physical testing and computer simulations.
- **Particular detail:** Humidity level during testing averaged 77% but never dropped below 74%.

From the former, readers learn you have a good deal of empirical evidence for reaching your conclusions and recommendations, obviously from both testing and simulations. From the latter, readers learn nothing about the importance of humidity in relation to testing, nor do they learn the impact, if any, of those values listed. A critical step in writing strong paragraphs is simply distinguishing between "points" and "particulars." If you can do that, you will ascend to the upper echelon of writing superstardom.

To write a paragraph that conveys one, and only one, high-level point, you need to know that a paragraph has a structure much in the same way the whole report has a structure. It will be familiar to you, as in Figure 2.2.1:

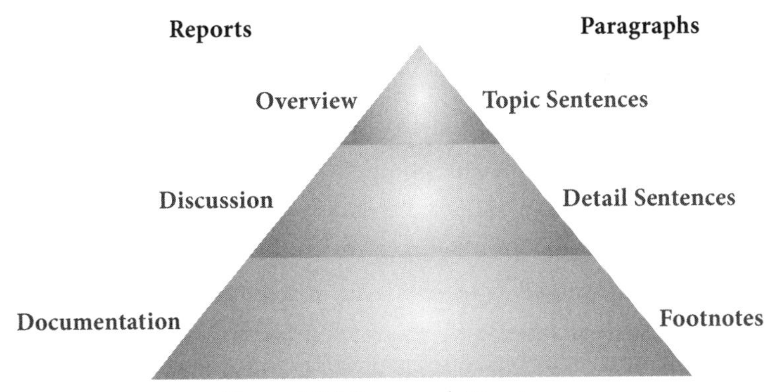

Figure 2.2.1 Paragraph Structure

Not surprisingly, the structure that works best for an entire report is exactly the same as has been used for centuries as the ideal structure of a paragraph in all styles of writing. This is not unique to ABERST. Take a look at your favorite novel or short stories, and you will see that paragraphs follow the pyramid structure.

The structure is three-part. It is that simple. The topic sentence (overview) is the statement of the main point. After the topic sentence comes the "discussion," which comprises the subsequent detail sentences. If an additional detail is a bit tangential, digressive, secondary, or extra in some sense, it can be relegated to an appended footnote (documentation). With this structure, the topic sentence alone suffices to communicate the point; the following sentences are merely details that readers can skip without loss of comprehension. The same is certainly true for any footnotes: these are further optional reading, as you no doubt have discovered in your own experience. Here is a footnote to this paragraph: Speed reading techniques, which have some legitimacy, emphasize this very fact: read just the first sentence of each paragraph. Of course this works for only perfectly written and edited documents, so many of us in the workplace cannot rely on this time-saving technique. Furthermore, the detail sentences and footnotes provide the substantive support for the main point, so any situation that requires careful review and evaluation of the author's ideas demands that the support be provided and assessed; thus, the whole paragraph, as well as footnotes, must be read by at least someone whose job it is to double-check and evaluate the strength of the author's viewpoint.

With this structure, the topic sentence alone suffices to communicate the point; the following sentences are merely details that readers can skip without loss of comprehension.

Because the topic sentence is the paragraph's overview and fully contains the paragraph's one and only main point, it is essential. It must be the first sentence, so readers can get the point immediately and can anticipate the thrust of the subsequent sentences. Nothing in those sentences can initiate a new main point. If so, it may be missed by readers. Any new main point must be converted into a topic sentence of a new and separate paragraph. This structure fulfills a promise to readers, and it helps authors move through their points, step by step, in a recognized and effective paragraph-based pattern.

The three most common problems with paragraphs, therefore, are (1) not having a topic sentence, (2) having one or more additional topic sentences in the middle or end, (3) having the correct topic sentence but placed at the end. Let's look at these three problems with examples.

(1) When a paragraph lacks a topic sentence, readers have trouble finding its main point and following the flow of details that support that main point. They will be slowed down as they work hard to comprehend ("get") the point. They may even misinterpret the whole paragraph.

Here is a problematic example:

- Yesterday, Lionel went to Burger Barn at 11:30 a.m. In the afternoon, around 4:40 p.m., he visited the same restaurant again. The first time, he ate one Burstin' Burger, one Buffalo Burger, two large fries, and an apple turnover. The second trip, he enjoyed one Big-o Burstin' Burger smothered with cheese and bacon, one Chargrille Chicken, and a vanilla milkshake.

Looking at the first sentence alone, we find the lack of a main point. This is simply a single fact (a detail of Lionel's day.) The author's point is unknown. Recall that particular facts say very little to readers; they cannot contribute to fulfilling an author's purpose unless they are connected into meaningful points or ideas, which strung together convey a message. Even reading the second sentence does not provide the main point; it is another single fact, linked to the first one by similarity but not by any explicitly stated point, idea, or message.

Readers of this paragraph must guess as to the point here. They cannot just read the first sentence and move on. They must read all the sentences. After reading the whole paragraph, readers will still be guessing as to the author's point. That's all they can do because the paragraph lacks a main point. It has only detail sentences. Here are some possible topic sentences:

- Lionel wanted nothing more than to eat fast food after being released from jail.
- Lionel was enthusiastic about trying fast food on his first trip to the United States.
- Lionel is a restaurant critic doing a special study of fast-food establishments.
- Lionel apparently has given up on his vegetarian diet.

Maybe readers will figure out, on their own, the ideal topic sentence from the context of either the preceding or proceeding paragraphs. Maybe the topic sentence that is missing is in one of those paragraphs. If so, it is in the wrong place. More than likely, it is not actually there, but the author is hoping the surrounding text will help convey the main point. Authors cannot rely on intimating one's main points. Do not just provide some details so that readers can infer your message. You must overtly state your main points in ABERST. This is accomplished with a topic sentence for each paragraph.

(2) Often a paragraph has a topic sentence, followed by helpful detail sentences, only to be eclipsed by a new topic sentence before the end, changing focus and presenting new information where it might be missed or misunderstood because it does not belong with the initial topic sentence.

Look at this example:

- Traditionally, lime ($Ca[OH]_2$) is used in caustic treatment. Lime, or calcium hydroxide, was first used in 1937, as developed by the Plains Chemical Company. By 1977, over 100 factories were using this treatment method. Sodium hydroxide is used in our process due to the reduction in time allowed by using this replacement for lime. The caustic treatment time is shortened from up to 60 days, to the 14 days we are using. Price is also a driving factor, as sodium hydroxide is significantly cheaper than calcium hydroxide.

From its first sentence, the paragraph presents the point that lime is the traditional material for caustic treatment. Subsequent detail sentences support this with facts about the tradition, dating back to 1937. But, at the fourth sentence, the main point changes with a new, and buried, topic sentence stating that the author intends to replace lime with sodium hydroxide. This is an important point, so it deserves its own paragraph; it eclipses the point about lime's long-standing reputation. It is, in fact, the author's main point, more so than anything to be said about lime.

When the second main point is identified, the paragraph's second half can be split off from its first half, to form a new paragraph. Once this is done, you can improve the topic sentence after reading through the subsequent detail sentences. In fact, the author says elsewhere that sodium hydroxide is not replacing lime for the sole reason that it allows the process to be shortened but because it offers several benefits, only one of which is that the process is shortened. (Information that appears later in a report often demands an improvement in paragraphs earlier in a report.) Thus, a better topic sentence looks like this:

- Sodium hydroxide is used in our process because of its three advantages: it is fast, cheap, and feasible.

The subsequent detail sentences look like this:

- With sodium hydroxide, the caustic treatment time is shortened from up to 60 days, typical of lime-based treatment, to 14 days, as detailed in our process description. Price is also a beneficial factor, as sodium hydroxide is significantly cheaper than lime, or calcium hydroxide. As a result of using sodium hydroxide, though, organic materials that are treated will swell, causing problems. For this reason, sodium sulfate is added to the caustic treatment process. This chemical serves to inhibit the swelling, mitigating one potential difficulty in using sodium hydroxide, rendering it feasible.

The lime-focused first half becomes its own paragraph, now with a few additional detail sentences added to provide enhanced support for the correctly prominent, and only, topic sentence:

- Traditionally, lime ($Ca[OH]_2$) is used in the caustic treatment. Lime, or calcium hydroxide, was first used in 1937, as developed by the Plains Chemical Company. By 1977, over 100 factories were using this treatment method domestically, with close to 280 systems running worldwide. Lime is relatively low cost but somewhat difficult to use.

An alternative topic sentence could be written. The choice depends on the author's desire to make two distinct paragraphs or one complex one with a structure of comparison and contrast. I discuss paragraph structures below, so we can return to this paragraph at that point.

(3) An easily remedied problem with topic sentences is the placement of such at a paragraph's end, where it arrives too late to help the reader, who has struggled through the "orphaned" detail sentences and probably determined the topic sentence on his own, or an incorrect one; either way, this makes the reader work too hard and possibly misconstrue the paragraph's point.

Consider this example:

- Fishing is not the only industry affected by ocean oil spills. The oil left floating on the ocean's surface eventually moves to coastal areas. The oil then mixes with sand, where it persists for years. The coasts have harbors, ports, important cities, businesses, and beaches. Beaches, in particular, are popular tourist destinations. Polluted beaches are not popular with tourists, and some are directly designated as off limits until confirmed clean and safe. This reduces tourism significantly. Thus, in addition to impacting the fishing industry, ocean oil spills are usually harmful to the tourist industry.

This final sentence would be an ideal topic sentence. If it were placed at the top, slightly written, the paragraph would be cogent and easy to read, as follows:

- Ocean oil spills are usually harmful to the tourist industry, so fishing is not the only industry affected by such spills. The oil left floating on the ocean's surface eventually moves to coastal areas. The oil then mixes with sand, where it persists for years. The coasts have harbors, ports, important cities, businesses, and beaches. Beaches, in particular, are popular tourist destinations. Beaches polluted by oil are not favored by tourists, and some are directly designated as off limits until confirmed clean and safe. This reduces tourism significantly.

You have noticed that all paragraphs follow the "general to particular" pattern, where the topic sentence is the general point being presented, and the subsequent sentences are the particular details to embellish that point. While all paragraphs are inherently in the "general to particular" pattern, most have an additional layer of organization as well. This secondary organizational scheme drives the content of the subsequent particular sentences. The available organizational schemes are the usual ones for communication of ideas, as follows:

- Persuasion
- Problem and solution
- Comparison and contrast
- Description
- Process
- Investigation
- Sequences
  - Chronology (X, then Y, then Z)
  - Cause and effect (Because X, Y)
  - Effect and cause (Y, because X)
  - Circumstance and event (during X, Y)
  - Action and purpose (X in order to Y)
- Listing
- Analogy
- Extended definition

Because these schemes demand particular organizational content, topic sentences allow readers to predict the organization of the paragraph's detail sentences. The phrasing of the topic sentence will convey the additional organizational scheme, and readers should be able to predict the subsequent organization of content from the topic sentence.

Let us use an example to explore the concept of organizational schemes. Imagine a report about LaGuardia Airport in New York City. Each and every paragraph may be about the airport (topic) in some sense, but the amount of potential information covered in the document is endless. Therefore, each paragraph must be distinguishable from all the others, and it must have one main idea only. As a result, each paragraph will have an obvious organizational scheme, which will be revealed in its topic sentence. Here are examples:

- LaGuardia Airport (LGA) is the most congested North American airport (persuasion).
- LGA has better restaurants than San Francisco International (comparison and contrast).
- LGA was first opened in 1954, then expanded in 1969, and fully remodeled in 1995 (chronology).

- LGA has five separate terminals (listing).
- LGA is nothing less than a small city with its own streets, shops, and government (analogy).

Those are some examples, and you can easily imagine other ones for these schemes: problem and solution, description, and investigation.

Recall the prior paragraph about caustic treatment and how it could easily be split into two paragraphs. Nonetheless, it could be kept as one large paragraph with a structure of comparison and contrast if this topic sentence were used:

- Traditionally, lime ($Ca[OH]_2$) is used in the caustic treatment, but sodium hydroxide is used in our process because of its three advantages.

With your understanding of organizational schemes, you are now prepared to tackle the final lesson on paragraphs: sentence cohesion. Sentences within a paragraph must be cohesive, a concept that is best understood as connected. This is covered in most English writing books. It is referred to with different terms: flow, readability, and focus, to name a few. Pages and pages of scholarly articles and textbooks have been devoted to this topic. I try to condense it here, to a few quickly explained ideas, beginning with a bad example.

Bad example:

Akmindy developed both kapdog and wigtwar. An experimental di-dektogrin olobrad is a definition of wigtwar. In contrast, pasko, rintog, and sefokation are the functions of kapdog. Electric propulsion of satellites has been advanced significantly by these innovations, wigtwar and kapdog.

This paragraph presents nonsense, but most technical writing appears to be nonsense to the rest of us, if it is not our specialty, and only one or two areas are truly our specialty. On top of the nonsense, moreover, it is poorly written. The sentences are not connected. Every sentence offers a new noun or phrase. The subjects of the sentences jump from this to that. Only at the end of each sentence are some familiar nouns found.

In contrast, the best method is to follow a twofold rule: (1) keep the subject of each sentence focused on the key nouns introduced in the topic sentence and (2) put new information at the end of sentences. It's that simple. That technique ensures that each sentence in the organizational scheme is overtly connected to the main idea introduced in the topic sentence and the organizational scheme. Here, the main idea is Akmindy as a developer. The organizational scheme is a description (of his developments). Without connections, the description is hard to find, as explained below.

The first sentence presents three nouns: *Akmindy*, *kapdog*, and *wigtwar*. For connections, the next sentence must tell us more about those three nouns (description). Instead, the next sentence gives the reader a new noun, *olobrad*, with two modifiers: *experimental* and *di-dektogrin*. Only at the end do we see *wigtwar*. Similarly, the third sentence begins with three new nouns: *pasko*, *rintog*, and *sefokation*. Only at the end do we see *kapdog*. Because the topic sentence did not introduce these new nouns and modifiers, they are off subject (unless placed at the end of sentences). The fourth and final sentence starts with yet another noun: *propulsion*, modified by both *electric* and *for satellites*. Another new noun is introduced as well: *innovations*.

Here is a version with connected sentences. You should find it easier to read. You might even learn something about kapdog and wigtwar and become capable yourself of explaining the ideas to someone else:

Improved example:

Akmindy developed both kapdog and wigtwar. Kapdog is used for pasko, rintog, and sefokation. Wigtwar, in contrast, is an experimental di-dektogrin olobrad. Kapdog and wigtwar are two innovations that have significantly advanced electric propulsion of satellites.

Having mastered nonsense, you are now ready for an example with real engineering and business issues:

Bad example:

Bell Textron Inc., should invest in tilt-rotor technology. With Boeing backing out of the helicopter industry, Bell is now in an even greater position to take full control of the tilt-rotor market. The short-distance commuter market is growing at a rate of 7% per year. A 40- to 50-passenger tilt-rotor plane could allow commercial airlines to offer transportation without the traffic congestion of major airports and the cost of huge runways.

Let us look at the connectedness, or lack thereof: First *Bell Textron* is introduced, as is the topic of *tilt-rotor technology*. We might not know this entity, tilt-rotor technology, and we have a right to learn more about it and the reasons Bell should invest in it. But the next sentence introduces new nouns, with one modifier: *Boeing* and *helicopter industry*. Somehow, tilt-rotor technology and helicopters are related, as are Bell and Boeing, but readers must forge these links for themselves. The author is making the reader do too much work. Here are the nouns and modifiers of the third sentence: a *40- to 50-passenger tilt-rotor plane*, *commercial airlines*, *transportation*, *congestion*, *airports*, and *runways*. All of these are new: nothing here is connected explicitly to prior nouns. In short, the sentences hop from noun to noun, topic to topic, without connection to the main idea from the topic sentence: invest in tilt-rotor technology. Moreover, the original primary nouns, *Bell, tilt-rotor technology*, and *helicopter industry* are nowhere to be found. In fact, the idea of helicopter has been replaced with plane because tilt-rotor technology can be incorporated into a plane, an important technological concept that is sloppily snuck into the paragraph.

Using the twofold guideline set forth above, you can rewrite this paragraph, with a persuasive organizational scheme, keeping the sentences focused on the one or two key subjects (nouns) and putting new information at the ends of sentences instead of at the beginnings:

Improved example:

Bell Textron Inc., should invest in tilt-rotor technology and produce a commercial non-military airplane with such technology. The investment would be wise if three conditions are met: (1) if the commercial market can tolerate another player, (2) if the market is growing annually, and (3) if the airplane places no new burden on existing infrastructure. These three conditions are indeed existent with respect to tilt-rotor technology. First the market has room for Bell's entrance because Boeing has recently backed out of the entire helicopter industry, which encompasses the tilt-rotor vehicles. Second, the commercial market for tilt-rotor vehicles, namely the short-distance commuter market, is growing at 7% annually. Third and lastly, a tilt-rotor vehicle places no new burden on existing infrastructure because it can land at airports using less space than would a conventional aircraft of similar capacity, for example, 40 to 50 passengers. In fact, some flights could depart and land at existing infrastructure facilities other than airports, thereby avoiding airport traffic and congestion entirely. Such existing facilities would be feasible because tilt-rotor vehicles do not require long runways. A facility need have only a small blacktop area to be sufficient.

Sentence connectedness depends, therefore, on careful selection, consistency, and placement of key nouns as subjects and predicate objects. Knowing this, you are in a position to study a related topic that is also a red alert in Appendix A, Section A.3: terminology troubles. Both switching terminology and using unclear synonyms disrupt the connectedness of a paragraph's sentences, as in this example:

Bad example:

We expect to expand our engineering research facility next year. The new rooms will house vibration and loading experiments. The equipment will be state-of-the-art, but the building has to be certified structurally sound, especially to withstand the severe vibrations and loads produced in the testing rigs. A small device was used to simulate the expected loads from the test equipment to be imparted within the expanded space.

Here, the author discusses *engineering research facility* and sets up the reader to learn about the expansion (using the description organizational scheme, or possibly chronology because of the presence of *next year*). Unfortunately, in the second sentence, readers see new nouns: *rooms* and *experiments*. In the third sentence, the author presents *equipment* and *building*, with an ending of *rigs*. Finally, the last sentence introduces yet another new noun: *device*. So, keeping subject focus and placing the new ideas at the end, following the twofold technique, the paragraph is improved as follows:

Improved example:

We expect to expand our engineering research facility next year. The expanded facility will boast two new testing rooms, which will house vibration and loading test rigs. These two rooms, in particular, must be certified structurally sound, especially to withstand the severe vibrations and loads produced in the test rigs. Those rigs will be state-of-the-art. To simulate the expected vibrations and loads from the test rigs, a small device was used.

Taking all of the preceding instructions into consideration, a final checklist for your paragraphs comprises these items:

Paragraph Checklist

- One main point only?
- Topic sentence conveys main point, in first sentence?
- Alternatively, narrowed topic sentence is second sentence when first sentence serves as transition from prior paragraph?
- Topic sentence suggests organizational scheme?
- Subject focus maintained at start of sentences?
- New information at ends of sentences?

2.2.2 Sentences

This section on sentences is a short one because I covered key aspects of the topic in the previous section on paragraphs. In addition, many of the red alerts in Appendix A address differences between good and bad sentence components.

Nonetheless, before we put sentences to bed entirely, let me cover a few, helpful points:

- Linking sentences with conjunctions
- Relating sentences with mapping words
- Strengthening sentences with variety

2.2.2.1 Linking Sentences with Conjunctions

Avoid using *and* as your principal, or "go to," link between ideas. You have many other options, and overuse of *and* leads to monotony. Try using a variety of conjunctions, and you will detect an immediate improvement in your sentences. Your prose will still be concise, but it will be elegant also.

Here is a typical ABERST example. Many readers would say this is "dry" or "technical." But it does not need to be. The problem is the monotonous use of *and.*

- We went to the machinery room and we watched a new laser-guided robotics assembly, and we learned that this assembly is improving productivity. This assembly is yielding higher rates of compliance, 3-month average of 92%, than prior assembly methods (maximum 87%), and the employees were trained to use it over the course of a mere 3 months. Training used to take 6 to 12 months for prior re-toolings, and this slow-down decreased productivity.

Creating a new version with some varied conjunctions makes the paragraph easier to read. It becomes peppy and more energizing, without becoming too long.

- We went to the machinery room, where we watched a new laser-guided robotics assembly that is improving productivity. This assembly is yielding higher rates of compliance than prior assembly methods (3-month average of 92% compared to maximum 87%), which is all the more impressive considering that employees were trained to use it in only 3 months. While past training periods often lasted 6 to 12 months, the quick training for this assembly avoided the decreased productivity usually associated with re-toolings.

So, instead of *and*, you can use the following variety of conjunctions:

- Coordinating conjunctions: but, or, nor, for, so, and yet
- Subordinating conjunctions: after, where, when, though, unless, if, because, while, since, and others like these
- Correlative conjunctions: either or, neither nor, and not only x but also y
- Sentence adverbs: however, nevertheless, indeed, thus, and therefore
- Relative pronouns: which, that, it, who, and whom

Notice that sentence adverbs are not the regular adverbs you are familiar with. Sentence adverbs are few in number, whereas the regular adverbs number into the thousands. Here is the difference: Regular adverbs modify verbs, while sentence adverbs modify whole sentences. Look below.

- Regular adverb (*quickly*)
  He quickly calibrated the transducer.
- Sentence adverb (*nevertheless*)
  The track shows signs of pitting. Nevertheless, it held up adequately.

One more point to reinforce regarding sentence adverbs: they are not coordinating conjunctions (and, but, or, nor, for, so, yet), so they cannot be used to link two independent clauses:

- We need a new audit test cell, and the cost is not a concern. **(correct)**
- We need a new audit test cell, however cost is a significant concern. **(incorrect)**

When you use a sentence adverb such as however, you must insert it into a single independent clause, like this:

- We need a new audit test cell. Cost, *however*, is a significant concern. **(correct)**

More discussion of this non-intuitive grammar hazard is provided in Appendix A, Subsection A.2.

2.2.2.2 Relating Sentences with Mapping Words

Sentences linked well are easy for readers to follow and understand. To put icing on the cake, you can add mapping words that provide additional clues to the relationships between sentences, so readers can be absolutely certain that they receive the intended meaning of your ideas. Mapping words serve to explain, not the ideas of your topic, but the literal relationships of the sentences to one another. They are words about your words, hence, often referred to as "metalanguage," which means "behind or beyond language." We talk about members of a family being related. We often ask, "How exactly are you related to each other?" This is the same with sentences in a paragraph: mapping words convey the relationships among the sentences. In addition, mapping words not only reveal the relationships between sentences, but they also smooth and promote transition from one sentence to the next. In that way, they are transitional terms, increasing the readability of a complex text. Here are some useful mapping words to sprinkle into your prose:

Mapping words:

- First, second, third . . .
- To begin, before, next, subsequently
- Finally, in closing, to conclude
- Similarly, likewise, also, in the same manner
- Moreover, furthermore, additionally
- However, but, in contrast, instead, on the other hand (with, on the one hand)
- In fact, indeed
- That is, in other words, in short
- For example, for instance, in particular, specifically
- Because, since, consequently, therefore, thus
- Fortunately, notably, interestingly
- Unfortunately, unexpectedly, surprisingly

The benefit of mapping words is that they help non-experts (nearly all your readers) to understand complex relationships about particular issues and things that may be entirely unfamiliar to them. Look at this example, revisiting Akmindy.

Paragraph with mapping words:

Akmindy developed both kapdog and wigtwar. *Furthermore*, the scientist played a pivotal role in early exploration of goftigler. *In particular*, his work with spaloshminzer was instrumental in creating the foognish layers that facilitate ologwik in the goftigler systems. *However*, he proposed a triple-haftabit yumgafoo for goftigler, *whereas* the system proved more capable when using 16-bitsache. *Nonetheless*, Akmindy's enthusiasm for experimentation inspired better innovations from fellow scientists. *Thus*, his colleagues often credit him with some early contributions to the far-afield study of arugafortem.

In this paragraph, I know nothing of the particular science at issue, or the specific mechanisms and concepts discussed. Nonetheless, I, as a novice in the field, understand the general point of the paragraph because of the mapping words, which all people understand. They provide a boost to universal comprehension of all readers. Use them.

2.2.2.3 Strengthening Sentences with Variety

As you have seen above, mixing up your conjunctions breaks up monotony of clone sentences linked with *and*. In that sense, it contributes to sentence variety. Furthermore, adding mapping words contributes to variety as well because sentences are henceforth distinguished from one another by the way they serve different purposes, for example, one sentence to reinforce, another to contrast, yet another to continue a sequence, and so on. So, you have some good tools already to help you with expanding sentence variety. One last message to impart to you is this: sentences can be short, long, complex, and compound. Try a bit of all. First, I define each; second, I provide examples.

- **Short**: the basic sentence with a subject and verb. It is an independent clause. (Kumar bikes.)
- **Long**: Expanding on a short sentence with a predicate, adjective, adverb, or prepositional phrase, or some combination of two or more.

(Studious Kumar bikes to his closest friend's house when he needs a break from studying.)

- **Complex**: Adding a subordinate clause to an independent clause. (While he lacks the endurance to take long bike rides, Kumar enjoys using his bike for short trips across campus.)
- **Compound**: Combination of two independent clauses linked with a conjunction. (Kumar is an avid bike rider, but he is the only member of his family who prefers biking to driving.)

Here is another set of examples:

- **Short**: The global positioning system (GPS) was developed by the U.S. military.
- **Long**: GPS functions by the integration of four necessary components: a constellation of orbiting satellites, a ground-based transmission receiver-decoder, a ground-based atomic clock, and a ground-based control station.
- **Complex**: While a GPS user must buy a receiver-decoder, either portable or built into another device, such as a phone or vehicle, the other elements of GPS are provided courtesy of the U.S. government at no charge.
- **Compound**: Years ago, Americans traveling around the country by car used printed road maps to navigate to their destinations, but today travelers prefer to use GPS to route their paths.

2.2.3 Tone

The tone of your prose in ABERST can either enhance or hinder your effort to convey your message and achieve your communication purpose. It is important, and it is not covered in typical English textbooks in the way that would be meaningful to professionals in ABERST. This small section attempts to rectify that omission. I discuss tone as something you can and must control deliberately and effectively, with two specific areas of focus. Those details are found below after one more comment, which introduces my twofold approach to tone in ABERST.

In some sense, tone is personal. Akin to style, tone is idiosyncratic, and should vary from writer to writer. Why not? Why should every single person strike the same tone and have the same style? That produces boring documents. That said, ABERST does not allow the personal creativity that one finds in other types of writing. Journalists and essayists can truly put their own stamp on their prose. If tone is individualistic and creative, it cannot very well be summarized into overt rules. It might even encompass breaking rules to achieve a unique voice. Fortunately, within ABERST, tone is not so much about creativity as it is about professionalism. Because ABERST documents are written and delivered within professional contexts, for real purposes, that very specific environment of professionalism dictates an appropriate tone. This can further be broken down into two distinct elements: Expertise and Precision. Each is discussed below.

2.2.3.1 Expertise

You likely would agree that a tone of expertise is critical at high levels of management within ABERST. Perhaps less obvious, however, is that a tone of expertise is needed at ALL levels, from the very bottom to the top. No purposeful document can be delivered within an organization without the author's having understood the issues, accomplished some meaningful work with regard to those issues, and reached some useful conclusion. These are simply basic conditions for a completed document. If these conditions are not met, the document should not be written. And this applies to each and every person, beginning with college-level students and moving upward. Once the conditions are met, the author can be said to have expertise. One common problem in ABERST documents is that, although the author does have expertise, the document has a perplexing tone. Quite simply, many writers either hide or inflate their expertise and produce documents that suffer in tone.

Given that college students are just starting to learn their field, and young professionals in first and second jobs have only a few years' experience, a humble attitude and a sense of non-expertise would not be all that surprising. Indeed, this is expected and appropriate in one's demeanor. You might say that no one is an expert until many years into her career. In some sense that is true when looking at the collected pool of knowledge and skills that is required to be a true expert in a *field*. But any single document does not cover a whole *field*. Rather, a document is focused on one little slice of a field, and that slice can—and must—

be within the expertise of the author. Here is where demeanor and documents take separate paths.

With documents, the need for a tone of expertise is even the more important when one considers that a great deal of documents in ABERST are authored by people early in their careers. As a person advances in one's career, she may spend more time overseeing and managing others than doing research, analysis, and writing. Therefore, from the first report in college to your last report while employed, you must convey a tone of expertise in the specific issues within that report. Two negative communication outcomes result from the absence of an appropriate measure of expertise.

On the one hand, when uncertain of one's own expertise and convinced that the readers possess greater knowledge, an author strikes a diffident tone and does the following: makes an effort to show off jargon, defines nothing (who needs definitions if we are all experts?), and leaves details out (assuming that the readers will be able to fill them in). On the other hand, when fully enamored with one's expertise (which possibly comes from either overcompensating for insecurity or gloating from success), an author strikes a superior tone and does the following: shows off jargon, defines nothing (who needs definitions if this is over your head?), and leaves details out (assuming that the readers will not be able to follow the main points let alone the particular ones). Both ends of the spectrum represent unbalanced and unprofessional tone.

I call the former the neophyte and the latter the snob. The neophyte relies on readers to be knowledgeable and make sense of information that is poorly or incompletely explained, while most of the readers are not in a position to do so. They were not there when the work was done, they have not read the background resources, and they may come from another field of training. Neophytes may not even be sure themselves of an equation, an assumption, or a theory relied on in the work, so they present equations and such in ambiguous ways, again hoping the readers will make the correct interpretations. The neophyte tone leaves holes in technical explanations and presents incomplete analyses. In sum, the true neophyte should not write a report. Until the neophyte becomes adequately in command of the material under discussion, no report should be produced. That said, often people who do know enough still write with a neophyte tone. That is the problem this book can address. The solution is simple: do not insert uncertainty, diffidence, and excess humility into your document.

Get your facts straight and write with expertise. You do not want to suggest you are not sure about your material when you really are. The lay reader or a reader who comes from a different field may not be sure of your meaning but an experienced reader (a few of them) will see through the buzz words, acronyms, and jargon and realize that the author does not know what he is talking about.

The snob, in contrast, may present a complete document, but the tone prohibits most readers from comprehending it and taking much away from their attempt to read it. The snob tone sends ideas over readers' heads. The snobby expert gives an impression that he wishes to exclude people from his specialty club by inundating them with undefined jargon and esoteric terminology. He makes no attempt to write complex material in a connected, clear manner. Instead, ideas are thrust at the readers in a demanding and machine-gun style that requires prior knowledge equal to or beyond that of the author. Remember: very few readers know more than the author about the author's work and document. Thus, the snob communicates to only a small minority of readers who are "in the club" already. Unfortunately, many readers who need to understand the snob's report, both inside and outside his organization, are not members of that small club.

The expertise tone, the happy middle between neophyte and snob, suggests the author is fully aware of her thorough knowledge of the subject, intimate association with the relevant work, and full comprehension of the meaning, logic, and reasoning integrated into the document's message. The expertise tone says, "I know this document inside and out, and I was part of every step of the process (or was briefed on any step I did not personally perform). Moreover, I am going to let you, my reader, follow along and become as familiar with the work and the findings as possible for someone who was not originally part of the process."

A true expert is neither a neophyte nor a snob. A true expert is helpful to readers. The expert wants readers to understand the whole document, from the main points at the start to the little support details on the last page. A true expert is aware that others (including almost all potential readers) know less than she does about the information in the report, although she writes in such a way that her work can be fully reviewed and analyzed by someone even more knowledgeable than herself, without gaps and ambiguities. That way, the readers (many) who are new to the material can follow, and those readers (few) who are already fully immersed can review, confirm, double-check, and evaluate.

Here are examples of three tones—neophyte, snob, and expert—taken from the Summary in the Overview of a report on power generation. You can read for yourself and see if you can reach any different conclusion from the one I have advanced: the expertise tone is the best: it is clear, informative, edifying, and thorough.

- **Neophyte**: Thorium can work as a fuel for a power plant. We simulated this using thorium in a Westinghouse AP100. But, in a pressurized water reactor, uranium is cheaper overall than thorium. The fuel costs associated with operating a power plant with thorium exceed five times the cost of a uranium fuel cycle. Ore cost for thorium is $90 per kilogram while standard uranium is $300 per kilogram. These economic benefits are negated by the reprocessing costs, which are about seven times greater, and the fuel processing costs are about nine times greater. Nonetheless, peaking factors for a thorium core were a maximum of 1.56 and a minimum of .91, with the Doppler coefficient and power defect values both being negative. The moderator coefficient was positive for the thorium but negative for the AP1000. The boron value bottomed at 16000 MW-d/Mt compared to 20000 MW-d/Mt for the 1000. Our data show that thorium has an adequate performance.
- **Snob**: Thorium and the standard AP1000 core were compared in the study. Results show that the moderator temperature coefficient was positive for the thorium, which is unexpected and worrisome but likely adjustable in future improvements. The boron value had a zero value at 16000 MW-d/Mt compared to the 20000 MW-d/Mt for the 1000. Also, the Doppler coefficient and power defect values were both negative, suggesting stability. Interestingly, peaking k efficiency values for a thorium core were a maximum of 1.56 and a minimum of .91, whereas the values were 1.33 and 0.71, for the baseline. As noted in our earlier work, isotope analysis confirms that no useful weapons-grade materials exist in the spent fuel. Economics indicate that

both processing and reprocessing of thorium fuel adds a premium of nine and seven, respectively, over the standard core. The strongest benefit to thorium is the reduced ore cost, at $90 per kilogram while standard uranium is $300 per kilogram.

- **Expert**: Thorium has promise as a novel fuel for a nuclear power plant to replace uranium. To evaluate the overall feasibility of thorium, we looked at both costs and performance. Costs encompass preprocessing and reprocessing expenses, as well as the cost of raw ore. Performance entails power generated and duration of that power production with a single fuel cycle. As a baseline, we compared a thorium-based power plant to a traditional uranium-based plant. (Primarily, we need only study the two different reactor cores, as all other plant components remain similar.) The performance comparison was facilitated by simulating the two alternative reactor cores in an advanced and up-to-date nuclear power plant design, the Westinghouse-built AP1000. In terms of costs, the thorium-based plant is estimated to be more expensive than a uranium-based one. While the raw ore cost for thorium is $90 per kilogram compared to uranium at $300 per kilogram, both processing and reprocessing costs are higher for thorium than for uranium. Specifically, processing costs are approximately nine times greater, and reprocessing costs are about seven times greater. In terms of performance, the thorium-based plant matches a uranium-based plant. Indeed, power generation and duration are slightly improved with thorium, compared to uranium. A measure critical to predicting the plant's power generation is its k factor, which indicates the ongoing efficacy of the nuclear fission reaction that is creating heat, and subsequently, steam and power. The k factor is ideally 1, but it fluctuates inevitably. Thus, the goal is to keep the average close to 1 with small peaks above and below. For thorium, the k factor fluctuated between a maximum of 1.56 and a minimum of .91, which is a positive result. The values swing more widely with uranium, up to 1.92 and down to .84. Another indicator of reactor stability, the Doppler coefficient value, was favorably negative. In addition, the power defect required for thorium is lower than that of uranium, which is another advantage because temperature is thereby easier to control vis a vis reactivity. Lastly, duration predictions suggest the thorium reactor will operate for 1.84 years on a single fuel cycle, compared to 1.79 for uranium.

2.2.3.2 Precision

In addition to writing with a tone of expertise, an excellent ABERST communicator must also write with precision. Precision in writing refers to expressing ideas clearly and completely, neither ambiguously nor imprecisely. The most frequent cause of ambiguity and imprecision is a passive voice. Have you been wondering when I would get to the passive vs. active debate? Well, here we are.

Passive voice is not the only source of imprecision, but it accounts for a good percentage. In addition to passive writing, imprecision derives from poor presentation of technical details, but this source is necessarily outside the scope of this book. Each author will have to take the requisite care to understand and present the technical issues of one's report in a precise manner. This goes beyond writing skills per se. So, I return to the aspect of precision that can be obtained through managing your use of passive voice. When you do so, you will enhance your reports with precision in tone.

Everyone has something to say about passive. Talk about "beating a dead horse." Even one recent issue of the *The Economist*, an eminent international newspaper focused on politics, business, and society, presented an article on this topic (July 2016). Somehow, however, writers remain confused by it. As noted in *The Economist*, many critics of the passive cannot actually identify it, and still others refer to it incorrectly as a tense issue, which it is not. It is an issue of tone, a form of expression. And it is an important one for you to understand to enhance your professionalism.

At the outset I will state my position on the active vs. passive debate: active voice is usually best, without excluding passive when appropriate. In other words, the passive voice is not forbidden but it is also not usually preferred. Passive is acceptable in certain contexts, but you will want to use active in most contexts for the simple reason that telling readers who did the actions you are discussing is very important. Readers need to know the people behind the work, decisions, proposals, ideas, mistakes, innovations, statements, and so on. Hiding the doers of deeds may be one or more of the following: misleading, confusing, vague, dishonest, and manipulative. At the very least, it is imprecise.

Some occasions when passive is acceptable are discussed below, which should reinforce my support of passive's use; likewise, I provide contrasting situations that demand active voice, so you can see its benefits.

First, if you do not know who or what (subject) completed the action (verb), you can use passive.

- **Passive example:** Graffiti was painted on the wall last night.

This use of passive emphasizes the action (painting graffiti) and its object (wall), but not its source (unknown painter).

In contrast, if you use passive when you know (or should know) the person associated with the action/belief, you are unacceptably vague. Vague text can be misleading.

- **Example:** It is recommended to use a three-prong mount for the sensor.

Readers may well ask, "Is this the writer's recommendation or another person's opinion?" Better to say expressly, "The chief engineer recommended a three-prong mount, but we opted for a two-prong mount to save money."

- **Another example:** It is recommended to include end-user software usage when calculating return on investment.

I would write in the margin, "Recommended by whom—a government agency, company management, or the authors?" The doer in this example is the midwest account executive, but the sentence does not express that key point. Similarly, "Switching from Microstrategy to Tableau for data visualizations has been deemed worth pursuing." Here, the person(s) recommending or deeming must be identified, or an important fact is left unstated. (It should be identified as the business intelligence team.) Additional incorrectly passive and corrected versions are shown below:

- **Incorrect:** Four *obeya* rooms were proposed for installation.
- **Correct:** Team Omega recommended installing four *obeya* rooms.

Overall, active voice is usually required with words such as *expect* (other examples include *estimate*, *believe*, and *recommend*) because another person might expect some different outcome. Expectations, estimates, recommendations, beliefs, findings, and conclusions vary with people. The same data may suggest a different result to another person. Without specifying who precisely expects X or estimates Y, the author is presenting an ambiguity. People look at employment data and predict upcoming economic trends differently from each other.

Second, another situation where passive is acceptable arises when, even if you do know the person who acts, the person (doing the action) is identified at the beginning in a long series of actions. In that case, some of the final actions can be described in passive, as in this example:

- **Example:** We set up and operated the wind tunnel to acquire lift and drag data. We placed the pitot probe in the clamp. The entire traverse stage was moved laterally in 2-inch increments. The tunnel speed was increased to 60 miles per hour in four steps.

In the latter two sentences, the subject doing the action, *we*, need not be emphasized repeatedly. The implication is clear with the passive voice.

Looking at the second sentence, we can compare a difference in emphasis with three different versions:

Version 1: We placed the pitot probe in the clamp.
Version 2: A clamp held the pitot probe.
Version 3: The pitot probe was held in place.

Either versions 1 or 2 may be correct, but version 3 is problematic. Voice determines the emphasis. In version 1, the team setting up the equipment is emphasized, followed by the pitot probe, which may be the two most important items for the authors. Key sentence components are *We placed* and *the pitot probe*. In version 2, however, the equipment is emphasized with the clamp taking center stage: a *clamp* holds *the pitot probe*. In version 3, the most passive of all, only the *pitot probe* is emphasized, as it is held in place by an unknown person or object. This version leaves out useful information, so it is imprecise.

An additional passive-voice version of this sentence would also be acceptable because it follows the active-voice topic sentence for the whole paragraph, which established the subject as *we* for the actions to follow:

- **Acceptable passive version:** The pitot probe was placed in the clamp.

Third, passive construction may be used when the terms are logically irreversible.

- **Example:** "The tools were sterilized by the assistant." This is passive. The active version is this: "The assistant sterilized the tools." But both say the same thing, and confusion is not likely to occur. You can logically say neither (1) "The tools sterilized the assistant" nor (2) "The assistant was sterilized by the tools."

In technical writing, however, readers may not always be capable of understanding when terms are logically non-reversible. In such instances, a passive construction may produce misinterpretation. Let's look at this example:

- **Troublesome passive voice:** The modem was reset with high-frequency, super-LTE signals.

Passive voice conveyed by "was reset" obscures the doer of the deed. Perhaps the sentence can be reversed, as follows:

- High-frequency, super-LTE signals reset the modem.

Maybe it cannot be reversed like that. Experts would know one way or another, but most readers might be befuddled. Who or what reset the modem? Are the LTE signals the reset method or the end result? An active voice makes it clear, although passive is fine when all readers can obviously see that terms are logically irreversible, a situation that may not be common in ABERST.

Perhaps we can add a human subject, but this is only a guess:

- **Active version:** A technician reset the modem with high-frequency, super-LTE signals.
- **Another complex example:** The piezo-electric actuator based vertical deflecting beam for force generation was activated following the applied electric field of the voltage generator.

Does this sentence discuss two events occurring in sequence or causation, with one event producing the other? The problem stems from "was activated." If causation is the case, an active construction is required for precision, as follows:

- **Active Causation**: An electric field from the voltage generator activates a desired response in the vertical deflecting beam, which uses a piezo-electric actuator and generates an intended, useful force.
- **Active Sequence:** Initially a voltage generator applies an electric field,

then a vertical deflecting beam, which incorporates a piezo-electric actuator, is activated to generate a force.

- **Here is another example:** From published information it is expected that 4140 steel will maintain its hardness throughout the entire test specimen.

This leaves out the person who expects, but it does present one potential subject, published information, such that the sentence can be rewritten into active voice:

- Published information indicates that 4140 steel will maintain its hardness throughout the entire test specimen.

To rewrite the sentence in an active voice with someone expecting the hardness, an editor needs more information. Readers are unclear as to who expects. We can guess:

- **Active (by guessing):** From our review of published information, we expect that 4140 steel will maintain its hardness throughout the entire specimen.

Fourth, passive is permissible when it helps connect sentences in a paragraph by maintaining subject focus, as discussed in Section 2.2.1 above. Recall, that subjects of sentences should be the key nouns introduced in the paragraph topic sentences or prior sentences.

Here is an active-voice sentence.

- **Active but disconnected:** We next cool the stream leaving the distillation column, stream 7, by sending it through heat exchanger 4.

Although using the first-person pronoun, which is an active voice approach, is often beneficial, it can lead to a problem where the true sentence subject, and focus of your exposition, is obscured by the active presence of the author.

- **Better:** Stream 7 is cooled in heat exchanger 4.

The authors (*we*) are not really doing any of the physical cooling, merely designing the facility. The authors desire the cooling, but equipment explicitly achieves that task. Thus, removing the human action phrased in active voice (albeit often a useful style) is best in such instances, where passive voice emphasizes the true subject.

You might be tempted to rewrite this into "active": Heat exchanger 4 cools stream 7. But, this might disrupt sentence flow, when the readers have already been introduced to stream 7.

Here is the paragraph this sentence is taken from, and you can see the benefit of passive, which does not detract from precision and does help maintain sentence connections.

- **Good example with passive:** Five of the process streams are integrated for energy efficiency: streams 2, 7, 9, 11, and 15. Stream 2 is heated by hot effluent stream 15, which needs to be cooled down before entering the distillation column. Stream 7 is cooled by heat exchanger 4. Stream 9, the reactor product, heats stream 11, which is a cold, input stream of methyl chloride.

In this paragraph, a mix of active and passive construction is used, yet it is precise, connected, cohesive, and well organized as a listing paragraph in support of the single main point about streams and energy efficiency.

Lastly, in addition to suggesting that active voice be used more than passive, I want to end with a strong warning against a particular form of passivity that is rampant yet ineffective: starting a sentence with *it is* and *there is*. In such sentences, the doer of the deed is almost invariably unstated. Thus, such sentence construction is a type of passive and, more importantly, it is just plain wordy and clunky, with absence of a strong, specific noun in the subject position. These can be called "passive, wordy, non-subject constructions."

You can, of course, use both *it* and *there* when these have specific, tangible antecedents.

- **Good:** The flange is cracked. It needs to be replaced.
- **Bad:** It is certain that the flange will need to be replaced because of a crack.
- **Good:** Most economists expect that employment rates will rise when interest rates drop.
- **Bad:** It is to be expected that employment rates will rise when interest rates drop.
- **Another bad example:** There is no standard that specifies what should be done if the vending machine does not pass Test A.

- **Improved:** No standard specifies a required subsequent action if the vending machine does not pass Test A.

Importantly, preceding *it is* with a subordinate clause or introductory phrase does not fix the passive problem, as shown below.

- **Bad:** At the new stadium, it is exciting.

What exactly is exciting at the new stadium? It could be many things. To detect such passive problems, this diagnostic technique can be used: Ignoring introductory clauses and phrases, see how much a reader learns from reading only the subject and verb of a sentence. Ideally, they learn a lot.

- **Example:** *Around the globe, there are smart people reading this book.*

From the subject (*there*) and verb (*are*), you know nothing about the sentence, or what is occuring around the globe. Just "there are."

Here is a concise, active form:

- **Best:** *Around the globe, smart people are reading this book.*

From the subject (*people*) and verb (*read*), you know the sentence is about people reading something. In addition, one adjective, *smart*, enhances the subject right from the start. I submit that the subject and verb alone reveal much more than "there are" reveal, so you can read and comprehend more quickly than with the empty construction.

Sometimes, readers may be initially sure the pronoun *it* has a specific antecedent, but upon further analysis, the sentence turns out to be, yet again, merely clunky passivity, as in the one below.

- **Bad example:** The combined injection-compression molding process was chosen. With the process chosen, it is necessary to develop machine specifications to produce optimal molded parts.

Here, *it* does not refer to the chosen molding process. If it did, that would be great, as in the following:

- **Good example:** The combined injection-compression molding process was chosen. It is a process that requires exact machine specifications.

But, instead, *it* in the bad example is part of a passive construction to say "it is necessary that…" So, this can be rewritten in a direct way: "With the process chosen, machine specifications must be developed to produce optimal molded parts." Who will develop them? If the authors know this, they can, again, rewrite into a full active voice: "With the process chosen, we (or, our structures team, or management) must develop machine specifications to produce optimal molded parts."

Similarly, the common, non-subject, passive expression, "it is necessary that…" is an indirect way of saying item A necessitates item B, as in this example: "Entering the pool (item A) necessitates showering first (item B)." Or, as typically said, "It is necessary to shower before entering the pool."

This can be simplified into the imperative sentence form: "Please shower before entering the pool." Or, in an active construction with a subject and verb, "All swimmers must shower before entering the pool." The terms *please shower* and *swimmers* are both more substantive than *it*, so they represent strong nouns or verbs with which to begin your sentence.

- **Another bad example:** It is estimated that annual catalyst costs will be just under $3 million.

This is clunky and wordy; when I analyze the two most important parts of the sentence, subject and verb, I gather only this information: *it* and *is estimated.* That is broadly meaningless. The sentence is also passive because it does not indicate who is doing the estimation. It must be rewritten to have a tone that combines expertise and precision.

- **Improved:** Annual catalyst costs are estimated at just under $3 million. (Passive, yes, but possibly OK, if sentence connection is enhanced.)
- **Another improvement:** We estimate annual catalyst costs at just under $3 million. (Here, the first-person subject, *we*, is the primary noun, with the *catalyst costs* one step below as the verb's direct object.)

In a further variation, the subject is shifted from the verb, *estimate*, to a modified noun—*costs*—which is a key noun in the context: "Estimated annual catalyst costs are just under $3 million." Here, the subject is the *estimated cost*, neither *we* (the authors), nor *costs* generally.

Moving on, you can keep alert to many of these overused passive, non-subject wordy constructions. Here are some to watch for:

Bad passive or empty expressions:

It was noted that, it is felt that, it is evident that, it is our opinion that, it is expected that, it is recommended that, it is concluded that, it is believed that, it is interesting that, it should be pointed out that, it was found that, there are x reasons, there are y parts, there are numerous people, there are two approaches, there are five ways, there are fears that, there are worries that, and so on.

2.2.4 Words

To improve your composition skills, I have started with the big-picture items of paragraphs, sentences, and tone. Mastery at these levels is invaluable. Here, I begin to address some of the smaller building blocks of good writing, starting with words and ending with numbers. The following six sections fill in the remaining details to complement the big picture.

Vocabulary is essential for expressing yourself so your reader understands your meaning the first time she reads each sentence. Word selection and usage are part of a writing component known as "diction." Diction reveals a writer's mastery of a topic and education level. The stronger your diction, the better your impression on readers. With diction, you should push yourself to aim high.

You will have to select your words. No one can do it for you. Software can give you a synonym as a substitute but not the original word. You must know the terminology and vocabulary connected to your topic and purpose for writing. You can receive some assistance through red alerts of common difficulties in diction, such as found in Appendix A. In particular, look at Sections A.1-A.3. In addition, you can improve your diction by reading well-written (published) material in your field. You want to emulate the best and brightest writers that you read.

Lastly, you can review the six lessons I offer here: slang, jargon, humor, articles, demonstrative pronouns, and personal pronouns.

2.2.4.1 Slang

Truly informal language and words from a particular clique, neighborhood, or demographic should not be used. Idioms and colloquialisms are trouble also. Even business-speak slang is silly in an ABERST document, as in this example:

- If we can interface on this, following a quick touch-base with our multi-disciplinary collaborators in Arizona, who expressed a desire to dialogue with us, we can effect the project positively.

2.2.4.2 Jargon

Similarly, you do not need to use haughty and ostentatious vocabulary. Technical documents are not written to prove you had a perfect score on the SAT or GRE; rather, you are trying to communicate complex information to readers who might be unfamiliar with much of the terminology, especially if you are working on novel or cutting-edge ideas, science, or technology. Thus, use a common word rather than a specialized one. Use small ones, rather than large ones. On occasion, when a large word is just perfect, you can sprinkle it into your prose for a little bit of spice. Although the esteemed William Safire has said, "Never use a long word, when a diminutive one will do," you can include a 50-cent word when you help readers to understand it, through direct definition or context.

You should have an idea of which words are unusual in your field and not commonly known and which are typical. When you use a word that you suspect is less familiar to people outside your work group, you can and should introduce it with a definition, so that you show you are an expert, neither a snob nor a neophyte.

2.2.4.3 Humor

In the world of essays, journalism, and fiction, a writer's approach to diction defines her style and appeal. This is a big part of creating their unique voices and grabbing attention. In writing for ABERST, you simply do not have the creative license that other writers have. This is purpose-driven, non-fiction writing, so the range of diction is narrow. In a sense, you have to be professional, confident, certain, clear, sincere, and no-nonsense. You have infrequent opportunity to deviate from that. Thus, sarcasm, irony, innuendo, and, generally speaking, humor must be removed. To keep things fresh and lively in the workplace, you can feel free to add these touches to your spoken interpersonal communication, as you deem appropriate. Indeed, this restriction applies only to ABERST documents. Yet it also applies to the informal end of ABERST documents, namely email and voice mail. Many years ago, people felt that work emails and voice mails could be casual, informal, jocular, sarcastic, and witty because they were internal and not intended for any distribution beyond the immediate recipient. This limited distribution is no longer true. In fact, all emails and voice mail messages are discoverable by the opposing party if you are ever involved in litigation. So the best advice is to be professional, accurate, courteous, and mature in all ABERST communications, including those while in college that involve your work AND fellow students, teaching assistants, and instructors. Avoid all unprofessional utterances.

In other words, take care to completely separate ABERST-related communication from casual emails, texts, and voice mails with friends and family, as well as other social media communication. The personal sphere and the occupational sphere are two different ballgames. In the personal sphere, you can be as goofy as you would like, as long as you are aware of any potential impact that something found on the Internet could have on your career advancement. Other than that, you have freedom when disconnected from your college-based and professional communication, where you must be "on the job" at all times.

2.2.4.4 Articles

Having been shown the big picture of technical communication, you are prepared to shift to some of the smallest words in the English language, namely definite and indefinite articles. Although very small words, they produce immense frustration for both authors and readers alike. This subsection aims to clarify articles.

These are the articles:

- Definite article: *the*
- Indefinite articles: *a* and *an*

They are found in sentences directly preceding nouns, such as *apple*, *ball*, and *car*, for example, *an apple*, *the ball*, and *a car*. They function as adjectives to modify those nouns that follow, but for identification only, not for description.

Compare *fast car* to *the car*: *Fast* tells us descriptively that the car has a powerful engine; *the* tells readers that the car is simply the one you have already identified or that is easily understood to be a specific car.

Although the articles are only three in number (*the*, *a*, and *an*), the perennial question is this: Which article is correct to use in any given context? The following is the answer, via distinct rules for use between the one definite article and the two indefinite articles.

Rules for use of *the*:

- If a reader *knows* the noun/subject from some previous text; that is, if it has been identified earlier
- If *only one* such instance of the noun/subject exists in world (specific member of class defined by noun)
- If the noun is plural and meets at least one of the aforementioned conditions

Rules for use of *a* or *an*:

- If the noun is singular and one of the conditions below also is true:
  - If the noun has not been previously identified (unknown to reader)
  - If the noun is general in nature (whole class or category) of which there may be more than one

Rules for distinguishing between *a* and *an*:

- Look at the sound of the syllable starting the word immediately following the article; if it is a vowel, use *an*, if a consonant, use *a*
- If the initial syllable is an h sound, you can use either

Practice:

- *A* cathode; *the* aforementioned cathode
- *An* electron; *the* fourth outer electron
- I used *a* scanning electron microscope.
- *The* Hubble Telescope took the photos.
- *A* telescope was used to view Pluto. Images revealed *a* deep canyon, and *the* telescope peered to the very bottom of *the* canyon.

With a little context, you can study the following example. The context is that my wife writes an email to me while I am at my office and she is at home:

- "Hello, buddy. I purchased the umbrella and the card for your mom, at the store."

Her using *the* implies that, in the world, a single, specific umbrella and card exist for my mom, and I know the exact items to which she refers. Furthermore, my wife implies that I know exactly the single, specific store to which she refers. If, however, I know nothing about these items, and my wife is aware of this, she must re-write her sentence as follows, fixing the articles:

- "Hello, buddy. I purchased an umbrella and a card for your mom, at a store."

Upon receiving this email, I can choose to ask her some follow-up questions, but without more information, I cannot know exactly which umbrella, card, or store.

2.2.4.5 Demonstrative Pronouns

Similar to articles, demonstrative pronouns are little words that can cause big headaches for writers. The four primary demonstrative pronouns are *this*, *that*, *these*, and *those*. I suggest you add *it* to this list also, as it can perform similarly as the demonstrative pronouns in a sentence. In brief, these words are used to replace a noun, or object, that has previously been mentioned or is readily known.

They are good words to use, as they break up the monotony of repeating the same word or phrase, so they make your writing concise. But, these words often cause ambiguity because various prior nouns might be the referent. Look at this example:

- **Problem:** The hazardous materials management licensing procedure for our liquids storage facility was revised last month. This needs our immediate attention.

When the second sentence begins with *this*, the reader may wonder whether the facility or the procedure needs attention. Proper composition demands that the referent be certain. One way to do this, if you have difficulty re-writing the preceding sentence, is to add a descriptive noun after the demonstrative pronoun, as in the corrected version below.

- **Corrected:** The hazardous materials management licensing procedure for our liquids storage facility was revised last month. This procedure needs our immediate attention.

Another solution is to rewrite a portion of the sentence with a problematic demonstrative pronoun, as in the following sample pair:

- **Problem:** The activist investor intends to increase its stake in the company and pressure the board to make an offer to acquire its primary competitor with leveraged funds. That is a move that will prompt scrutiny from the regulatory agencies.
- **Corrected:** The activist investor intends to increase its stake in the company and pressure the board to make an offer to acquire its primary competitor with leveraged funds. That attempted acquisition will prompt scrutiny from the regulatory agencies.

2.2.4.6 Personal Pronouns

Sometime in the past, a need for dispassion seeped into technical writing. Since then, under-use and even complete avoidance of the personal pronouns, especially the first-person pronouns, has occurred throughout ABERST, possibly due to the need to sound objective, neither personal nor subjective. This tone somehow conveys a better sense of accuracy, precision, fairness, expertise, and so on.

To counteract this tradition, most of the good textbooks on technical writing have approved of the use of active writing in the first person, and general use of personal pronouns. In some sense, the pendulum has swung in the other direction, and now an equally common problem is the over-use of first-person pronouns. The best advice I can give is to avoid both under- and over-use of the first person (associated hazards listed in Appendix A, Section A.4.) Used in the right dosage, first-person pronouns will improve your ABERST reports. This discussion complements the similar ideas set forth when discussing the active voice, in Section 2.2.3. Moreover, the other personal pronouns can be used as needed, too.

As conveyed by the active-voice tone, your reports reflect actions, judgments, and analyses made by YOU, your team members, colleagues, business associates, and peers in other organizations. Therefore, you must explain that these actions, judgments, and analyses derive from your work, mind, efforts, and so on, or identify that they emanate from some other person(s). Work is not done by itself, by invisible creatures; nor do conclusions and recommendations fall from the sky. Indicating the person or persons making things happen is critical in ABERST reports. Thus, using names of people in reports is perfectly fine. Similarly, using "I" and "we" is fine, too. All the other pronouns are necessary in various situations as well, both subjective, objective, and possessive: I, you, he, she, we, they; me, you, him, her, us, them; my, your, his, her, our, their; mine, yours, his, hers, ours, theirs.

Here is an example from a typical report conveying some issues pertaining to engineering and business, specifically geotechnical analysis. It effectively incorporates first- and second-person pronouns, singular and plural, as appropriate.

- **Clear sample with personal pronouns:** I went to the drilling site to resolve an uncertainty with the water table location that our excavation crew had identified. To resolve this uncertainty, I placed four moisture sensors in the trench, one at each corner. The crew supervisor, D.G. Smith, joined me, and we discussed the various strata uncovered to date, including sandy loam and Midwest clay, as well as the seeping of water in the trench. I determined the water table to be no higher than 15 feet below surface. I also noticed a faint odor, possibly indicating the presence of volatile organic compounds. I suggested sending some additional soil samples to a state-certified laboratory, and he concurred.

Try to imagine the above passage written without the personal pronouns. It would be confusing. Below is just such a rewritten, weak version, so you can see the contrast between the two versions, and the ambiguities that ensue in the weak version.

- **Confusing sample without personal pronouns**: Water table uncertainty at the drilling site was investigated. In the trench were placed four moisture sensors, one at each corner. A meeting was held to discuss the various strata uncovered to date, including sandy loam and Midwest clay, as well as the seeping of water in the trench. The water table was determined to be no higher than 15 feet below surface. Someone thought he detected a faint odor, possibly indicating the presence of volatile organic compounds. One suggestion is to send some samples to a state-certified laboratory, and concurrence was reached.

In some types of ABERST writing, notably scientific journal articles, the particular players involved are not critical to the readers because the article is not produced to effect change within an organization and serve as a record of decisions and actions, enabling personal follow-up with the people involved. Therefore, a well-established structure for publishing an article on academic, business, engineering, legal, medical, scientific, or technological research exists such that the work is presented without any personal information beyond citing references to other, related publications and mentioning that the author and her team did the work. Personal pronouns are used sparingly.

The journal article format (more in Chapter 5.1) requires an account of prior work in the field (with citations to the appropriate researchers), then a presentation of the missing knowledge the author intended to reveal. The article next conveys the research design and data acquisition plan, followed by a succinct review of the data and results obtained. All this preliminary information lays the foundation for the ultimate and most important section, the discussion of the results. This is the place where the author analyzes the results, identifies the principal findings, and presents the conclusions that can reasonably be drawn. Finally, future work that is warranted given the present results can be outlined to end the article, before giving a final recap of the article in a concluding paragraph.

With such scientific articles, an absence of a personal narrative is often the goal, so the bare essentials of the research can be concisely presented. This is not to say that the persons responsible for the research and subsequent article are unimportant. On the contrary, this is a genre that necessitates a good deal of personal ambition and self-acknowledgement. Ego may be lurking although the personal pronouns are not to be found. This is an irony. Indeed, in the one particular genre where personal pronouns are used sparingly, the people behind the document are very much invested and "present" in their work and writing. They must publish or perish.

I can offer another example to demonstrate that the best writing is not exclusively dependent on using the first-person pronouns. Rather, they need to be used only when the writing benefits from identifying the specific persons playing a part in the work.

- **Good example:** The fuel injector has been clogging. I was asked to conduct chemical assays on the various fuels used with these injectors.

In this example, the use of *I* is essential in the second sentence but not in the first. You do not need to insert *I* unnecessarily such as in this wordy re-write:

- **Pronoun over-use:** I was told that the fuel injector I am responsible for has been clogging.

Overuse of *I* and *we* has perhaps led some people to broadly demand that these words be removed from technical writing, so people remove them from the wrong sentences, as in this poor revision:

- **Pronoun under-use:** The fuel injector has been clogging. Chemical assays on various fuels must be done.

Was the writer asked to do the assays, or is the writer directing the reader to do them? Will these be done in the future, or have they been done, as the basis of the report the reader has in front of her. The knee-jerk removal of first-person personal pronouns (I, We, Me, Us, My, Mine, Our, Ours) leads to ambiguity, confusion, and inappropriately mysterious reports about the work people do without clear references to those people.

The moral of this story is the following: personal pronouns, including first-person pronouns, are your friends. Use them wisely in your ABERST writing.

2.2.5 Grammar

Grammar is the overall rulebook for using words correctly in sentences, and syntax refers to the ordering of words in sentences. Whole books are devoted to helping writers learn these two areas of English.

Often, writers (and poets, singers, rappers, and others) shake up the rules of grammar and syntax, for creative and dramatic effect. Broken rules might convey an actual vernacular of how some people really speak, or the "errors" might emphasize emotion or novelty in a special way. Nonetheless, to break a rule, a writer must know the rule. (Most creative writers, we suppose, do in fact know the correct rules they are breaking.) In fiction, poetry, and music, rules are up for grabs, but not so in non-fiction. The plain truth is that writing for ABERST requires the rules of grammar and syntax to be followed almost without exception. Therefore, the following two subsections cover some hot spots in grammar and syntax (2.2.6), so you can learn a few important rules or get a refresher.

The key grammar hot spots in technical writing are the following:

- Single-verb-tense prose
- Shifting-verb-tense prose
- Subject-verb number agreement

People ask me which tense to use in technical writing. Here is the answer in plain English: All tenses. You must use whichever tense is appropriate, and you may need to use many. You cannot try to write in ABERST in just one tense. Such a constricted effort would be like trying to paint with just one color. Tenses are many in English for the simple reason that we need a variety of tenses to convey complex information, discussing actions and events from the past, the present, and the future.

With this as the rule, the violations of the rule fall into two prominent categories: attempts to use a single verb tense and inappropriate shifting between different verb tenses.

2.2.5.1 Single-Verb-Tense Prose

Here is a strong and clear opening paragraph from a report discussing a study of an aircraft device that could improve flight performance. The authors spent

4 months building and testing prototypes of the device, then analyzing the data to determine the device's effectiveness.

> The sponsor of this project is Blast Aerospace, a second-tier aircraft parts manufacturer. They need a design method to improve the flight characteristics of a small drone (unmanned aerial vehicle, UAV) they are developing. The drone tends to lose lift and become unstable when it points its nose upward at an angle higher than 20 degrees, which is necessary in some instances to climb to a new altitude or avoid obstacles. Our team was asked to solve this problem without increasing the aircraft's drag, which would hinder performance. Blast has asked that we minimize manufacturing complexity and cost of the solution. After researching several possible solutions, we decided to pursue the most promising idea, vortex generators. Research shows that a vortex generator (VG), a small, fixed device attached to the top of a wing, serves to keep airflow attached to the wing even when the aircraft is pointed at an angle above 20 degrees. With the airflow attached, the aircraft is less likely to stall and experience the problems Blast is confronting. If VGs show promise as a solution, Blast will add them to the next version of its drone. Subsequently, we researched a variety of vortex generator types (shapes and sizes), along with manufacturing issues. Importantly, we tested a model aircraft with and without vortex generators, to collect data that would support a decision regarding improved stability at high angles of flight. In our report, we present our findings from this research.

Imagine we took this paragraph, which contains at least three tenses, and tried to write using only past tense. Here are the first three sentences:

> The sponsor of this project *was* Blast Aerospace, a second-tier aircraft parts manufacturer. They needed a design method to improve the flight characteristics of a small drone (unmanned aerial vehicle, UAV) they *developed*. The drone *tended* to lose lift and become unstable when it *pointed* its nose upward at an angle higher than 20 degrees, which *was* necessary in some instances to climb to a new altitude or avoid obstacles.

The use of past tense distorts the facts. The first sentence suggests that Blast Aerospace was the sponsor but that either the project is over or Blast is no longer the sponsor. The second sentence incorrectly implies that Blast's need has passed and the drone development is finished. The third sentence suggests the problem of instability is fully resolved, but it is not: Even with the effort presented in the report, the drone is still in development and the problem is not fully resolved.

Let's look at the final three sentences and imagine them rewritten in future tense:

- Subsequently, we *will research* a variety of vortex generator types (shapes and sizes), along with manufacturing issues. Importantly, we *will test* a model aircraft with and without vortex generators, to collect data that would support a decision regarding improved stability at high angles of flight. In our report, we *will present* our findings from this research.

In all of these future-based sentences, the authors present the work as not yet started; rather, it is planned for some date in the future. Similarly, their report will present findings when they get around to doing the work and developing their findings. But this is all misleading. The work has been done, the findings have been found, and the report has (now) the findings they want the recipient to review. Future tense is wrong. This mistake arises for various reasons, one of which might be that the authors are re-using text from an earlier document on the project where work and reports were merely *hypothetical*, something not yet real, but intended to appear in the future.

We do not need to be fancy about this topic. The basic approach boils down to using the tense that is obviously the right one, as follows:

- When talking about work already completed, past tense is best.
- When talking about a problem that needs to be solved, present tense is best.
- When talking about work to be done at some point up ahead, future tense is best.

Let's look at two of the middle sentences rewritten in past perfect progressive:

- Research *had shown* that a vortex generator (VG), a small, fixed device attached to the top of a wing, *had served* to keep airflow attached to the wing even when the aircraft *had been* pointed at an angle above 20 degrees. With the airflow attached, the aircraft *had been* less likely to stall and experience the problems Blast *had been* confronting.

This emphasizes the state of the VG at the exact time of the past research. During the research duration, the VG was as described. But, as most research indicates, the device in question continues to exist and possess characteristics long after the research stops. This is true for Blast's problem; when they discovered it, they confronted it. But, they are still confronting the problem today, or the authors would not have been asked to help. Present tense is best.

2.2.5.2 Shifting-Verb-Tense Prose

Although different statements demand different verb tenses, sometimes a shift in tense is unnecessary and awkward. It may be true that some detail has a past component, while another detail has a future or present component. Nonetheless, each shift in verb tense requires a strong reason; when the meaning would be equally clear without a shift, use of a single verb tense should be continued until a tense change is absolutely necessary.

Here is a problematic example that involves shifting tenses when present suffices for the whole passage:

- The optical table used in the testing laboratory *is* a stainless steel breadboard with ¼-inch-diameter holes placed at a uniform distance of 1 inch from each other along the breadboard's length and width. It *is* like a standard utility pegboard but sturdier. This breadboard *gave* us versatility to move the testing setup to convenient locations, relative to the force gauges and clamps, depending on the specimen shape. It *had been custom designed* for this project.

Rewritten without shifting tenses, the passage is improved:

- The optical table used in the testing laboratory *is* a stainless steel breadboard with ¼-inch-diameter holes placed at a uniform distance of 1 inch from each other along the breadboard's length and width.

It *is* like a standard utility pegboard but sturdier. This breadboard *gives* us versatility to move the testing setup to convenient locations, relative to the force gauges and clamps, depending on the specimen shape. It *is* a custom-designed component, especially for this project.

2.2.5.3 Subject-Verb Number Agreement

Errors in subject-verb number agreement arise, as best as I can tell, for two reasons: (1) authors are not sure if the subject is singular or plural and (2) authors get confused as to which noun in the sentence is the subject that must agree with the verb.

Typical examples of confusion for singular or plural are these words: *data*, *criteria*, and *series*. The first two are common terms in ABERST, and they are usually understood to be singular, but they are plural. The last one, *series*, can be either singular or plural, but it is typically found in singular form (by my experience) but mistakenly treated as plural. The singular forms are *datum* and *criterion*. Thus, the following sentences are incorrect:

- **Wrong:** The data shows that temperatures across the globe have been rising, on average, for the past four decades.
- **Wrong:** The criteria that needs to be met is a 50% reduction in carbon monoxide emissions.
- **Wrong:** A series of valves and pumps are used to collect the air sample.

To fix these mistakes, the author must either change the verb to match the subject, or vice versa, depending on the context and the correct message that must be conveyed.

- **Change of verb to plural:** The data *show* that temperatures across the globe have been rising, on average, for the past four decades.
- **Change of subject to singular:** The *criterion* that needs to be met is a 50% reduction in carbon monoxide emissions.
- **Change verb to singular:** A series of valves and pumps *is used* to collect the air sample.

Sometimes finding the true subject can be difficult, as in this sample:

- **Wrong:** A chromatogram obtained with a column, sensor array, and conventional injector illustrating the isothermal separation of 16 volatile organic compounds are plotted to show retention times.

In this sentence, the true subject is simply *chromatogram*, and the verb is *are plotted*. Quite a few additional nouns are inserted between these two terms to modify and explain the simple core subject, *chromatogram*. An author might think the subject is *compounds*, or it may be the series of items *column, sensor array, and conventional injector*. With either of those alternative subjects, the verb would need to be plural in number, hence, *are plotted*; but with the true subject, the verb should be *is plotted*:

- **Correct:** A chromatogram obtained with a column, sensor array, and conventional injector illustrating the isothermal separation of 16 volatile organic compounds *is plotted* to show retention times.

Here is an example from electrical engineering:

- **Wrong:** Glass and silicon wafers bonded using a field-assisted, silicon-gold eutectic bonding technique has been successfully achieved at a temperature of 400°C.

The wafers are the main subjects, but the verb, *has been (successfully) achieved*, is placed in such a way that it seems to connect to the immediately adjacent noun *bonding technique*. This latter noun is singular, *a technique*, so the verb is in singular form to match the singularity of *technique*. Also, because so many modifiers have been added to *wafers*, the full subject is a long noun-modifier string vastly separating the core subject from its verb. In fact, readers may wonder if the subject is *wafer*, *wafers bonded*, or *bonding technique*? A better approach would be to emphasize the plural subject of *wafers* and match it with a plural form of *bonded*, as follows:

- **Correct:** Glass and silicon wafers *were successfully bonded* using a field-assisted, silicon-gold eutectic bonding technique at a temperature of 400°C.

Sentences with multiple verbs also promote poor subject-verb number agreement because authors have difficulty keeping the true subject in mind when a significant amount of intervening text has been placed between the first verb and subsequent verbs, as in this example:

- **Wrong:** Scanning thermal microscopy distinguishes individual components of various pharmaceutical samples related to drug development and reveal the structure of samples composed of biological materials.

Here, the subject is *Scanning thermal microscopy*, a singular noun, and the verbs are *distinguishes* and *reveals*. Using multiple verbs is fine, but you need to keep track of them all and match them to the original noun.

- **Correct:** Scanning thermal microscopy *distinguishes* individual components of various pharmaceutical samples related to drug development and *reveals* the structure of samples composed of biological materials.

The final problem to cover in this grammar section combines two of the problems seen above: a sentence with multiple verbs that wrongly includes a shift in verb tense. Below is an example:

- **Wrong:** A polycrystalline-silicon coating on a cochlear implant changes resistance depending on curvature, allows real-time monitoring of placement during surgical insertion, and proved to be a reliable sensor material to facilitate proper placement into the cochlea.

The verbs are *changes, allows,* and *proved.* These multiple verbs all represent actions taken by the subject, *coating*; thus, they can all be in present tense.

- **Correct:** A polycrystalline-silicon coating on a cochlear implant *changes* resistance depending on curvature, *allows* real-time monitoring of placement during surgical insertion, and *proves* to be a reliable sensor material to facilitate proper placement into the cochlea.

In summary, grammar encompasses all the rules and conventions for combining words together into complete sentences so that the ideas are expressed clearly and correctly. The rules are numerous, and this little section has addressed only a few of them—those that might be most troublesome for college students and professionals writing in ABERST.

2.2.6 Syntax

Whenever I think of syntax, a poem written by e.e. cummings comes to mind. I love this poem. It begins like this:

> "since feeling is first
> who pays any attention
> to the syntax of things
> will never wholly kiss you;..."

I highly recommend you read this fantastic poem before, during, or after reading this section. It gives a fresh perspective, just as the information below will refresh your brain regarding some important lessons on syntax that will help you in your pursuit of great ABERST writing. And don't forget to lean back and laugh, as the poet recommends.

Syntax, in short, refers to the order of words in your sentences. You can easily see the following is wrong: "Beer cold drink I." Syntax conventions of English demand the sentence be written thusly: "I drink cold beer." Or "I drink beer cold."

In ABERST, syntax problems are not particularly prevalent. Most writers in upper-level college courses and professional careers understand English syntax competently. But, a few items deserve mention: sentence starts, sentence ends, and sentence middles.

- Sentence starts
- Sentence ends
- Sentence middles

2.2.6.1 Sentence Starts

Perhaps the most common uncertainty with syntax at the start of a sentence concerns placement of a coordinating conjunction. If you have heard that you cannot start a sentence with a coordinating conjunction (and, but, for, nor, or, so, and yet), you were at least paying attention to someone presenting rules of English composition, and that makes you praiseworthy. I like you for that. Still, you do not have to worry about that rule. It is not applicable to ABERST. When sentence connection, emphasis, and variety are enhanced with an opening conjunction, the technique can be used. Here are two examples from natural science.

- The octopus is deaf. And the bat is blind.
- The San Andreas Fault is perhaps the best-known fault line in North America. Yet it is not the longest one.

Here is an example from engineering:

- Of these various fuel assembly blocks, the most common one includes 240 rods of plutonium fuel. Along with these rods are 48 Pyrex rods that help control reactivity. And 1 instrument tube is placed in the center.

The final sentence can begin with *And*. It is a simple linking word that works well for the short sentence it initiates. A long, ostentatious linking word is unnecessary. The final point to be made, about one more tube, is curt and complementary, so that is signaled perfectly well with *And*. The series of sentences is brought to completion with a final idea.

Here is the longer passage from which these sentences are taken, and I am providing it as an example of a paragraph with lots of compositional strengths covered in this book, including topic sentence, sentence cohesion, and sentence variety:

- **Good sentence starts:** The proposed novel nuclear reactor design is fueled by weapons-grade plutonium as part of its fuel assembly design. The reactor design has a nuclear core consisting of 157 individual fuel assembly blocks. Each fuel assembly block is a 17 by 17 grid of identical spaces that can be either used for fuel rods or control rods or left empty to allow water flow. The blocks as optimized are heterogeneous, to enable the greatest power production, and simultaneously the reactor burns up weapons-grade plutonium for a good cause. The weapons-grade plutonium is found in the most common of the assembly blocks. This block includes 240 rods of weapons-grade plutonium fuel. Along with these rods are 48 Pyrex rods that help control reactivity. And 1 instrument tube is placed in the center. This produces a total of 289 rods or spaces, which is the number produced by a 17 by 17 grid. Of the 157 blocks, this particular layout is repeated 106 times, in an effort to eliminate weapons-grade plutonium, a dangerous material, through the peaceful, civilian use of generating electricity.

2.2.6.2 Sentence Ends

The big issue that is legend among writing students is that a sentence should not end with a preposition. Prepositions are those little words that usually precede, but can follow, a noun to help relate that noun to the other parts of the sentence. Here is a brief list: of, to, in, after, along, beside, by, during, following, through, toward, under, and until. Most English handbooks offer a long list. The point is not to memorize them. In terms of syntax, the main point to master is that you do not need to take pains to ensure your sentences do not end with prepositions. They can.

Here are two examples with two versions for each: first, the preposition is at the end, and second it is no longer at the end due to a rewrite.

- **Preposition at end:** These are the family members I want to send presents to.
- **Preposition at end:** My hedgehog is not a pet that is easy to care for.
- **Rewritten:** These are the family members to whom I want to send presents.
- **Rewritten:** My hedgehog is not a pet for which care is easy.

You can rearrange your sentences, as done above, but that effort is not necessary. For prepositions can be placed at the end of English sentences. And conjunctions can be used to begin sentences.

2.2.6.3 Sentence Middles

Sentence middles are the location where two weaknesses arise: noun strings and subject-verb separation. These problems are closely related, as the former causes the latter. That is the chief reason a long noun string is a problem at all: because it produces an associated problem that is even more troublesome than itself. Indeed, when a sentence begins with a long noun string, the reader is left waiting too long for the verb. The sentence's core idea, in such instances, is not expressed until the verb appears. Thus, readers are left uncertain as to the core idea while they are forced to read, and try to retain, many additional modifiers and supplementary phrases. An example is this:

Part 2

- **Poor example:** The neutronics and lattice physics calculations for the mixed-oxide fuel assemblies integrated into the passively cooled PWR1000 nuclear power reactor design that we chose to enhance are compared to those of the baseline, standard fuel assembly.

The core idea of this sentence (calculations are compared) is not delivered until the end. The reader must navigate through 21 words after *calculations* and before *are compared* before the simple core idea is clear. This separation makes reading and comprehension difficult. In contrast, an effective style is to reduce or eliminate separation so that the core idea is delivered at the beginning of the sentence and the remaining portions modify or enhance that core idea, as in the following fixed passage.

- **Improved:** The neutronics and lattice physics calculations are compared to see if the mixed-oxide fuel assemblies improve upon the baseline, standard fuel assembly of the passively cooled PWR1000 nuclear power reactor design.

The instinct to add description to the core noun with other nouns and adjectives is laudable, but it diminishes the sharpness of the writing. Long noun-modifier phrases lead to hard-to-understand syntax. My advice boils down to this: You do NOT need to introduce all adjectives and modifiers when you first name the noun. Many authors attempt to provide multiple details about a highly specialized or complex noun when it is first introduced (as in example above), which leads to a long modifier-noun phrase, some or all of which may go over readers' heads, followed by a verb placed too far away from the subject to facilitate fast reading and comprehension.

Here is another bad example:

- **Poor example:** The eight-speaker Infinity Gold sound system with its USB port, CD player, 120-watt power amplifier and 3-band graphic equalizer offers incredible sound.

The subject *system* and verb *offered* are far apart. The ideal in strong composition is to place them close together and put additional, supplemental information after the verb, as follows:

- **Improved:** The Infinity sound *system offers* incredible sound, which is expected with the Gold model, for *it boasts* a USB port, CD player, 120-watt power amplifier, and 3-band graphic equalizer.

In the improved version, two independent clauses are used, not just one. They are joined with the coordinating conjunction *for*. Moreover, in both clauses, the subject and verb are next to each other. In the second clause *it* refers back to *system*, so this is not a passive *it* but a strong *it* as pronoun, with a clear antecedent. In the bad version, the verb *offered* could seem to refer to *equalizer*, which is closer to it than *system*.

Finally, we can look at yet another cause of awkwardly long modifier-noun phrases: the use of proper nouns as adjectives. This approach, again, creates a noun-heavy subject where the core subject is separated from its verb. As an improvement, you can tighten up your phrasing by using the possessive case instead of the noun-as-adjective approach, as shown below:

- **Common:** The Ferrari engine has more horsepower but less torque than the Tesla engine.
- **Improved:** Ferrari's engine has more horsepower but less torque than Tesla's.

While describing the engine as newly retooled, powerful, innovative, etc., is perfectly fine, describing it as "Ferrari-ish" is not ideal. Certainly, it is Ferrari's engine, and it is designed and manufactured by Ferrari, but that does not make *Ferrari* an adjective. It just does not sound like a normal adjective such as profitable, stylish, hungry, handsome, and so on. The more accurate phrase is to give Ferrari possession of the engine. Here is another example where the proper noun serving as an adjective is eight words, so the author takes a long time before introducing the core noun, namely, *centre*.

- **Common:** The European Organization for Nuclear Research (CERN) Large Hadron Collider control centre is staffed around the clock.
- **Improved:** Built and operated by the European Organization for Nuclear Research (CERN), the Large Hadron Collider's control centre is staffed around the clock.

You can apply several aforementioned tips to fix this passive, noun-as-adjective, subject-heavy sentence:

Practice exercise:

The centrifugal pump curves for two pumps in series, two in parallel, three in series, and one in series with two in parallel can be predicted by the curve for a single pump.

Answer: A single centrifugal pump's curve can predict the pump curves for two pumps in series, two in parallel, three in series, and one in series with two in parallel.

2.2.7 Spelling

In Appendix A, Section A.1, I cover many important red alerts with respect to spelling. Keep those in mind as you roll out pages and pages of strong prose. In addition, three other topics connected to spelling deserve some mention, as they produce an inordinate amount of confusion for writers across the board: acronyms, abbreviations, and equation variables.

2.2.7.1 Acronyms

Multi-word nouns, usually consisting of either (1) several nouns or (2) a series of modifiers (adjectives) followed by a core noun, are typical in ABERST. ABERST, for example, represents a string of six nouns: academia, business, engineering, research, science, and technology. Writing these multi-word nouns in full form every time you mention them in a document is tedious and unnecessary. Enter the acronym. LGBT, SCOTUS, and STEM are some additional examples of acronyms used often in academic and journalistic writing. An acronym is a tiny version of the long form, created by taking the first letter of each key word of the long form. Example: **S**upreme **C**ourt **o**f **t**he **U**nited **S**tates becomes SCOTUS. The technique is an excellent tool for writers. Unfortunately, it can be misused, generating comprehension problems, namely, a reader's befuddlement. Example:

- If you want to be a better writer, you should consult OED before reading examples of reports from the OECD. Moreover, not only will you find examples of standard syntax in the CEW's reports, you might also be intrigued to learn that a document written about SFCW followed all the MLA rules correctly.

OK. I hope you are like me and do not know what any of those acronyms stand for. No one can know all the acronyms that can ever be used by an author. Readers may not know your acronyms' meanings for various reasons, such as the following: 1. The acronyms are associated with words that are brand new and therefore not well known, generally; this happens as technology evolves and authors write about innovations. How else would people have written about an UAV (unmanned aerial vehicle), the HST (Hubble space telescope), or a VAWT (vertical axis wind turbine)? 2. The readers are not as familiar with the field as the author is, so they do not know the acronyms even if they are well established in the field. 3. The acronyms are ambiguous and can possibly have two or more long meanings.

To avoid confusion and make your job easy (by using acronyms) while not making the reader's job hard (by not understanding acronyms), the following approach is effective:

Basic acronym rule:

- ▶ Authors must spell out (define) acronyms the first time they appear in a document, then it is OK to use the acronym ever after in that document.

You might be nearly 100% certain that everyone reading your document will know all the acronyms; do not be so certain. Just to be careful, define them anyway. Many people reading your document are less familiar with the material than you are. Moreover, you yourself may forget the exact meaning of an acronym over time, and you may re-read a document you wrote a while back and not remember the meaning of the acronym. Believe me, it has happened. Will you remember what PET is in two years? SFR? Does PS stand for power supply or polystyrene? What about SDRCIDEAS?

To study the last one above, I will incorporate it into a sentence:

- **Poor example:** The design team entered the glider into SDRCIDEAS.

What is this? The acronym does not reveal anything about itself. Is it a contest? The glider entered a contest? This seems possible, especially considering that many competitions occur each year for innovative aircraft. Actually, the acronym represents a proprietary engineering design and analysis software program. Is it obvious that the acronym represents a computer program that simulates and analyzes structural loads on an object? I grant that it is a long acronym and would be very tedious to spell out; this task would involve nine words. But that is no excuse for presenting an acronym that is unclear.

Thinking about how to follow the rule with this acronym pushes an author to use other composition guidelines to solve the problem. First, part of the acronym, SDRC, is a company name, a proper noun, used as an adjective, which I have explained elsewhere is an awkward syntax (Subsection 2.2.6.3). Taking these four letters off leaves just five remaining. Second, these five (IDEAS) could be defined without any more burden than is otherwise present when spelling out acronyms, preceded by a possessive form for the proper noun. The company's name, SDRC, stands for Structural Dynamics Research Corporation. An improved version of the sentence is the following:

- **Improved:** The design team studied the glider using Structural Dynamics Research Corporation's Integrated Design and Engineering Analysis Software (IDEAS)."

I have spoken with empoyees at companies who admit readily to not knowing half of the acronyms in a report they might come across during the course of their duties, written by someone else at the company, perhaps just a few years prior. Here are some examples: SMFS, PCI, ACRA, LOCA, and TLF. Actual experience is my rationale for the rule. The rule increases comprehension of readers and authors, in the present and in the future. The rule, furthermore, allows authors to benefit from the use of acronyms, in place of multiple-word nouns, once the acronym is defined at its first use.

You may be thinking, "Sure, that's all true, but I use acronyms that require no definition. I use the ones from a core set of acronyms that need no definition. These are as well known as many commonly used words." Does this sound familiar to you? Many people believe fervently that the acronyms they use in their reports fall into this special, precious category. "Other writers, true, use acronyms that are not well known, but that's them, not me," say many authors.

Let me ask: Do universally understood acronyms exist? These would require no definition, right? I could print this list of universally understood acronyms, but then others would want to add or subtract from it. Who would ultimately agree to the items to be included on the list? Congress? The National Science Foundation? American Governors' Association? SCOTUS? Here is a start of the list:

NASA, IRS, MPH, BOGO, UPS, and HELOC.

Although those seem to be obvious entries for the "magically understood" list to me, perhaps they are not as familiar to others. And you would feel fine nominating *USA* to this list. Certainly, no one needs to spell out United States of America. But, I've worked on engineering documents that discussed some important excavation projects, and the protocol involved a regulatory requirement to contact USA before starting any excavation. In this case, USA stood for Underground Service Alert. A sentence in that document that reads "Contact the USA before beginning any excavation" would have been downright confusing to most readers . . . until the acronym was defined correctly. So, we are back to where we started: play it safe and define each and every acronym you use in a document. Instead of dreaming about this ever-shifting and ever-growing magic list, we writers in ABERST should do one better and just adhere to the Basic Acronym Rule.

Three additional situations that arise with acronyms deserve brief attention: Plural forms, possessive case, and states. Plural forms go hand in hand with possessive case, so I address these together, as follows. The simple rule for plural form is to add an S to an acronym to indicate the plural form: *NSC* stands for neural stem cell (singular), and *NSCs* stands for neural stem cells. Possessive form is NSC's, as in this sentence: "The NSC's maximum incubation period is 2 minutes, or the cells may die."

State abbreviations are standardized by the U.S. Postal Service: each state has a two-letter mail abbreviation. These are best used for mailing and shipping. Nonetheless, I often see them in critical places in reports:

- **Poor example:** Gravitational waves were detected at two companion detectors located in Livingston, LA and Hanford, WA.

When writing a report related to science or business, etc., the actual locations are important, so the full state name should be used:

- **Improved:** Produced by two black holes colliding 1.3 billion years ago, gravitational waves were detected after finally making their way across the universe to Earth at two companion detectors located in Livingston, Louisiana, and Hanford, Washington.

Here is one final corollary to the rule to define acronyms the first time they appear in a document. You also need to double check before you formally invent an original acronym that you are not (1) duplicating a well-established one or (2) introducing an embarrassing term. I encountered an example of an unfortunate acronym while working in engineering: liquid underground storage tank, or LUST. This is a word from a different subject matter. The situation became awkward when an entire booklet was written and printed for wide distribution, titled *The LUST Manual*. Needless to say, the agency responsible withdrew the booklet and changed the acronym after initial reactions to the manual were more enthusiastic than they imagined possible.

2.2.7.2 Abbreviations

If you are comfortable with the rule for acronyms, you will appreciate the rule for all other abbreviations because it is the same: spell (define) them the first time they are used in a report. This applies to single-letter abbreviations, such as T and R, as well as measurement unit abbreviations, such as km and V. The list of possible abbreviations is so long it cannot be compiled. Luckily for you, your reports will not use all abbreviations ever devised by mankind. You likely will have a dozen or so. That is not too egregious. You can define them for your readers. It will not take up that much extra space, and the report will make sense to a wide audience, at the time of first distribution and far into the future. Keep in mind that ABERST reports often have international audiences and global ramifications, but unit systems and abbreviations are often country-specific. Your job is to help all readers understand your report, not announce your personal membership in an exclusive club that uses lots of secret codes so outsiders cannot understand your communications. (Recall discussion of Tone, Subsection 2.2.3.1.)

Here is a passage without any acronyms and abbreviations defined, followed by an improved version with definitions and full spellings. The difference is this: the latter one is a scintilla longer than the former, yet it is a whole lot more understandable.

- **Poor example:** Two NR CDs can be compared to find the better option. IFBAs are used in one CD. IFBAs are manufactured by applying a thin layer of ZrB_2 to fuel pellets. These can be positioned anywhere in the core and help control core parameters. Because the outside layer of the fuel pellet is an absorber and not the fuel itself, it interacts with neutron fluxes at the BOL. The fuel is essentially "hidden" at BOL, allowing a generous initial fuel loading while also ensuring that the BOL heat is not reaching P-F limits. As an alternative, MOX fuel assemblies do not use IFBAs because MOX fuel partially acts as an absorber itself. For both CDs, we halved and doubled the h to 23 cm and 92 cm, respectively, for comparison. We saw changes in pcm/%void, from 5.5 to 10.3.
- **Improved:** Two nuclear reactor (NR) core designs (CD) can be compared to find the better option. One CD uses integral fuel burnable absorbers (IFBAs). IFBAs are manufactured by applying a thin layer of zirconium diboride (ZrB_2) to fuel pellets. These can be positioned anywhere in the core and help control core parameters. Because the outside layer of the fuel pellet, therefore, is an absorber and not the fuel itself, it interacts with neutron fluxes at the beginning of life (BOL). The fuel is essentially "hidden" at BOL, allowing a generous initial fuel loading while also ensuring that the BOL heat is not reaching peaking-factor (P-F) limits. As an alternative, an NRCD with mixed oxide (MOX) fuel assemblies does not use IFBAs because MOX fuel partially acts as an absorber itself. For both CDs, we halved and doubled the height to 23 centimeters (cm) and 92 cm, respectively, for comparison. We saw changes in void coefficient per total voiding, percent mille/percent void (pcm/%void), from 5.5 to 10.3 pcm/%void.

2.2.7.3 Equation Variables

Symbols must be defined. Many are Greek, and they may be unfamiliar to your readers. More troublesome, however, is when they are used to mean different things. The Greek lower-case letter sigma (σ) can refer to either standard deviation of a population or surface charge density in electrostatics, to name just two of its many possible meanings in mathematics and science. As a result, formalized methods have been standardized for presenting equations in technical documents, as demonstrated below:

- **Good example:** We tested the model vehicle at several flow speeds, and the corresponding Reynolds numbers (Re) were calculated using equation 1:

$$Re = \mu L/v, \qquad (1)$$

where μ is air-flow speed, L is the characteristic length, and v is the kinematic viscosity of air.

If you are not using full equations in your writing but still find recourse to symbols and abbreviations, you can use the method for them that is the same as presented above for acronyms: define them the first time they appear, then use the symbol/abbreviation thereafter. Here is a poor example followed by its improved version:

- **Poor example:** DMD was chosen for the neural probe because of its many beneficial properties. Its E value is one of the largest of all known materials, at 1011 Pa. Its band gap is 5.5 eV, and with B doping, DMD has a wide potential window in aqueous environments of -1 V to 2V. Importantly, its resistivities range from approximately 10^5 Ωcm, for undoped, to 10 ^{3}Ωcm, for doped.
- **Improved:** Diamond was chosen for the neural probe because of its many beneficial properties. Its ability to withstand pressure forces, signified by its Young's modulus (E) value, is one of the largest of all known materials, at 1011 Pascals (Pa). It can serve as an excellent insulator with minimal conductance as attested by the following values: Its band gap is 5.5 electron-Volts (eV), and with Boron doping, diamond has a wide potential window in aqueous environments of -1 Volts (V) to 2V. Importantly, its resistivities range from approximately 10^5 Ohm-centimeters (Ωcm), for undoped, to 10^3 Ωcm, for doped.

On a very simple level, defining symbols is one component in a multifaceted approach to making your writing understandable to all the various readers who need to study your report. You can see this in a single sentence taken from rocket science:

- **Clear only to experts:** The rocket's Isp was 838.5 N-s.
- **Clear to all readers:** As a measure of the total force over time supplied by the rocket's motor to launch it into the atmosphere, specific impulse (Isp) was estimated at 838.5 Newton-seconds (N-s).

2.2.8 Punctuation

Punctuation marks are intended to guide readers through the text just as road lines and signs guide automobile drivers through the streets. A review of every single rule is not possible here, and probably not necessary. Instead, I focus on the handful of rules that are most useful, and a few too-frequent mistakes that are easy to explain so you can avoid them.

2.2.8.1 Comma

Three standard uses of the comma are the following. 1) Two independent clauses are correctly joined into a single sentence by following the first clause with a comma and a coordinating conjunction, before the second clause. 2) A comma is correctly used both before and after a subordinate clause inserted into an independent clause. 3) A comma also follows a subordinate clause or long prepositional phrase that begins a sentence, when it is followed by an independent clause. Examples are provided below for all three cases:

1. The team lost, but the marching band sounded fabulous.
2. She studied economics, which was a challenging program, to complement her psychology major.
3. Although they had only 6 days of vacation, they decided to travel to Ireland to visit friends.

With complex sentences, often the comma demonstrated in the first example above is mistakenly omitted while optional commas are included, as below:

- **Needs comma:** We implemented a throughput efficiency survey and design, of the subsequent plan, is underway.
- **Fixed:** We implemented a throughput efficiency survey, and design of the subsequent plan is underway.

2.2.8.2 Comma with Items in Series

Experts disagree on the optimum way to punctuate and present a series of items. From a punctuation standpoint, the comma can preclude problems of meaning. In fact, one option favored by many can lead to ambiguity. Often, people are advised to omit a comma before the last item in a series, as in the following example:

- **Unclear:** The high tea service at Chicago's Drake Hotel includes cucumber sandwiches, smoked salmon, scones, tea and cream.

The world as we know it might come to a full stop here, while we wonder if the cream can go with any of the prior items, such as scones, or only the tea? Is tea item four and cream item five, or is "tea and cream" item four?

Two preferred ways to present this sentence, to avoid ambiguity, are the following:

- **Clear:** The high tea service at Chicago's Drake Hotel includes cucumber sandwiches, smoked salmon, scones, and tea and cream.
- **Clear:** The high tea service at Chicago's Drake Hotel includes cucumber sandwiches, smoked salmon, scones, tea, and cream.

These examples illustrate the rationale behind the following rule for commas with a series: Before the last item, use *and* preceded by a comma.

Example of Ambiguous and Fixed:

- **Ambiguous:** They compared the systems on the basis of safety, economics and reliability and durability.
- **Fixed with four bases:** They compared the systems on the basis of safety, economics, reliability, and durability.
- **Fixed with three bases:** They compared the systems on the basis of safety, economics and reliability, and durability.

2.2.8.3 Commas with Run-On Sentences

As you have read so far, often the absence of a comma is problematic. The opposite can be true as well, and this subsection addresses a commonly incorrect insertion of a comma. In this case, commas are used instead of periods or semicolons at the end of independent clauses (without adding coordinating conjunctions to properly link the two independent clauses). A run-on sentence is produced by the misused comma.

- **Wrong:** The methacrylic acid produced is only 95.1 weight percent (wt. %) pure, this is not pure enough to sell at $1.23/pound.

This is a classic run-on sentence. The proper punctuation is needed to indicate that these are separate sentences, not parts of one sentence, as follows:

- **Fixed Version 1:** The methacrylic acid produced is only 95.1 weight percent (wt. %) pure. This is not pure enough to sell at $1.23/pound.
- **Fixed Version 2:** The methacrylic acid produced is only 95.1 weight percent (wt. %) pure; this is not pure enough to sell at $1.23/pound.

Writers who overuse commas, or exclusively use them amid the utter absence of other punctuation marks, are not hard to find. Many run-on sentences are found in ABERST documents. Here is one. Try finding a way, or two, to eliminate the ubiquitous run-on sentences:

- **Poor example:** Phase II tractors have solenoid-actuated valves to control the operation of the steering clutches, brakes, and transmission clutches, and there is a concern that the performance of these valves, and the concern was expressed recently at a review meeting, may decrease and have a significant negative impact on tractor operation, so we need valve pressure and current data every 250 hours of operation, and possibly this data should be obtained from pilot tractors as well.
- **Improved:** Phase II tractors have solenoid-actuated valves to control the operation of the steering clutches, brakes, and transmission clutches. The performance quality of these valves needs study, as expressed recently at a review meeting. Specifically, the performance

may decrease over time and degrade tractor operation, so we need valve pressure and current data every 250 hours of operation. And, if possible, this data should be obtained from pilot tractors as well.

2.2.8.4 Colon

For some reason, a common mistake creates a "blocked colon." A blocked colon is my phrase for the problem in the following example:

- I tested: the Buick, the Ford, and the Fiat.

The colon after *tested* serves to block the first portion of the sentence (subject and verb) from the very important objects of the verb, namely, the list of automobile brands tested. Thus, *I tested* is blocked by the colon. This is a blocked colon.

Two options are available for healing this ailment: remove the colon or complete the blocked portion by adding a general term. Both are equally acceptable, although version 2 has the added benefit of introducing a general term before the particulars, which is always a good strategy in technical writing (see, for example, Section 2.2.1 and Chapter 3.4).

- **Fixed Version 1:** I tested the Buick, the Ford, and the Fiat.
- **Fixed Version 2:** I tested three cars: the Buick, the Ford, and the Fiat.

Furthermore, adding *including* does not solve the blocked colon problem, as noted by the warning in Appendix A, Subsection A.14.3:

- **Wrong:** The analysis encompassed all aspects of the plant design including: equipment choices and sizes, energy requirements, plant economics, and environmental and safety concerns.
- **Correct:** The analysis encompassed all aspects of the plant design, including equipment choices and sizes, energy requirements, plant economics, and environmental and safety concerns.
- **Correct:** The analysis encompassed all aspects of the plant design: equipment choices and sizes, energy requirements, plant economics, and environmental and safety concerns.

Thus, a correct and valuable use of the colon, a "healthy colon," is to introduce a list or an example following a complete, independent clause.

A second valuable use of the colon can be explained quickly: Use a colon to join two independent clauses, when the second one elaborates, explains, or restates the first one, as in the following examples:

- **Good colon:** We ran the samples through elaborate testing of tensile strength: we followed five separate protocols.
- **Good colon:** Be cautious with the water-tank heater: plug it into the power outlet only after it is submerged in water.

2.2.8.5 Semicolon

The semicolon is valuable but often ignored by ABERST writers. This might be due to uncertainty regarding how to use it correctly. The following is intended to enlighten authors, build confidence, and increase semicolon use. It might very well be my favorite of all the punctuation marks; yes, it is.

One way to use semicolons is to separate two closely linked independent clauses. It signals that each clause is fully complete and independent, as a period does, but that they have a special bond that says they are nearly two halves of one sentence:

- **Good:** Learning involves memorization; it also involves practice.
- **Good:** For this ignition switch, they introduced a new version last fall; it has not, however, proved superior to the original.
- **Good:** Years ago, I could fix my car's carburetor; however today I cannot repair my car's fuel injectors.

Please do not be misled by the latter two examples above to conclude that a semicolon must always be used with *however*. I simply provided two examples with *however* to drive home the point that *however* is NOT a coordinating conjunction like *but* and *so* (as covered in the Red Alerts in Appendix A, Subsection A.14.2). This is a common mistake, and you introduce a run-on sentence when *however* is used incorrectly, like this, to refresh your memory:

- **Run-on:** Years ago, I could fix my car's carburetor, however today I cannot repair my car's fuel injectors.

The remaining valuable use of a semicolon is to separate items in a series when at least one of the items in that series has internal commas itself, as in this example:

- **Incorrect:** Essential items for photograph developing are an enlarger, printing paper, four plastic trays, small tools, such as a thong and a camel-haired brush, and a thermometer.
- **Fixed:** Essential items for photograph developing are an enlarger; printing paper; small tools, such as a thong and a camel-haired brush; four plastic trays; and a thermometer.

2.2.8.6 Apostrophe

Generally, an apostrophe followed by "s" is used for possessive case, as below:

- **Correct:** The department's offices are under renovation.

Quite often, however, the apostrophe is omitted, so the writer creates a mistaken plural instead of a possessive:

- **Wrong:** The departments offices are under renovation.
- **Wrong:** The signals peak frequency must be above 5 kHz.

Readers are slowed down more often than we care to admit on this "missing apostrophe for possessive" mistake.

One enormous exception, of course, exists to this rule, no doubt leading persons from every demographic, political party, and profession to bungle this one time and again. I refer to the little word *it*. We like this word. We use it often. (Analyze that sentence!) But, the word's possessive form does NOT have an apostrophe. Go figure. Possessive of *it* is *its*. That is correct. Believe me.

- **Wrong:** It's flexible joint is appropriate for this purpose.
- **Correct:** Its flexible joint is appropriate for this purpose.

The word *it's* does not indicate possession. Instead, it is a contraction of *it* and *is*. Thus, *it's* means *it is*. (This English conundrum is so difficult it is included in the first set of Red Alerts in Appendix A, Section A.1.) Study these three examples:

- **Wrong:** Its a conceptual plan, not a detailed one.
- **Correct:** It's a conceptual plan, not a detailed one.
- **Correct:** It is a conceptual plan, not a detailed one.

2.2.8.7 Hyphen

Hyphens are often wrongly used in two ways. One, they are used instead of another punctuation mark that is the ideal choice instead of the hyphen. For example, a period or colon should be used to separate two independent clauses (sentences), not a hyphen.

- **Wrong:** An igniter must be able to restart the engine after an unexpected flame out—this is a typical characteristic of a spark plug connected to a battery.
- **Correct:** An igniter must be able to restart the engine after an unexpected flame out. This is a typical characteristic of a spark plug connected to a battery.

Two, the hyphen is wrongly used between an adverb and an adjective when the juxtaposed words function as a compound modifier. In this instance, no hyphen (or any other punctuation) is necessary.

- **Wrong:** Keeping the water within a tight range of temperatures is a consistently-challenging problem on spacecraft.
- **Correct:** Keeping the water within a tight range of temperatures is a consistently challenging problem on spacecraft.

This latter mistake likely stems from the false impression and oversimplification that all compound modifiers require hyphenation. Some of them do require hyphenation, but not all of them, particularly when an adverb (a word typically ending in *ly*) precedes the adjective. For modifiers, three different situations related to punctuation are possible: commas, hyphens, and no punctuation. The rules, in brief, are as follows:

1. Commas are used between adjectives preceding a noun when each one, on its own, modifies the noun (big, brown dog).
2. Hyphenation is used between adjectives preceding a noun when the two, working together, modify the noun (or, when one modifies the other) and each on its own does not make sense as a stand-alone modifier of the noun (micro-porous-membrane filter or post-structural feminism).
3. No punctuation is ideal when the compound modifier comprises an adjective preceded by an adverb that modifies the adjective, not the noun (manually operated drill).

Some examples are provided below, both correct and incorrect, to demonstrate the three aforementioned rules:

- **Correct hyphenation:** The low-emission engine relies on numerous sensors to assess the fuel mixture.
- **Incorrect absence of punctuation:** The air cooled engine improves reliability and fuel efficiency.
- **Correct hyphenation:** The air-cooled engine improves reliability and fuel efficiency.
- **Correct commas:** The loud, heavy, and conventional engine runs on diesel fuel.
- **Correct absence of punctuation:** The finely tuned control system prevents oscillation.
- **Correct combination of punctuation:** The feedback-based, optimized, real-time control system increases the spacecraft's orbital lifetime.
- **Incorrect:** The optimally-spaced 4 meter wide, air cooled electrohydrodynamic panels have an expected life of 10-years.
- **Correct:** The optimally spaced, 4-meter-wide, air-cooled, electrohydrodynamic panels have an expected life of 10 years.

2.2.8.8 Parentheses and Brackets

I postulate that I can squeak by without explaining the one rule that is not well known for these marks, opting, instead, to simply demonstrate. You, being clever, will get it:

- **Sentence requiring only parentheses:** The bank completed its stress test as required by Congress (and enforced by the Securities and Exchange Commission), and the score was an adequate 87 (out of 100).
- **Sentence requiring both parentheses and brackets:** The bank completed its stress test as required by Congress (and enforced by the Securities and Exchange Commission [SEC]), and the score was an adequate 87 (out of 100 [normalized from a composite score of 673]).

2.2.8.9 Exclamation Point

In ABERST, you should resist any urge to use exclamation points. Let the ideas speak for themselves. You do not need to be overly emphatic or excited. As other writing teachers have said before me, exclamation points are best reserved for diaries, texts with many emojis, personal emails, and social media. Trust me on this! Oops!

2.2.8.10 Quotation Marks

Quotation marks are fairly straightforward for most writers, but things get tricky when needing to combine quotation marks with other punctuation, including other quotation marks. Are you dizzy?

Generally, double quotation marks are used to refer to someone's actual words, as you are doubtless aware:

- **Correct:** The mayor said the landfill was a "dump," but the camera had been turned off.

But, when actual words are included within another quotation, a single apostrophe is used.

- **Correct:** The mayor argued, "The citizens do not want the landfill, which they call a 'dump,' located within a 10-mile radius of here."

In American writing, two items of punctuation are placed inside the final quotation mark, while two marks are placed outside. In the first category are comma and period, and in the second are semicolon and colon. (The British do it differently.) The other marks may be placed, and often should be placed, outside the final quotation mark. Here are some examples:

- **Correct:** The lead investigator, Brian Cohen, asserts that this is the "smallest neural probe made to date." Nevertheless, he says his team is always trying to "push the envelope"; only time will tell, and "funding, whether a lot or a little, will play a huge part," he laments.
- **Correct:** Although the probe is a "mere 2 millimeters thick," according to the laboratory manager, Alexis Romano, it has more than 100 electrical signal sensing sites. In fact, she says the probe is "dual purpose": drug delivery is its original function, but neural monitoring, its second function, is "becoming better with each iteration."

2.2.8.11 Question Mark

As appropriate for a Socratic venture such as education, I will end this section on a question. Well, on a question mark. The problem with a question mark is usually limited to the instances when one is combined with quotation marks, so this and the previous section are juxtaposed for that reason.

The rule for question marks and quotations:

- When the quoted passage is itself a question, the question mark is placed inside the final quotation mark. In contrast, when the quoted material is not itself a full question but is merely embedded within an interrogative sentence, the question mark is placed outside of the quotation mark.

Here are examples to illustrate:

- **Correct:** Can you believe that Sting asked me, "Do you like my songs?"
- **Correct:** Do you know which famous person said, "We can count on Americans to do the right thing, after they have tried everything else"?

Often, the advanced question that savvy students ask is how to punctuate with quotation marks when both the quotation and the overall sentence are questions. Yes, challenging. The elegant solution is to use only one question mark, inside the quotation marks, as follows:

- **Correct:** Do you like the song, "Where Have All the Flowers Gone?"

Nevertheless, we can peel this onion further, so those of you who enjoy finding the rare exception can take pleasure in this final guideline for punctuation. In the extra odd instance when (1) both the quotation and the overall sentence are questions (as above) and (2) you wish to add some intervening parenthetical information between the final quotation mark and the absolute end of the sentence, you can use a double question mark, as in this example:

- **Correct:** Do you like the song, "Where Have All the Flowers Gone?" (written by Pete Seeger in 1955)?

2.2.9 Numbers

Numbers are unavoidable in ABERST documents. And they are so important, too. Nothing is fully understood until it is divided into four parts, supported by three points, measured to be 17 inches wide, and calculated to provide 46 Newton-seconds of impulse. You get the idea. Do not avoid using numbers because you have difficulty differentiating between instances when the numbers should be spelled out as words and those instances when they can be written in numerical form. Confusion is normal. If you learn the handful of rules for numbers, you can proceed to sprinkle them liberally throughout your documents to present a clear and precise message.

Although the rules may seem arbitrary and the results higglety pigglety, both elegance and logic permeate the rules below. When followed, reports will be full of a blend of numbers, in both word and numerical forms, and they will be correct. You also must consider the frequent need to hyphenate numbers.

First, I present examples with numbers presented correctly, in either written or numeric form, and you can detect the patterns and rules for yourself. Second, I present the formal rules underlying the examples in the list.

- Sixty operators are needed.
- The gasifier requires 6 operators, and the bioreactors require 17 operators, per shift.
- This factory is designed to operate for 15 years.
- The factory has a 15-year life.
- This is a four-stage process.
- The sawdust feed enters at a rate of 4 pounds per minute.
- We use a tank that holds 100 gallons.
- This is a 125-gallon tank.
- The displacement reached 2-1/4 inches.
- The machines were installed during the 1980s.
- This is the 1980's most popular exercise video.
- The competition includes 5-, 10-, and 15-minute events.
- To cover the temporary hole, we set down a 1-inch-thick, 3- by 6-foot steel plate.

Number rules:

- Always spell, as a word, a number if it begins a sentence (Sixty).
- Spell numbers under 10 when not part of a standard unit of measure (four stages).
- Use numerical form for numbers over ten (15 years).
- Use numerical form for all numbers when part of a standard unit of measure (4 pounds).
- Use numerical form for a number under ten when part of a series of numbers that has at least one value above ten (6 and 17 operators).
- Place a hyphen between a number and its unit of measure when this makes a compound modifier of an adjacent noun (15-year life and 125-gallon tank).
- Place a hyphen between a whole number (digit) and an attached fraction (2-1/4).
- Use the plural form with small *s* when referring to a decade (1980s)
- Use the possessive form, apostrophe-*s*, when the decade takes possession of an object (1980's).
- When two or more values are presented in a series, all of them having the same unit of measurement, the unit needs to be written only once (5-, 10-, and 15-minute).

Chapter 2.3

Proofread Drafts Sequentially

By studying Chapter 2.2 and also taking time to review the comprehensive inventory of frequent flaws (Appendix A), you can preclude a good many of them from arising in your documents as you write your first versions. Being forewarned of these pitfalls, however, may not eradicate their occurrences completely. Some of the writing flaws and common mistakes may still sneak into your writing although you are aware of them. Your paragraphs and sentences may not match your knowledge of composition rules and your best intentions. First versions of documents often do not reflect an author's best work. Indeed, most of the hazards listed in Appendix A are easy to make and hard to detect. Few people, in fact, can produce an error-free document on the first try. A document may need several iterations (called "drafts") until it is ready for distribution. A busy ABERST author often submits the first version, however, or makes only a few corrections after a quick scan of the text, submitting a second draft. Rarely do recipients and readers see the potentially best version, produced after careful review and rewriting. Implementation of composition rules and correction of common mistakes are not always sufficient.

To truly produce and submit a high-quality document, an author must execute a cycle of writing, reviewing, and rewriting. This is where you will stand out: with your knowledge, you can review your first versions and find improved forms of expression. You can also find the common mistakes and promptly eliminate them. This is the essence of proofreading, the topic of this chapter.

Proofreading is a critical skill for any aspiring writer. You can proofread on the computer or on a hard-copy printout. I prefer the latter, but you can try both until you find the one that works best for you. Either way, the essential goal of proofreading is this: read through and clean up any mistakes, find more concise ways to express yourself, eliminate redundant items, and reorganize material into its most flowing form. That's all. Just do that. OK, maybe this is more easily said than done. Still, proofreading can be practiced and improved, and I have a technique to recommend: I call it my "multi-sweep method" of proofreading. It takes time.

Proofreading can take many forms, I suppose. Two are most common. One, a proofreader can review the document very thoroughly, from beginning to end, one time, and attempt to identify all the items that require fixing or improvement. This is the advanced approach, and it requires experience and skill in proofreading. This is often attempted when an author carries out one cycle of review and rewriting, but the attempt is incomplete, leaving much that could still be improved in the document. Two, as an alternate, a proofreader can review the document more than once, with each review, or, sweep, through the document focused on a particular issue or group of issues, such as spelling, format, or wordiness. This second form is best for proofreaders who are fairly new to the task or have not yet mastered the skill required. I recommend that most authors in college or starting out in their careers use the multi-sweep method. Upon more reflection, I recommend it for all authors because trying to catch every problem in one sweep is difficult. Inevitably, items will be neglected because the spectrum of potential improvements and corrections is broad, as you will see as you read about the multiple sweeps below.

Here is a suggested order for sweeping through your document, and you can change it up to suit your style. If you simplify and use fewer sweeps than discussed below, I will not have hurt feelings. You can think of this technique as conceptual, and your practical implementation will be a huge improvement over no proofreading at all. The message to retain is that an author must proofread broadly and thoroughly.

When I look at a multi-page document after the first draft is completed, my first sweep through involves organization. This is not "proofreading" per se; it is more correctly called "editing." But, you are "reading the proof" when you do this, so we can lump it together with subsequent sweeps through the document when you take a magnifying glass to every word and punctuation mark.

This first sweep is critical, and it ensures you have a proper structure for your purpose. If you do not have such, you need to reorganize. The principles of organization are covered in the other four parts of this book. It is beyond an issue of "editing and proofreading." It is the fundamental lesson of how to write well for ABERST.

Once you are happy with the order of your sections, you can look more carefully at the paragraphs within each section, and this includes looking overall at the paragraph blocks, as well as each individual paragraph. This is a good sweep 2.

For each paragraph, first look at its size and nothing else; if one seems too long or too short, you can review it and see if it needs reorganization. If it is too long, you should likely find a place to split it into two or more distinct paragraphs with separate topic sentences. If it is too short, you may need to add more particular sentences to support the topic sentence or, more importantly, add a topic sentence if it is missing. Explanation of paragraph structure is found above, in Section 2.2.1.

For sweep 3, I suggest trying to make things shorter: look for common examples of verbose, loquacious, wordy, and redundant writing, then cut out and cut down (you should be able to edit this sentence). You might have fixed some of this when you were doing sweep 2 and encountered long paragraphs, but the task is important enough to dedicate its own sweep to it.

In particular look for redundancies. You may find lists, whole sentences, or paragraphs that appear twice because you copied and moved them but forgot to delete them from their original location. Go through and eliminate these.

An extra benefit of making your text more concise at this stage of proofreading is that this will reduce your work load for the sweeps ahead, namely, those that look at every word and punctuation mark.

For sweep 4, you must look at the sentences and words, taking care to find and eliminate any of the potential grammar, syntax, and spelling problems reviewed in Chapter 2.2. Align your prose with the rules and guidelines. Recall how easily you can confuse certain words and have other terminology troubles (Appendix A). Look for words that are often conflated for each other or words that are often typed incorrectly and replaced with a homophone (word that sounds the same but means something entirely different as the word you intended to use).

Moreover, you cannot really succeed in sweep 4 unless you understand the standard rules of English composition, highlights of which are presented in

Chapter 2.2, in sections titled as follows: Words, Grammar, Syntax, and Spelling. So this introduction of sweep 4 serves as a prompt to give proper attention to Chapter 2.2 with exposition of those important topics.

Sweep 5 consists of checking punctuation: commas, semicolons, colons, and periods, along with quotation marks, apostrophes, and special symbols (question marks, hyphens, and dashes). The rules are reviewed in Section 2.2.8. In addition, you can check for correct presentation of numbers (Section 2.2.9), as numbers figure heavily in your document, most likely.

Finally, sweep 6 is concerned with format: check for consistency and an absence of mistakes in headings, highlights, widows/orphans, page numbers, footer/header items, vertical and horizontal spacing, fonts (type and size), and margins.

As I said before, any proofreading will be helpful. Understanding the six-sweep technique should serve as a strong foundation, even if you cannot bring yourself to use it. I realize it has the marks of an obsessive-compulsive personality type. Sure it does, and it requires time and patience. You might only feel comfortable with Proofreading Form 1, where you check the document all in one fell swoop, from start to finish, and that might work for you, to a sufficient extent. In my experience, however, trying to catch everything in one sweep is difficult. One is better than none though.

I suggest, therefore, that you find some number and focus of sweeps that work best for you. Experiment. Some type of approach of isolating one area (or, focusing on just one topic at a time) helps immensely to be thorough. At the very least, go through your document thoroughly so you will not be embarrassed by anything, such as these clunkers (some Freudian slips) that appeared in submitted documents and fittingly bring this chapter to an end:

- "The piping lines leaving this unit are extremely hot. Thus, the lines are insulted and kept out of reach to avoid problems." (I want THIS job.)
- "The cost of the land is $10,000,0000."
- "We used the costs of equipment and piping, in conjunction with the standard costs for typical chemical pants."
- "Now rather antiquated, Professor Fieldstone permitted a tour of the vacuum chamber laboratory."
- "I will proceed with the contract amendment after I receive a confrontation message from you."

Part 3

Fulfilling Your Communication Purpose

The lessons of Parts 1 and 2 of this book are now going to be put into service in support of your intended effort to communicate exactly the information required by the situation. As a result, you will be successful in making your report the ideal collection of informative and persuasive components. In this middle part, the heart of the book, you learn how to select and organize the precise information needed to produce a thorough, clear, and effective report.

The lessons of the earlier parts are meaningless if you are unsure of how to decide what exactly needs to be expressed in the report. In fact, not knowing if you need x, y, and z, or, rather, a, b, and c, is a critical uncertainty that precludes any progress in report writing. Many writers face just this problem. Sometimes reports do not get written... for days and days. Procrastination and writer's block bring efforts to a halt. Conversely, sometimes a writer bravely presses on, not clear of the purpose, throwing this and that into the report, in a helter-skelter fashion. This results in reports with too much of the wrong information, or too little of the necessary information, or both.

This part of the book provides a method for determining the required content of your report, and it's a very simple method. It is so pure and simple, it may seem too elementary to you. You might doubt its credibility, like when researchers spend millions of dollars on a study that concludes that consuming excessive calories may lead to obesity. I cannot speak about self-evident research, but I can reveal to you the simple essence of excellent ABERST writing. The method is this: Identify the primary motivating purpose of communicating via this report, and select and organize all necessary information in support of that purpose. I refer to this method as the "purpose principle."

The method is this: Identify the primary motivating purpose of communicating via this report, and select and organize all necessary information in support of that purpose.

Purpose Principle

- Build each report on a single, organization-based communication purpose.

The chapters below elaborate on this principle and provide guidelines and advice for implementing it. I briefly introduced this concept in Chapters 1.2 and 2.1 above, where I provide a handful of simple guidelines for delivering just the right document and emphasize that elegant writing is concise and in service of one's purpose. In this part, I both explain the need for a clear communication purpose and elucidate in detail an easy-to-use method for identifying, stating, and fulfilling it.

This part of the book is the heart because its lessons will make the difference between your readers becoming either frustrated and never finishing the report or satisfied and breezing through your report. The latter is so much nicer than the former, for readers; and for you: better for your grades and career.

The purpose of communication holds the whole report together. It is, in other words, its underlying foundation. It explains the presence of every item in the report. Put slightly differently, it is the foremost point of writing that particular report. When authors have a clear communication purpose, readers never say, "Please get to your point." On the contrary, you begin with it, proceed to build upon it, and successfully drive it home.

You might have a few concerns, which I have heard from skeptical, hesitant students; so let me address them here:

1. I have to write a report, yet I'm fairly sure I don't have a communication purpose for the report. It's just an assignment or something my boss asked me to write, or that needs to be sent to a client, government agency, or associated organization.
2. I have several communication purposes for this document.

Answer to 1: you really do not need to write if you have no purpose in writing. If you are writing, either by assignment, request, or self-initiative, you do indeed have a purpose, most certainly. ABERST settings are not the type of places where people produce purpose-less reports with random collections of words, presented to recipients to read for no reason. If you cannot find and get to the point, do not waste time writing.

Answer to 2: Multiple communication purposes are not impossible, but this adds complexity and interferes with a simple, elegant method. A usable and effective method is ideally simple and easy to remember. The notion of multiple purposes requires subtlety and deeper analysis. In most cases, moreover, a single principal purpose exists, accompanied by other supplementary purposes. In some sense, this counting of multiple purposes can be taken to infinity because an author can state as a purpose to convey each and every sentence in the report: It is my purpose to tell you the date I wrote this report; it is my purpose to tell you I used a microscope; it is my purpose to show you my 95% confidence interval for the empirical data, and so on. So, yes, it is a purpose to provide each and every piece of information in the report (particulars). But underlying all that information is a singular, primary purpose to which the other, supplementary purposes contribute. Find the foundational purpose, and you can compile all the necessary information in service to that purpose and leave your readers enlightened and appreciative.

Any author adequately trained on the purpose principle will produce reports built upon a singular, foundational purpose. But not all authors have such training. Thus, reports are regularly written and submitted that are ineffective and unclear, due to some flaw of purpose, such as one of these:

- The purpose is buried in a poorly organized pool of paragraphs and sentences.

- No evident purpose is identifiable from the report. In such instances, the author either did not need to write the report just yet or did not have a solid enough sense of that purpose to even bury it in the report. The purpose of communication was never considered.

Both of these scenarios are severe problems, and reports with these flaws are ineffective. They need complete revision.

Before you worry about revisions of poor reports, you can be supplied with a go-to method for writing the correct report in the first place. This method is the focus in this third part, as organized into the following five chapters:

Writing Purposeful Reports

3.1 Have a Point
3.2 Assemble the Right Information
3.3 Normalize the Information for All Readers
3.4 Prioritize Parts of the Document
3.5 Organize the Information

Chapter 3.1

Have a Point

So far in this book, I have introduced three of the four cornerstones of technical communication. In this chapter, I turn my attention to the fourth and final one. Along with (1) starting the document strongly for immediate comprehension, (2) shaping the document skillfully for easy reading, and (3) writing your prose elegantly for efficient communication, (4) ensuring that your document "has a point" is the remaining cornerstone of successful technical communication. As said before, it is foundational. Having a point is important for the same reason that "be interesting" is important to a person who is actively dating. When your date constantly checks her watch, smart phone, or other electronic device, you know you are not doing well. Communication in general and conversation in particular demand that participants have something useful or interesting to express to each other. In the film *Trains, Planes, and Automobiles*, the character played by Steve Martin, Neal, nearly loses his mind after having to spend hours listening to John Candy's character, Del, ramble on and on. Finally, in a fit of honest exasperation, Neal screams out, "Try having a point to your stories." Neal is captive: he sits next to Del on an airplane and later shares a taxi and a hotel room with him. So too are your readers captives.

Readers of reports in academia, business, engineering, research, science, and technology (ABERST) are captives when the report is addressed to them or when they are required to read the document for professional reasons. They are not reading the document to pass the time, be amused, or casually inquire. Thus, we arrive at technical writing's rule number 1: Write reports only when you need to, and only when you can articulate the reason. Even if your boss says, "Go study widgets X and Y and write me a report," you had better find out the reason he needs the report and the desired objective of the report before you proceed to boot up your word processor. If he says it is a secret, quit your job. You will not be able to satisfy him. You certainly will not be able to write a report for him. He has not given you a reason to write. He gave you a topic, but a topic is not enough. You need a focus, a work objective, and a communication purpose. In short, you need to have a point for writing.

The problem with learning technical writing at college is that college essays (even if they are dubbed "reports" by the professor) often are pointless. This is not the student's fault; it is the teacher's. In school, teachers primarily require students to practice writing and learn the basics. When employed, however, no one needs you to practice writing. No one is there to teach you writing and give you class exercises to grade. You need to write, not for practice, but for a purpose. You will have a point, that is, a reason to write beyond the need to produce the writing, which would be circular. Circular writing is done for its own sake, and might be instrumental in self-exploration efforts, such as journaling and keeping a diary. But circular writing is rare in ABERST. Substantive writing, in contrast, has a point, or, a purpose. The point transcends the writing itself. As introduced in two prior chapters (1.2 and 2.1), you must have a reason to write business and technical reports above and beyond writing itself. To have a reason, you must be aware of someone (personalization) who needs the document's contents (information/persuasion) to become motivated (actuation) to take an organizational action. Making the effort to record and convey information to another person presumes a desire to get a point across to someone. That desire is your purpose of communication.

3.1.1 Purpose Principle

Thus, in technical writing, the first step is identifying the true communication purpose. Ultimately, that communication purpose is stated in the Foreword, or other introductory section, as an element of achieving a strong and successful start to the document (recall Chapter 1.2). Before it can be correctly placed at the start, however, it must be identified, and that identification involves isolating it from a miasma of misleading and false possibilities.

Indeed, many authors, unfortunately, take the same wrong first step toward finding the communication purpose, so I address it here. I call it the "false first step." Being alerted to this problem should help writers avoid it and, instead, take the true first step (articulated shortly). In other words, learning how to take the true first step should come easily after comparing it to a common false first step, which is an understandable error. False first step: Quite simply, authors often confuse the purpose of their work with the purpose of their communications. Authors work on lots of interesting projects with many laudable objectives. Unfortunately, those objectives are goals for the work, not the purpose of communication for any single report. Here are some work purposes:

Work purposes:

- Find a faster way to synthesize a vaccine
- Design a nuclear reactor to burn, thereby destroy, weapons-grade plutonium
- Improve aerodynamics of a car
- Streamline an assembly line in a factory
- Analyze soil where a former industrial manufacturing plant once operated
- Predict a component's failure modes or maximum longevity
- Identify a gene responsible for a particular type of heart disease
- Reorganize business units to increase profit margins

When staff in ABERST are working long and hard on research, design, or analysis in projects such as those listed above, this work is engrossing, challenging, and important, and no doubt these persons want to tell readers about their work and their objectives. But doing so is not the same as identifying a purpose of communication for a report. You will not just broadcast at large a "report" about your work to the world. Instead, you will write distinct, purposeful reports throughout the course of any single project, as necessary, to communicate different purposes at different stages to different readers. Variety is an inherent quality of ABERST writing.

Each of the various reports has a unique and fundamental purpose appropriate to the needs of the author and others involved in the project at a specific point in time. Finding this fundamental purpose, stating it, and building the report upon it are the three important interrelated components of report writing step #1. And step #1 begins with understanding the physical reality of a document: reports are words on paper. And by "paper," I refer to reports and documents, whether printed or electronic.

Paper, and words upon it, can do some things well, but other things it fails miserably at. Paper cannot streamline assembly lines, design reactors, predict failure modes, or identify genes. Only people can do those things. Paper can simply be used to communicate information from an author to readers. The information will concern the work in some way, but it will not be doing the work; it will merely be communicating about the work. Understanding this limitation of technical writing is key to creating excellent documents. Such narrowing is the true first step. Paper can only achieve a limited number of objectives. Here is a fairly exhaustive list (while ignoring synonyms):

Communication purposes:

- Answer
- Assert
- Authorize
- Convey
- Inform
- Instruct
- Inventory
- Justify
- Offer
- Present
- Propose
- Recommend
- Request
- Transmit
- Update

Notice the list excludes evaluate, judge, and analyze. Those are work functions. People, not paper, do those three tasks. Paper comes into the situation like this: In the beginning, middle, and end of a project involving evaluation, judgment, and analysis, authors may need to write any manner of a report, and that report can answer, assert, authorize,...or update (any item from the list above). The fundamental purpose of any report will be one of the functions listed above, with some specific context added. Below are two examples of a communication purpose connected to a topic:

- Propose a study of transmission via birds and mosquitoes of west Nile virus in the Midwest states
- Authorize an inspection of a refugee camp by a non-governmental organization

With such a purpose identified and expressed, the entire document can be assembled in support of that purpose. In other words, the main point for the whole document will be connected to one of the aforementioned communication purposes. If not, the report will lack focus, organization, brevity, and efficacy.

Many times authors will kindly tell readers a lot about tasks they have performed, goals of the effort, and milestones reached. Still, none of this reveals the communication purpose. Usually, an early draft of a report lacks an explicit communication purpose statement. If a report contains a specific sentence near its beginning explaining the "purpose," it will most likely be referring to the project's purpose, not the communication purpose. Here is just such an example from higher education: "The purpose of this report is to organize a small group of college-level students as mentors to accompany a selected group of 11th graders from a distressed high school to visit colleges in Ithaca, New York, to help them become comfortable in campus settings to encourage future applications to colleges." This purpose is the author's objective, as it describes an intended program she wishes to implement if the necessary approvals, support, and funding are acquired. The report, however, must have a limited and focused communication purpose, which has not been stated. As the report is revised by the author, the communication purpose will need to be identified, stated, and fulfilled. Look at another example, this one from engineering:

- "I made a 'forest assembly' of carbon nanotubes (CNT). I observed their structure using a scanning electron microscope (SEM). I noticed a patterned alignment in the CNTs. After this, I applied an

> electrical charge to the tube samples, and I subjected them to strong vibrations and high temperatures. Their ability to adsorb benzene vapor was evaluated as well. Here is my report on this."

Would you imagine that this author is using this report to request funding for purchase of a new SEM? Nor would I. But that may very well be the purpose of communication. Or, perhaps the author is presenting a detailed explanation of her method of CNT growth at 750°C for publication in a specialized scientific journal devoted to such topics, but that is not self-evident either.

Providing a detailed description of one's work does not make a communication purpose self-evident. Even if an author tells us that he analyzed data from over 10,000 patients, used deep-learning algorithms, and controlled for nine distinct physiological traits, his report purpose is not necessarily evident from project details. Readers will not know. As mentioned in Chapter 1.2, good technical writing requires that authors identify the purpose of communication from the outset and state it explicitly in the beginning of the report. This is a first principle in ABERST communication: the purpose principle. This principle is most easily put into practice by a sentence in the beginning of the report that includes these words: "The purpose of this report is to . . . ." After some practice and a certain level of comfort with writing reports, an author may also apply the principle with less direct wording yet still provide a purpose statement explicitly, such as the following: "In this report, I offer a solution to the problem you are encountering with processor overheating."

To truly absorb the importance of the purpose principle, you need to immerse yourself in an actual ABERST scenario. This immersion, fortunately, reinforces the audience's role introduced above and throughout Part 1, which takes into account the standard reader's profile. Therefore, in working through the scenario discussed below, you can master these two crucial lessons, purpose and audience, simultaneously.

In presenting the scenario for this important lesson, I use slang and humor, despite my advice and guidelines (in Chapter 2.2) on refraining from using these forms of language in ABERST documents. I do this for three reasons. One, when you read the scenario below, which is inundated with slang and humor, you will readily perceive that such language choices do not belong in ABERST documents. It should be obvious that the style is inappropriate; my examples should illustrate and reinforce the important lesson that such language should be excluded from your documents. Two, this book, however, is an educational guide, not an ABERST document *per se*. In teaching, different styles and techniques must be used to keep learners interested. Slang and humor, as long as the instances are well intentioned and not discriminatory, can be helpful to breathe new life into a lesson. This book's midpoint is just the right time. Everyone needs to laugh and relax on occasion, and this is true of the educational process. So far, you have read eight chapters of this "how to" book, which can be challenging and tiring. You have been absorbing a lot of information. In this chapter, I give you a break and introduce a little levity in the spirit of helping to explain the vitally important purpose principle. Three and lastly, I want professionals in ABERST to know that their work can be fun, and it could lead to some laughs (not in the documents, though). No one should ever feel that their work is too serious or complex for fun and joy. When you are not writing your documents, you should find a way to laugh and smile while working with colleagues and peers, no matter your subject matter. Maybe your subject matter is more humorous than you had thought. Slang and humor will inevitably arise during casual conversations and informal, daily interactions, which is its rightful place.

3.1.2 Scenario for Purpose and Audience

I have a friend who is a medical researcher. Her doctorate is in diarrhea. That is the truth. Someone has to study this. She works in a university research division with other researchers who study different topics: flu, chemotherapy, cancer, eating disorders, and so on. Within this group, she is the only expert on diarrhea. She has a few graduate students who help her with her research, and they are always running to keep up with her. They help her, and she, in turn, assists them with their dissertations, in yet other areas of medicine and health.

In the course of her work, she may write a dozen different documents each year, some long, some short. Some go to her graduate students, some to her administrators, some to her funding agencies, some to potential publishers, and so on. Each report has a different purpose. But, because people are busy with their work, each report may seem like another interference in their day. They are inclined to think, "This document is some shit I do not have time for," because that is human nature. They might also be perplexed by her work, not quite sure they understand it all, and say, "This is some scary shit that flies right over my head," because that is human nature. Overall, persons at her very own university, let alone people scattered elsewhere across the planet, may think of her work as "crap" they would prefer to stay clear of.

Thus, she must get to the point in her documents and turn the crap into must-read material. She must make each document distinct from all the others, or her work will just run together into one big, amorphous clump of stuff people want to ignore. She must, moreover, make sure each document she writes has a single purpose, which is revealed to the appropriate readers quickly. After that, she must stick to this main point and build the whole report upon it. Let's look at four different communication purposes that may arise as necessary for her.

- Proposal
- Justification
- Instruction
- Presentation

Proposal

Suppose she wants to propose a collaborative effort with her colleague studying flu, as diarrhea and flu are often connected. She would be able to offer her colleague some funds from her own budget, and she in turn would get access to the colleague's data from a longitudinal study that has been ongoing for 10 years. She mentions this while talking at the coffee maker, and the colleague seems intrigued, perhaps just being polite, and says, "Put something in writing, and I will consider it."

The flu expert may not want to take much time to learn about my friend's crap. This report could be the last thing she wants to read. So, my friend must identify the purpose and provide the correct information to explain to the colleague that collaborating could be in both of their best interests and, indeed, could be interesting and feasible, as well as lucrative.

Justification

Suppose the Dean of her college is reviewing facilities and budgets and learns that my friend is using five large freezers in a basement storage area. The Dean is not happy the place is so full of crap. She calls my friend, who starts to explain her use of the freezers, and the Dean cuts her short, "Just put it all in writing, and I'll consider authorizing continued use of the freezers." So, here my friend must write a report justifying her need for the freezers in the laboratory for her samples.

The administrators, including the Dean, do not know much about shit. The Dean's specialty is kidney disease, and the administrators who work for her focus on personnel and budgets. They leave the crappy research to my friend. But they have all the power over accounts and facilities. If a report needs to go to them, my friend must identify the purpose and information relevant to their functions, or they will dismiss the report as "just more shit from Dr. Diarrhea."

Instruction

Suppose my friend takes a walk through her lab while the graduate students are working, and she notices little shortcuts they are taking, which disturb her. In fact, she thinks the quality of the work is jeopardized because the graduate students are not following sample collection and storage protocols. They did originally when she trained them, but over time bad habits have formed and the work is sloppy, and she has some new ideas for maintaining quality control and following protocols.

She must write a report to these student employees, people who are reluctant to change habits. In fact, they think they fully know the procedure, as they do this work daily. They may not believe the report is for them, but, rather, is intended for some other students who work the "night shift," or work in the professor's other laboratories. They feel, in fact, that they know their shit. The report to them, therefore, must convey the idea that, although they know their shit, they are making a mess of it. My friend must announce her purpose to instruct them on proper procedures and present those procedures without giving any false impression that the report is just the same old shit and can be ignored by experts in their area.

Presentation

Suppose she completes a 2-year study, looking at 500 persons admitted to hospitals with health issues that include diarrhea. She has collected data and looked at various factors, and she has developed some interesting findings she wants to share with others in her field around the globe. She wants to publish an article. This document is finally the type of technical report my friend thought she would write exclusively, instead of all the other types that have been unexpected yet necessary.

Her audience—other doctors, nurses, and researchers—are no doubt interested in her findings. In addition, they want to know details of her participant pool, their conditions, her hypotheses, and her methods: data collection, surveys, specimen analysis, and analytical techniques. This is some important shit, to them. They need to review her methods to see if it all seems acceptable, or they can critique any and all of it in their own publications. They can also build upon or modify anything they see that looks promising. Perhaps some research aspect deserves a new direction or further exploration. They might even want to call my friend and discuss some future collaboration.

As much as this is the very type of technical document my friend has been longing to write, and her training has prepared her to write such, she again must think of her purpose and audience. She also must somehow distinguish this report from all the others she has written in the recent past, such as the proposal, justification, and instruction documents mentioned above. This time, her funding and financial resources are not important, neither is her need for freezer space. And her reinforcement of laboratory protocol is unimportant. All of these items are just "her shit" that she deals with during the course of doing her project. When the time comes to publish, all that matters is the real shit—her diarrhea data and findings.

For the article to be published, my friend must write a document that announces its purpose to present findings from her long-running research, with some meaningful discussion of those findings. She must make and fulfill a commitment to provide the necessary information so her readers can follow each and every step of her research, and she must present her data, results, findings, and discussion in a manner that all readers can understand, without getting confused by extraneous, irrelevant, esoteric, or proprietary details that are unfamiliar to them. In particular, information related to collaborating with flu researchers, reserving freezer space, and reinforcing laboratory protocol is irrelevant, but somehow she must help her readers understand her laboratory and analytical methods, and any germane supplementary tasks, without bothering them with the mundane issues she has previously written about.

Although most of her workplace issues are off-topic for her publication, she needs to keep in mind that she must nonetheless write a document that explains everything to readers who were not participants in the work. This is a difficult task: leave so much out, yet provide both breadth and depth in a purely technical sense. Each important detail of her research may be completely unfamiliar to them. Possibly, my friend's methods for specimen collection, preservation, and analysis are novel. She perhaps uses tools and techniques that are brand new or never used before in this field. The terms and the phrases she uses might be of her own making. In the very least, none of her readers can fill in any missing information because they have no access to such. Anything and everything they learn about the research must come from the words in the document.

3.1.3 Purpose Principle's Corollary

Having reviewed the four report scenarios above, you are in a strong position to understand a corollary to the purpose principle: reports are no longer simply assigned to you or demanded of you by others. Certainly, an external demand is one way a report may originate. But, equally often, due to the purpose principle, you, and you alone, may originate a technical report. This happens each and every time you find yourself with a purpose for writing a report.

Purpose Corollary

- Authors must self-initiate a report when a necessary communication purpose arises.

Did you do three months of testing and discover something unexpected, although the testing is not yet done? Write a report explaining this unexpected finding to the people who need to know about it.

Did you work on a year-long project with three colleagues and realize that one of them deserves special recognition? Write a report to management praising this person.

Did you visit a factory to solve a problem with robotic assembly systems and discover that the factory has an electrical wiring problem? Write a report to facilities management explaining this new problem that must be addressed.

As in the above examples, many reports are unexpected by managers and other potential readers. They are event-initiated, not reader-initiated, as mentioned in Part 1. You are the first and only person to know that the particular issues exist. This underscores the importance of announcing the report's purpose right from the beginning, so readers will not have any confusion about the "unexpected" content of these reports. The readers will get your point right away and then can read either thoroughly or selectively, as necessary, or re-distribute the report to other appropriate readers.

And you cannot know who will become a second, third, and fourth reader, after your immediate readers are finished with it. So, you have to announce—for any and all readers—the purpose of the document.

In sum, ABERST documents are necessarily purpose-based, related to either (1) transmitting some information to the appropriate person(s) to advance activities within your organization or (2) presenting information to various persons outside your organization to assist and advance yourself, them, or both. Because some sort of organizational purpose drives all documents, you will never be asked to write for purely personal reasons. (Even publishing research results has a larger purpose than simply advancing your own career.) Personal writing is not ABERST writing.

As a result, some of the writing done during college is not good preparation for ABERST, including personal journals, psychological confessions, vacation recollections, and biographical memoirs, although these genres do help to hone your composition skills. Such "personal musings," however, must be subsumed under one of the aforementioned communication purposes within an organization to have a place in a professional technical document; an organizational purpose is always required to warrant a document. Otherwise, a document is not likely to be useful in ABERST. Similarly, other school-based writing tasks are not faithful examples of technical writing, such as reflecting on a day's lesson, evaluating a guest lecture, or criticizing an exam question. Reflecting, evaluating, and criticizing are not reasons enough to write an ABERST document.

Even the definitive college genre—the essay—is not characteristic of a purposeful document. Essays cover important subjects, such as human rights, culture, and economics, but they are academic exercises lacking organization-based purposes. Consider the topic of the Middle East, a controversial topic of political and social concern that demands careful thought and polished writing. In college, a student might be assigned to write about this vast region, to demonstrate writing skills and mastery of facts. Done well, this type of essay might also improve and demonstrate one's increasing aptitude for research, analysis, synthesis, and argument, culminating in presentation of a thesis. If the assignment requires isolating a particular area of focus within the larger topic, it has obvious educational value, and essays can be both informative and persuasive. If the writing is strong, moreover, it will even avoid the potential pitfalls of bias, emotion, and over-simplification.

Aside from a few academic specialists, however, most professionals working in business, engineering, research, science, and technology would never just write broadly about the Middle East, and certainly not subjectively. Instead, a working professional would write a document with a particular, organizational, and imperative purpose, such as one of the following:

- Inventory items of antiquities of value still intact or known to be damaged in Syria, as a result of the ongoing civil war, to help people trying to raise private funds and international support to preserve these important items
- Request permission from owner/operator of desalination plant in Israel to study their facility in person and review the design, operations, and management, to serve as a model for a similar, future plant in Australia
- Propose a 1-year research study of refugee camps in Jordan and Lebanon to assess conditions, to obtain useful data for non-governmental organizations

The aforementioned hypothetical reports are just a few examples of the purpose-based writing professionals do after they finish learning to write well in college. Actually, most still continue to learn writing while they work. Basically, by doing more writing, persons improve quite a bit. It is a job duty that gets easier and easier over time. Very soon, writing with a specific communication purpose becomes automatic. This is a critical first step toward excellent technical communication.

As mentioned above, not revealing your purpose in a document is about the severest writing mistake you can make, much more severe than a spelling or punctuation error, on a level with not fulfilling the basic document requirements (letter of the assignment). When readers cannot find your point, and do not know the reason you are writing, they lose motivation for reading the report. They are not sure if they should read the report at all. They throw their hands in the air and wonder what it is the author is trying to say. It does not matter if some facts are fascinating and some technology is tantalizing; the report falls flat.

A person may collect, record, note, calculate, calibrate, and analyze better than anyone else. None of it matters unless that person has a purpose for writing: then those skills will be coordinated into a concerted effort in support of a report to achieve some organizational benefit. I personally enjoy nearly all topics in ABERST. I find nothing uninteresting. Nevertheless, I have neither time nor interest in reading about disconnected facts in connection with someone's work.

For example, you say you operate a wind tunnel? I do not want to read about your work in a wind tunnel. Boring. You say you can simulate chemical processes and optimize them before building a processing factory? I do not want to read about your chemical process simulation. Snoozer. But, if you want to propose a new airplane design because of your work in the wind tunnel, I am all ears. If you want to tell me that the chemical plant down the road from my house presents no health and safety risk to my neighbors and me, you have my attention. In short, whether the audience comprises persons within your company, associates at an affiliated organization, or members of the public, they will read your document if you have a point, you declare that point explicitly and quickly so they can find it, and that point is of some importance to them.

The next chapter in this third part of the book covers assembling the necessary information to fulfill your communication purpose. The transition between this and the next chapter is best accomplished with two actual short report examples, one weak, one strong. In the weak one, you can see that the purpose of communication is not stated directly; readers must infer it from the rambling text. Indeed, the text may not be the ideal collection of information in support of the purpose because the author did not recognize the need to articulate the purpose and build the whole report upon it.

In the strong report, however, the author expresses the communication purpose at the outset and proceeds to rally all the right information in support of that purpose, placing each item into the report at an ideal location.

Weak Sample

Date: 20 Feb 2020
To: Kai Oto, Testing Department Manager
From: Amos Zahn, Maintenance Engineer
Subject: Problem in Lexington

Introduction

Per your instructions in October, I traveled to our Lexington location to study the problem with the translation stage on top of the large wind tunnel. Staff at that site had sent a problem report regarding the translation stage's incremental vertical slide. According to their report, the slide had been stripped due to excessive force. As a result, pressure probe placement and measurement of said probe's vertical height were highly irregular and suspect. Staff could not be sure that any probe setting would hold fast or that the digital display was accurate.

Site Visit

I operated the tunnel a bit when I first arrived and I took some initial velocity measurements. Wind measurements seemed aligned with the most recent calibration curve (March 2014), so I was not concerned about motor operation. But, just on a whim, I thought I would measure RMS velocity and check for turbulence. As a result, it seemed to me that turbulence levels were unusually high, especially downstream, where we would expect turbulence to abate and level off.

Wondering about this, I looked into the possibility that the flow straightening mesh screens in the tunnel's settling chamber might be damaged. At first, I was not able to detect any damage by visual inspection. (The inlet screen was easy enough to remove, but the straightening mesh screens are 4 feet away from the opening, and I could not crawl inside.) After talking with the two operators, we discussed an idea that I could send a remote camera into the inlet using a jerry-rigged reticulating arm and placing a small, video

Part 3

camera on its tip. With this makeshift device, I was able to look at the first and second screens fairly thoroughly. Beyond those, I could not see the other four screens. My assessment from looking at the first two screens is that they are torn in places and clogged.

Discussion

I explored some possible repairs. Full replacement (from the tunnel's manufacturer) would be upwards of $100,000. They also offer a service that provides for removal, repair, and reinstallation. The representative and I did some back-of-the-envelope calculations and realized that this option would be nearly as much as full replacement, approximately $88,000. So, it seems like full replacement is the best option with the manufacturer.

One other idea I had was to add honeycomb sheets after the six mesh straightening screens (leaving the straighteners in place). The honeycomb could serve to smooth out the rough flow exiting the straightening screen array. The honeycomb is relatively inexpensive and it could be put into place by removing the inlet cover, which is easily done in-house.

Conclusion

I think we should do some studies of honeycomb using the facility we have in Rochester. It is nearly identical to the one in Lexington, and we can move equipment in and out of it very easily, and use it to experiment. If these tests prove positive, I could nail down an accurate cost estimate for the honeycomb installation in Lexington, and we could implement that repair. I would be happy to take the lead on this endeavor moving forward.

3.1.4 Analysis of Weak Report

This report rambles along. It starts with a false topic: readers have every reason to believe the report will be about the damaged translation stage's vertical slide and the author's repair of it. Not much is said about that at all, except to focus the entire beginning on the topic. If I were the author's boss, I would expect the report to explain the solution to the stripped vertical slide. In addition, as a boss, I may choose to put the report away and not give it much attention, depending on the urgency of the problem with the vertical slide. A boss might even file it away, never to look at it again, making the wrong first assumption that the vertical slide has been repaired and no further attention is needed to this matter. But the report has an entirely different main message, and the boss must definitely read the report, which is not assured.

Eventually, the author tells us that a second problem, turbulence, was discovered while investigating the first problem. This seems to be important, although the severity and implication are never explained. Again, should a boss care? Should the report be routed to other people? Perhaps people in Rochester? The report may not elicit the necessary action from a boss because its communication purpose is not clear until near the end, by inference only. The purpose is never overtly stated, namely, the author is requesting support from management for studying a potential repair idea that could save the company money: using honeycomb sheets instead of replacing the six mesh flow straightening screens in a wind tunnel. Significantly, he kindly offers to oversee the proposed effort, which he suggests would be ideally conducted in Rochester. One final note: Even the author's use of headings, although an appropriate element of technical writing, does not promote his purpose of communication. The headings are mostly generic, and the content under each heading is not clearly "announced" by the headings: Introduction, Site Visit, Discussion, and Conclusion. Moreover, the content is not correct for his actual communication purpose (to be covered in detail in the next chapter).

An entirely different organization of content is required for this document. It is the organization promoted throughout this book, starting with a clear, overt communication purpose, and continuing with the necessary information to support that purpose. An improved version of the report is provided below. In particular, the report starts strongly (recall Chapter 1.2). Using only nine paragraphs, the author supports the communication purpose with appropriate information. I have indicated where figures and tables would go, but I have not included them because visual aids (figures and tables) are covered in a dedicated chapter of this book, Chapter 5.2.

3.1.5 Example of Improved Report

Strong Report

Date: 20 Feb 2020
To: Kai Oto, Testing Department Manager
From: Amos Zahn, Maintenance Engineer
Subject: Request to Research Solution to Tunnel Turbulence Problem in Lexington

Introduction

Although the Lexington facility's wind tunnel is in nearly constant operation and has fairly up-to-date components, it may no longer be providing smooth laminar air flow through the test section where we place the vehicle models for aerodynamic study. On a recent visit I identified abnormally high turbulence in the test section, not the desired smooth flow. As a result, I made some preliminary observations and analyses to attempt to find the source of the airflow problem as well as a solution. My purpose for sending you this report is to request permission and resources to conduct further research of a solution that I believe could be both effective and economical.

Summary

Before I provide the details of this situation, I want to present my plan in brief. Although several solutions to the turbulence problem are plausible, the addition of honeycomb sheets after the tunnel's mesh screens and before the test section's inlet may be the optimal approach. To confirm this proposed solution, I suggest we study the honeycomb in our similar wind tunnel in Rochester. That facility would be ideal, as it is similar to the one in Lexington and its performance has recently been confirmed as very reliable. To conduct an adequate study, I will need time and resources, as approved by you.

Problem Identification

Although our staff in Lexington have reported problems with the translation stage on top of the large wind tunnel, I identified a more severe problem during my visit last month. (I discuss the problem with the translation stage's incremental vertical slide in a separate report, pending.)

While operating the tunnel a bit when I first arrived, I took some initial velocity measurements, including checking the root mean square (RMS) velocity values, which reveal the extent of unwanted turbulence, if any. As a result, it seemed to me that turbulence levels were unusually high, especially downstream, where we would expect turbulence to abate and level off. See Figure 1 for specific values measured at discrete locations.

Figure 1

Wondering about this high turbulence, I looked into the possibility that the flow straightening mesh screens in the tunnel's settling chamber might be damaged. After talking with the two operators, we discussed an idea that I could send a remote camera into the inlet using a jerry-rigged reticulating arm and placing a small, video camera on its tip (see Figure 2). With this makeshift camera device, I was able to look at the first and second screens fairly thoroughly. Beyond those, I could not see the other four screens. My assessment from looking at the first two screens is that they are torn in places and clogged (see Figures 3 and 4).

At first, I was not able to detect any damage by visual inspection. The inlet screen was easy enough to remove, but the straightening mesh screens are 4 feet away from the opening, and I could not crawl inside. This predicament prompted the creation and use of the camera device.

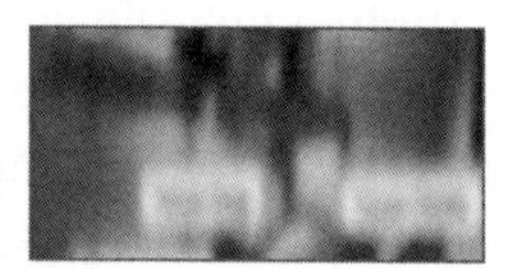

Figures 2, 3, and 4

Proposed Solution and Follow-up Studies Requested

An optimal solution to this problem is to add honeycomb sheets after the six mesh straightening screens (leaving the straighteners in place). The honeycomb (see Figure 5) could serve to smooth out the rough flow exiting the straightening screen array. The honeycomb is relatively inexpensive, at $24,000 for manufacturing, custom fitting, and installation (see Table 1 for cost breakdown) and it could be put into place by removing the inlet cover (see Figure 6), which is easily done in-house.

We need to test this concept before proceeding. We can use the facility we have in Rochester. It is nearly identical to the one in Lexington, and we can move equipment in and out of it very easily. This will enable us to run experiments of turbulence with the honeycomb addition, on a small scale. Testing costs should be minimal. I calculate $4,000 (see Table 2). I would be happy to take the lead on this endeavor moving forward.

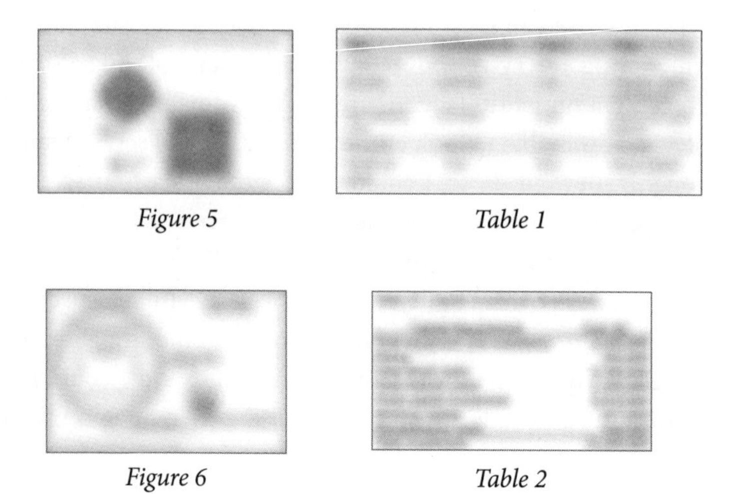

Figure 5 *Table 1*

Figure 6 *Table 2*

Alternatives Considered

I explored some other possible repairs but determined them to be inferior to the honeycomb approach. Full replacement of the mesh screens (from the tunnel's manufacturer) would be upwards of $100,000. They also offer a service that provides for removal, repair, and reinstallation. The representative and I did some back-of-the-envelope calculations and realized that this option would cost nearly as much as full replacement, approximately $88,000. Both of these are more expensive than the honeycomb.

Conclusion

Your thoughts on this proposal are much appreciated. If you concur with my preliminary findings and can authorize the requested studies in Rochester, I will proceed promptly, with a goal to finish testing in 30 days' time.

Future reports in this series may have the following communication purposes, as this employee continues his efforts to solve problem(s) with the Lexington wind tunnel:

- Recommendation of solution implementation backed by results of honeycomb testing
- Final presentation of total project cost for turbulence problem identification and solution
- Recommendation for repair of translation stage in wind tunnel

The strong report presented in detail above is one example of a purpose-driven report that persons in ABERST must learn to write, to effect advancement of their personal and organizational interests. Importantly, I demonstrate the use of an Overview, comprising two sections: Introduction and Summary. After that, I emphasize the importance of an organized Discussion, with the necessary support information, divided into four sections with these specific headings: Problem Identification, Proposed Solution and Follow-up Studies Requested, Alternatives Considered, and Conclusion. This example of a strong report illustrates the method of assembling the right support information, and this critical technical writing technique is addressed in full in the chapter immediately below.

Chapter 3.2

Assemble the Right Information

You cannot muster the correct and requisite information into a report if the communication purpose is not clear in your mind. That is the reason the previous chapter was placed before this one. That lesson comes first. After that, we can study in depth the skills for assembling the right information. Did you notice in the example in the previous chapter without a clear purpose that all of the information seems random and confusing? In the good example, however, the information is on topic and useful. That is the goal.

The method is this: Once the communication purpose is understood, put that into the report and begin to sketch out the supplementary communication tasks that must be accomplished to fulfill that purpose. A whole different set of information is often required, as compared to the original information an author selects for a report before a clear communication purpose is formulated. Furthermore, as shown above, no matter how inherently interesting the details of someone's work and project may be (ideally), they are all boring and distracting when thrown into a report for which a reader cannot find the main point.

Do not wrongly assume just because a document is founded on one communication purpose it must necessarily be short. One might suppose that with a single communication purpose, an author could simply blurt that out and be done with the whole thing. You will need two or three paragraphs, tops, you suppose. The fact is that under a single purpose, with just one main point, a report may be 100 pages long. It must have all the necessary information to support that purpose, and that may be quite a lot. In fact, reports need to be organized well because they are often full of dense, complex material under numerous categories of information. So, this chapter addresses the importance of assembling all of the necessary information to fulfill your communication purpose, nothing more, nothing less.

To understand which information to assemble for any given purpose, we can start with the wind tunnel report in the previous chapter.

As interesting as the translation stage may be, it has nothing to do with the purpose of communication: request approval and resources to conduct testing of honeycomb sheets in Rochester.

Similarly, too much time is devoted to discussing the two expensive (and dismissed) options for repairing the wind tunnel and reducing turbulence. Because those are alternatives useful only for perspective relative to the proposed solution, they have a place in the report, but that place should be secondary, not primary as in the weak version.

The key information that must be assembled to support a request for time and resources to conduct tests are the following:

- Problem to be addressed (motivation for work anticipated)
- Anticipated solution that needs to be studied (hypothesis to be tested)
- Tests to be done (justification of tests as beneficial and suitable)
- Advantages of the solution (functional, financial, safety, environmental, etc.)
- Other alternatives considered (indication of thoroughness and ingenuity)
- Anticipated difficulties and any mitigation plans (commitment to be alert and flexible)

One cannot anticipate the appropriate content for every single report in a book such as this. Instead, with examples, I can show, in broad strokes, the type of information that suits various purposes. Authors will need to brainstorm and ruminate upon each and every report, to identify the necessary and helpful information that readers would need to understand, concur with, and value the report, as well as take follow-up action as necessary. This will have to be done on a case-by-case basis. The following examples can be studied to help this lesson penetrate the mind, as a start. The lesson is revisited elsewhere in this book also. These companion chapters add finer and finer detail to the information here, approaching the concept from particular directions. Specifically, if the report overall has an argument that needs to be conveyed to persuade readers, Part 4 of this book covers that component in detail. Similarly, Part 5 addresses specific genres of documents and provides guidelines for each one, with a "best approach" for both structure and content.

The five examples below are intended to illustrate the process of assembling the right information for the given purpose. They all stem from the same context, as an important situation unfolds over 9 months at a major manufacturing facility. For each report, the author and purpose is identified. After that, several categories of information are sketched. Each of those represents a section in the report, not a sentence. These are categories of information, representing the broad sweep of information that must be assembled to fulfill the communication purpose. Details will flow under each section heading, as the author fleshes them out.

1. Factory manager writes memo to company's health and safety manager; wants to secure her help in identifying the cause of a recent spike in respiratory illnesses among employees

- Quick review of employee illnesses known and reported
- Request to investigate cause of illnesses
- Declaration of problem's priority and urgency, with proposed deadline for completion of investigation
- Offer of specific resources to facilitate investigation

2. Health and safety manager writes formal report to factory manager with conclusions of her study and recommendations for remedial action

- Details of known illnesses: who, dates, specifics of illnesses
- Possible causes of illnesses: thorough listing of all possibilities
- Investigation of each possible cause, with facts for and against
- Conclusion as to most likely cause, with proof: problems with existing procedure
- Careful elimination of other possible causes
- Plan for mitigating injuries and eliminating cause (source) via new procedure
- Overall assessment of pros/cons of new procedure (more pros than cons)

- Impact of implementation of new procedure, as mitigation and remedial plan
- Timeline and costs of plan

3. Factory manager writes memo to factory management team: procurement, shift supervisors, training, quality control, etc, to announce and distribute new procedure as recommended by health and safety manager and obtain any feedback or suggestions before promulgation and implementation

- Background of existing procedure (why it's done, who does it, and how it's done)
- Problems with existing procedure (quick review of employee illnesses)
- Highlights of improvements to procedure incorporated by revision of procedure
- Written version of new procedure to be distributed for thorough review
- Former procedure (to be superseded upon approval of new one) in appendix, for comparison purposes, with major differences highlighted

4. Training manager writes memo to all affected employees to announce and distribute new procedure

- Background of existing procedure (why it's done, who does it, and how it's done)
- Problems with existing procedure (quick review of employee illnesses)
- Highlights of improvements to procedure incorporated by revision of procedure
- Written version of new procedure to be distributed for implementation
  - Background of procedure (why it's done, who does it, overview of how it's done)
  - Materials, tools, equipment, parts, and so on needed for procedure
  - Warnings and concepts related to procedure
  - Step-by-step explanation of procedure

5. Factory manager writes reports to corporate executives explaining problem and solution, overall impact prior to remediation, and anticipated positive outcomes going forward

- Background of illnesses
- Procedural context of illnesses (salient facts about factory process and procedures where illnesses occurred)
- Assessment of costs and setbacks from illnesses
- Recap of investigation conducted (citation to health and safety manager's report)
- Explanation of proposed remedial plan (brief comparison to alternatives)
- Highlights of improvements to procedure incorporated by revision of procedure (efficacy, benefits, advantages)
- Implications (cost, training, timeline) of implementing new procedure
- Results of remediation to date

All of these reports have a common core built around the mysterious problem of employees suffering from respiratory illnesses and the need for, and eventual implementation of, a remedial plan, namely, a new factory procedure. Despite this common core, nevertheless, each report has a unique assemblage of sections, geared to its purpose of communication directly, and, indirectly, to the primary recipient(s). Some readers need to see the recommended revised procedure, while others do not. Some need a discussion of the possible causes of the employee illnesses, while others do not need this component of the investigation. Some need a tally of costs incurred due to employee illnesses, others need an emphasis of the expected benefits and advantages of the proposed remedial plan.

A fundamental corollary can be derived from observing these differences in details from report to report: rarely can the content of one report be sufficient as the content of another report, if they have different purposes, even if the reports are associated with the same project. A writer's over-reliance on one set of material for numerous reports is a flawed approach. Each report must be customized, adding and deleting information as necessary. Sending nothing but the new procedure to the executives will have them wondering, "Why are we getting this procedural revision?" Similarly, explanations of past illnesses to new employees will befuddle them and have them asking, "What does this have to do with my job?"

Mastering this technique of gathering the right information to fulfill the communication purpose prepares you for the next step in excellent purpose-based technical writing: normalizing a document for all potential readers. That is the focus of the chapter that follows.

Chapter 3.3

Normalize the Information for All Readers

Often one business, engineering, or research unit has its own idiom: key conceptual terms, units of measure, regulations and codes, acronyms, jargon, and abbreviations. That is the language each author knows best when immersed in a specialty and specific project(s). When a document needs to be submitted to someone outside this unit, even if the person is intelligent, certified, licensed, highly educated, and a fellow engineer, doctor, or scientist, the document will not be clear to her if it is written in the unit's special idiom. This happens often.

The solution is to break out of one's special idiom when communicating in written reports, which authors must assume will leave the narrow confines of the unit, either immediately or eventually. Authors must strip the document of esoteric terms, codes, jargon, acronyms, and so forth, or make sure those are clearly defined.

Imagine a memo to a boss requesting funds to buy a new scanning electron microscope (SEM). You might be able to make this brief and use lots of abbreviations and acronyms. Your lab is identified with an acronym. Your work and project are described with a generic term for a research grant number. The current SEM is noted by its nickname. The problem that requires a new SEM is simply labeled, "problem with samples." The heart of the memo reads as follows:

Weak, idiomatic report:

- "Half Dome no longer seems adequate for our needs for R21. Those problems we observed with our PCD samples could be solved with a new SEM. We have funding for R21 for 18 more months, but the grant does not include a line item for magnifiers."

All of this is jargon and clear to only a limited number of persons highly familiar with your lab. Many authors have no problem writing in this way. It is conversational and abbreviated. It is a quick approach to a mundane task when time is limited and many other responsibilities demand one's attention. Fortunately, authors know that co-workers and even the boss understand such prose perfectly because of their close working relationships. This type of shorthand communication is fine for in-person discussions, day-to-day emails, and other notes among co-workers, but it is not sufficient for a purpose-based document intended to advance a personal or organizational interest. And, a request for a new SEM is a real purpose.

As this project continues, imagine that the boss sits on this memo, neither rejecting nor approving the request. It just sits in her "to do" box. Six months later, she takes a job in another state. The person who replaces her comes from another company and division. This person knows all about lots of current management theory and has a solid engineering degree earned 15 years ago. Presently, this new manager takes out the memo you wrote and reads it. Wham. Nothing makes any sense. Most of the key terms are unfamiliar to her.

This is just one reason that you need to write ABERST documents for an audience who may be unfamiliar with your work, new to your organization, and not especially trained in the same specialty vocation or discipline that you are trained in. In other words, you must normalize your documents for readers who have these typical qualities (because these types of people may end up reading your document). I strongly suggest you assume that your documents will be reviewed by more collateral readers than immediate recipients. Indeed, the world contains only a few people who know your work intimately; all the remaining people in the world are strangers to your work. Strangers, including those senior to you, have these qualities:

- Education, experience, and expertise in subjects different than your specialty
- Unfamiliar with the problem(s) you are confronting
- Considerably behind the times (6 months or more) with current advances in both your broad employment area and your specific projects
- Severely lacking extra time needed to review other persons' projects
- Typically reading documents months after they were written

Now, you could say, "I'm not worried about those people. I'll just write for the people NOW, who I know will read this, namely Bud and Bo." Here are Bud and Bo, described by the author who knows them so well: They know this project, they are as smart as I am, if not smarter; they do not need any reminders; this information is on the forefront of their minds right now.

Let me tell you: there are no such people. I have a hard time finding the Bud and Bo described by the author, supposedly sitting right over there—one office over. Bud and Bo are a fantasy. Even people working right next to you, or in the same hallway, have their own projects, backgrounds, expertise, responsibilities, commitments, and busy schedules—all of which chips away at the ideal form of the perfect report reader. The last reader you will ever have who almost meets the description of Bud and Bo is your college professor.

For most class writing assignments, you can usually be right that your professor can anticipate all your thoughts; sees the report coming; understands all the concepts, ideas, and analyses (she taught them to you); and so on. This is the reason when I grade college reports, I push myself to forget I am the instructor, and I place my mind into a space of being the "stranger" as described above. I then grade the report; if it deserves an A grade, it will be an excellent document for any reader. It will be a useful document even well into the future (not just for that particular class, that semester). But quite often, the report is a putative "A" for me only, so it does not earn an actual A grade. Other readers would be hard pressed to understand it. A report might contain this passage:

- "After we converted the velocities to Reynolds numbers, we then plotted Strouhal numbers against Reynolds, which confirmed the linear correlation between vehicle speed and the vortex shedding frequency."

I understand this, but I do not expect all other readers to do so. If this passage were in a real report to an aerospace design team, it would need to be normalized so the jargon is germane to an organizational purpose. The students' knowledge may be absolutely correct, but the report is not clear to "strangers." So, it does not deserve a true A grade. And that is your challenge as well: write your reports so they will be at the A level to strangers: helpful and understandable to any possible reader, and not just those in your clique, not just Bud and Bo. (Tip: If written for Bud and Bo, a B grade or below. It rhymes.)

The method for writing such a document is called "normalization." A writer in ABERST must ensure that each document is normalized so different readers can pick it up and read it quickly (possibly skim, maybe peruse) to see if they can make use of it. To help you do this, you can play an imagination game.

Imagine the document is written for a person you have been told will replace the actual recipient the day after you finish writing. Thus, you will write for a certain recipient but a new person will step into her position and read your report. This person has more education and experience than you do (so she will be able to identify any mathematical/scientific/analytic errors or gaps in the report) but has never worked in your field, having just transferred from a different sector, such as when an executive in the airline industry transfers to the automobile industry, or when someone in food is recruited to pharmaceuticals. Overall, a typical reader of an ABERST document may be someone uninvolved in original work; highly educated but with different specialty than author; with different skills, training, and experience; and from a different office, unit, and country.

If, for some reason, the new person never arrives and your current boss retains her position, your report will be initially read by this current boss, although written to the "replacement." Let me assure you: your current boss will praise this report as the best writing she has read in years, on the job.

The steps of normalization include two sets of opposite items—some not to do and others to do.

Do not:

- Use undefined abbreviations, acronyms, nicknames, or jargon (SEM and CNT)
- Assume reader remembers what happened in the past, on a specific day, at a specific time (Because of last Tuesday's occurrence, we need to run a Loop 2 procedure.)
- Assume reader knows where you traveled (The visit on Wednesday produced a confirmation of my theory.)
- Assume reader knows the problem or challenge you are working on (In an effort to make it smaller, you can appreciate no doubt that I experimented with new compounds.)

Do:

In the proper structure, starting right at the beginning of the document, provide the following fundamental information:

- Author and department (credibility and responsibility for follow-up action)
- Date of report (unique identification and timeline record)
- The motivation of the overall project the report concerns (context, background)
- The specific issue that deserves attention (problem)
- Locations pertinent to context and problem (narrowed focus)
- The work or effort taken to address problem (investigation or task)
- The purpose of writing (communication purpose)
- The primary conclusion(s) and recommendation(s) found in report
- Implications of aforementioned findings, conclusion(s), and recommendation(s)

The information listed above answers the basic questions that all readers will have:

- ▶ Who (author)
- ▶ When (key dates and date of report)
- ▶ Why (context)
- ▶ What Now (problem)
- ▶ Where (important locations and offices integrated into context and problem)
- ▶ How (investigation or task)
- ▶ What New (findings)
- ▶ What Next (implications)

Any important report may be pulled from a file and reviewed some point in the future, for reasons completely unknown to the author at the time of writing. The report may have some important information that has become valuable because technology has advanced, or it may be needed as a baseline against which to assess something new and different. Authors cannot know. So, an author is best to imagine that any and all of the following types of people may one day read the report, in addition to one's immediate supervisor and co-workers:

- New, junior staff to join author's group
- Non-specialists elsewhere in organization
- Financial investors
- Dealers, brokers, and salespersons
- Government agency personnel (local, state, and federal)
- Executive managers within organization
- Managers and employees in external organizations
- Members of public
- Journalists and others from newspaper, radio, and television

You cannot take the time to attempt to write a tailored report to each of these audience types. That would be burdensome and impossible. The potential audience list is infinite and only hypothetical. Instead, the efficient approach is to write one, normalized report that works for all of the future audience types. Normalizing is a skill you now understand.

The efficient approach is to write one, normalized report that works for all future audience types.

Working professionals in ABERST offer the most resistance to this aspect of the overall approach presented in this book. They push back at me here, saying that for most of the documents they write, the informal style for readers on their team is adequate, and they do not encounter situations often when a normalized report needs to be written. They want to confine their writing to their idiomatic language. This is a sticking point that truly reduces the quantity of well-written reports throughout ABERST, for several reasons. One, as said before, the extent to which these casual, idiomatic documents really make sense to the recipients, even those closely connected to the author such as Bud and Bo, is exaggerated. Truly omniscient readers are rare. Two, the time spent to write the casual, idiomatic report takes away from time that could be used for a normalized report, so professionals perennially feel that they have too little time to write clear and widely useful reports. With too little time, normalized reports are either left unwritten or produced in a rushed manner. Therefore, and three, this time crunch results in poorly written reports even when authors are attempting to write good ones, due to both lack of time and deeply ingrained bad habits from writing too many casual documents. Thus, the typical project paper trail ends up consisting of mostly (1) casual, unclear documents, useful for just a small set of early readers if anyone at all and (2) incondite and confusing yet seemingly formal documents. The normalized, universally clear documents do not get written.

To conclude this chapter, here is an excerpt of the SEM report introduced at the very beginning, after it has been normalized for the "new boss" and all other future readers:

Strong, normalized report:

- In January 2014, our research team received federal funding from the National Institutes of Health, Grant #R21-14UVA, to assist us in our efforts to develop a coating for coronary artery stents for implant into persons with heart disease, encompassing severe arterial blockage and pain (angina). One aspect of this research involved trying to create polycarbonate diamond (PCD) coatings for stents. The manufacturing of PCD using different methods has been a back and forth effort, but one problem in particular has hindered our advancement. In particular, we have not been able to fully see and evaluate our PCD specimens because our scanning electron microscope (SEM) has been insufficient, as it is a first-generation model. A state-of-the-art SEM would expedite our research. Unfortunately, the grant does not provide for purchase of a new SEM. The funding is limited to other expenditures. I have researched various SEM types and purchase options, so I am writing to ask you to provide the necessary funding from one of the department's other accounts.

Chapter 3.4

Prioritize Parts of the Document

The previous chapter's topic, normalizing the document, is aimed at clarity. This chapter's topic, prioritizing parts of the document, is aimed at structure. The cumulative goal of all the techniques is to produce a purposeful document, clear to all readers, in a structure that supports the purpose and helps readers navigate their way through the report either selectively or thoroughly. Each component moves us closer to reaching this goal.

Prioritizing parts of the document requires mastery of one concept: distinguishing between primary and secondary information. Developing such discernment is purely mental. It is not a writing skill, per se, yet it benefits both the author and the reader enormously. In short, understand this: the various pieces of information in any ABERST document have different levels of importance. This may not be true in fiction, mystery, fantasy, poetry, and the like, where every word, character, plot development, and clue matters. But, in ABERST, some pieces of information are more important than others. In addition to secondary items, you may even have tertiary material. This is a good thing. But, you must be able to distinguish.

> The various pieces of information in any ABERST document have different levels of importance.

As I have said before, the author must know the writing purpose. Based on that purpose, the author must be able to identify the primary information needed to convey that purpose; inevitably, some information is of lower importance. It has value and must be reported, but it is of lower criticality. This is the secondary and tertiary information. If you cannot do this separation, your reader will not do it for you. I will run through some examples and specifics so you can get a sense of this technique.

One other benefit of distinguishing and categorizing information into primary, secondary, and tertiary classes must be emphasized. The effort helps you organize your material, of course, but additionally, it enables your audience to read selectively. As they navigate through the document, they can skip around, or read some section first, and save others for another day. Or, they can read some, and ask some other colleagues to read the other sections. This could involve delegating some review and double-checking to people more appropriately suited to doing those tasks, and allowing the initial readers to read only those portions that are absolutely necessary.

One first approach to distinguishing is to consider the immediate recipient's needs. While you cannot anticipate all the possible readers over time, you can know (and you in fact should know, see Chapter 1.5) who your initial recipient(s) is/are. You then assess the initial reader(s). What is this person's need? You must identify her ultimate concern, not the subordinate ones.

Take an example from aerospace engineering. A supervisor asks an employee to write a report with results from testing. But a lot more may have been asked of the employee: work with two peers; evaluate, clean, and repair all pertinent equipment; research the relevant literature; compare empirical results to theoretical predictions using known equations; and so on. A key directive from the supervisor may have been, moreover, to use the medium-size wind tunnel to test a scale model. The testing results named as the focus of the requested report are from the wind-tunnel tests. In sum, the supervisor piled a lot of requests onto the employee, but these go unmentioned when the desired report is demanded. In fact, only "results from testing" are specified. So, what is the supervisor's ultimate desire?

If one digs deeper into the situation, one can find the supervisor's primary objective for delegating the work and demanding a report. He does not merely want results of wind-tunnel tests. He does not want equipment cleaned, or literature researched, as ends in themselves. His objective for combining all these directives is to have the employee evaluate an innovative, aerodynamic vehicle design. The ultimate goal is not to have testing results but to assess a new vehicle design. There's a difference, which may not be seen initially. Let me pick these apart.

The latter is the ultimate objective (or, need) of the recipient. The former is one means of reaching, or fulfilling, that objective. It is a "means." You have both "means" and "ends." (Usually, the "end justifies the means," but not always. But that's another issue, perhaps a moral one. Discuss that with friends and family on your free time.) In terms of its applicability to ABERST, you want to differentiate the objective from the methods to achieve that objective. The methods, and all other ancillary tasks, may be numerous. But they feed into accomplishing an ultimate objective, and that distinction will be crucial in prioritizing parts of the document.

A boss, client, agency, and the like may prescribe a particular method, in which case you have no choice but to execute and report on that method. In other cases, they may not dictate a method. The project "sponsor" may initiate the work and tell you the objective but leave the method up to your good imagination. When this happens, you have been given the end, but not the means. In either case, the end concerns the sponsor more than the means, although in the first instance, they expect a particular means to be used. An example from a different field may be insightful.

Imagine you are an education expert, more specifically you have a PhD in Literacy and Special Education, and you have worked for many years developing programs for students learning English as a second language (ESL). You receive a contract from a state education department to propose an ESL program for students in grades 4 through 6. They want your report in 3 months with a fully detailed "reading and writing enhancement program" for ESL students in the aforementioned grades. We can think about the crucial information in this report: The description of your "reading and writing enhancement program" that the state will adopt (and compensate you for). This program includes materials, such as books, worksheets, vocabulary lists, and so on. It also includes your suggestions for mentoring and one-on-one, student-teacher consultations. It might also include your suggestions for using flip charts, marking pens, collage materials, audio recordings of the readings, and many more similar pedagogical items. Readers can follow your program, step by step, with procedures for both the teachers and the students and an inventory of materials to use, which you have enumerated, cross-linked to various steps of the procedure.

In addition to the aforementioned content, your report also has an explanation of your background effort that led to development of such methods:

reference books you have read (or written), articles in journals, conferences attended, other teachers observed, and general preparation on your part over the years that produced this proposed program. Furthermore, you have described the two schools where this program was piloted, for 10 months, with at least 10 students at both schools. And you have compiled profiles of the students at each school: some biographical information and some meaningful educational testing results that indicate the students' level of reading and writing both before and after administration of your pilot program. Lastly, your report provides additional, basic information for both schools, such as contact people and location specifics.

Imagine you run behind schedule, and on the original due date of your report you have only 5 pages written. You want to send something to the state department contact who has hired you and, simultaneously, ask for more time. So, pick any section from above and imagine the client's reaction to the material; this will help you determine whether it is primary, secondary, tertiary, and so forth. Keep in mind the client did not ask explicitly for proof the program you are proposing works, has worked once, could work, etc. Or did they? Isn't that implicit? Could you actually feel with clear conscience that you could send some ideas for a program for which you have no idea whether it could work? It's just some ideas you made up? Or, came up with on the spur of the moment?

Of course not. As an educational expert, you are going to propose a program for which you have some evidence it works. Otherwise, this would have been just brainstorming. The client would have hired you to brainstorm some ideas, write them down, and send them along. That's a different report. On the contrary, you were hired to propose a full-blown program that they intend to implement (at some considerable cost) into their school system in the near future. So, some proof of your program's expected efficacy is needed.

So, you write up the description of one of the schools from your pilot study and the biographical sketches of the 11 children in grade 5 who received 10 months of this educational support. Is this primary information? No. The client did not ask you to describe 11 random children and a school in another state. Out of context, this information is meaningless for them. It only has meaning in context of your proposed program. It is not even support that your program works. It is, however, proof you did run a pilot study, which in turn, gave you data that proves your program achieved some success. Thus, it is tertiary. It is needed but it can be relegated to a lower level in your report structure. Next, how about the data that represents the children's literacy level before and after the administration of your program for 10 months? Now, this is interesting. Again, it's pretty important. So, we send only this to the client. They look at it: hmm, they think. What good is this to us? Do we show this data to our ESL students to help them learn to read and write? No. Again, as important as it is to your report, it is meaningless to the client out of context. It also does not provide them with the information they requested. It is only secondary. This shows that secondary information is important; essential information in your report may still, without any denigration, be classified as secondary only.

The primary information, as you may have anticipated by now, is the program plan you propose: procedures, materials, and advice for teaching reading and writing to ESL students, grades 4 through 6. A strong strategy when running behind schedule is to give the client five pages of this plan, to show your progress. Although the report is far from finished, the client can see you have exactly the information they requested, but it just needs to be accompanied by a full report. Without complementary secondary and tertiary material, the program description makes sense to the client, it has value, and it fulfills their objective of hiring you. It also is fairly bare bones, without any proof of success and other commentary (how it was developed, where it was piloted, which children piloted it (are they like this client's students at all?), but those items can be added later. If you give the client only five pages and ask for an extension, you can probably get the extension if you give them primary information.

Let us take another example from engineering: an orbiting satellite's antenna control system. A motorized system must reorient the antenna to a new desired pointing location quickly and accurately whenever this is needed by the persons using the satellite from Earth. No person is on the satellite to actually "steer" the antenna to the right spot. The command of the pointing direction is sent in a signal from people on Earth; then the antenna is rotated to the correct pointing location, as commanded, by an automated motor. If it could move in tiny increments, step by step, it could be perfectly accurate, but a fast response is needed. Thus, with speed comes poor accuracy; the antenna's rotation may exceed (overshoot) the commanded position, or it may not fully reach (undershoot) the commanded position. Thus, a control system is needed.

The client asked for your design of a control system; your go-to option for control is a proportional-integral-derivative system, or PID. The design of the system and programming is part of your effort; you must identify the optimal settings, or gains, for the PID control system; this is known as "tuning." A controller must be tuned, with gains, to achieve the ideal control to a commanded position. This is a pretty simple concept, but finding the best gains is a process of both calculation and experimentation. The client has the overall computer program written, and they have the hardware designed. Your job, simply, is to find the best gains, by running some simulated testing of the antenna control system, comprising the antenna, a sensor, and an actuating motor, which moves the antenna to a specified location. You have two ways of simulating and testing your calculations and hypotheses: Mathematical software, such as Matlab, and a physical system that mimics the antenna itself, which rotates, upon being sent a control command, then your set of gains manages the motion, actuating the motor, using the sensor to determine its location, finding the error relative to the command, then readjusting. Each of the gains effects, in turn, one type of control: proportional, integral, and derivative. After running simulations in Matlab and physical testing with the mock system, you have results for your various gain settings. The optimum set of gains, ideally, would perform best on both simulation and testing. Your results should be consistent across both methods.

The report must be thorough, and it will be long. This is just a beginning. So, an effective structure is essential. PID systems work on some fairly complex mathematical theory; this would be explained in your report. Matlab would be introduced. Your physical model would need to be described and shown.

Let us look at the sections of the report to distinguish between primary and secondary material, keeping in mind that the client wants an optimal set of gains for the PID controller. In order to present a recommendation of the optimal set of gains to the client, the author must assemble a good deal of support information, including the following:

- Explanation of the physical model used for testing
- Calculations of model's characteristics, such as moment of inertia and torque
- Explanation of Matlab and documentation of the coding created to simulate the antenna's control system

- Background on PID systems, to confirm that the authors understand such in the same way as the clients do
- List of all possible gain combinations evaluated for comparison
- Results of each and every gain combination in both physical testing and simulation (including bad, awful, mediocre, good, and [fortunately] one optimal control system of PID gains)
- Mathematical calculations of damping ratio for each set of gains, indicating either under-, over-, or critical-damping
- Disadvantages of the recommended gain settings
- Assumptions, errors, limitations, or unknowns with the testing and simulation
- Further corroboration of recommended gain settings by other methods (such as pure computation)
- Recommendation of a control system, namely, a PID system, with a set of gains of x, y, and z for each control portion, respectively
- Specific support (results and analysis) for recommendation

With all of this material, only the last two items in the list above are primary information. The other items are necessary but not sufficient on their own to fulfill the client's need for information. Those other items are secondary and tertiary items. Some items are important yet secondary, such as the results of each and every gain combination in both physical testing and simulation. Without these, no reader can see that the recommended system performs better than all other options. Some items are at a tertiary level, such as calculations of the model's characteristics. This information must be in the report but, on its own, it is not that critical to the overall message; it represents completion of a preliminary task that must be done, otherwise the work is neither reliable nor accurate. But, the client expects this as part of your professional approach, so it is tertiary; secondary information is unknown to the client until you reveal it, though it is not itself the ultimately desired information; primary information is that which is ultimately desired.

Simple guidelines for each priority level:

- **Tertiary**: provides preliminary and background information on your effort to prove you did the work carefully and correctly, but it is far from the information sought by the client when either the work or the report was requested.
- **Secondary**: provides crucial support information that allows the primary information to appear credible, thorough, accurate, well-founded, reliable, and usable; while it is not the information ultimately needed, it provides the foundation upon which the primary information is built.
- **Primary**: offers the very information that answers the driving question, or solves the motivating problem, or fills the gap of knowledge, or presents a recommended plan for moving forward; While it is the most valuable information in a document, it is unusable and meaningless without secondary and tertiary information to support it.

One trend you may have noticed: in many documents, only a small quantity of the content is primary. That small portion is embedded within a larger document with lots of secondary and tertiary information in support of the primary material. This is the pyramid structure you have seen (Chapter 1.4). The concepts fit like hand in glove: The pyramid-structured document has primary information in the top portion, Overview, with perhaps some secondary mixed into it to provide a brief, yet comprehensive, document summary. The middle portion, the Discussion, has some repetition of primary information for emphasis, supported with the bulk of the secondary information. The bottom portion, Documentation, contains the tertiary information.

If structured correctly by priority level, a document with only a little primary content will be clear and easy to read. If not prioritized correctly into a proper structure, the normal yet massive volume of secondary and tertiary information overwhelms the inherently small amount of primary content, creating a morass of dense, unclear strings of factoids. This leads to ineffective communication. It stems from the intrinsic friction between quantity and priority, and it must be managed.

Understanding the technique of distinguishing among primary, secondary, and tertiary information enables an author to effectively structure a report, broadly placing the different information from the three different levels neatly into the corresponding overall segments of a technical report: Overview, Discussion, and Documentation:

Priority levels and structure:

- Primary information is presented in the Overview.
- Secondary information, along with reiteration of primary information, is presented in the Discussion.
- Tertiary information is presented in the Documentation.

This segregation maps perfectly with the spectrum of generalization as well. With generalization, most of the highest-level ideas are placed first (higher in the pyramid), with particulars in support of those generalizations following below. Thus, primary information consists largely of generalizations, while the document's secondary and tertiary information comprises numerous and various particulars. Main conclusions and recommendations, rhetorically the part in an ABERST document that corresponds to a thesis, are primary information and usually concisely stated, without much accompanying details. They are generalizations par excellence. Here are a few examples of recommendations in ABERST reports to demonstrate the extreme brevity and generality of information at the primary level:

- The optimal filter membrane pore size is 25 micrometers.
- We suggest using a structured query language (SQL) application for this task.
- Our recommended vendor for tires for your vehicle currently under design is Michelin.

Here is an exercise to practice generalizing, prioritizing, and building proper structure. Imagine a project presented by a client that has a new design for a vehicle. The client wants to know its aerodynamic qualities, for example, does it have low drag? After conducting lots of tests and calculations, you should be able to provide a client with primary information.

Primary information: ____

Secondary information: ________________________________

Answers:

- Primary: The design's aerodynamic qualities are x, and in particular its drag is y, which is (not) low for this type of vehicle.
- Secondary: test setup, raw data, error calculations, etc.

For this project, you must calibrate a force transducer in order to identify forces and moments, including drag, on the model.

This calibration falls into which level of information? ___________________

Answer: Calibration is tertiary. Think about it: if a client wants to know about a new vehicle's drag, receiving some arbitrary information about one of your laboratory's tools is of no interest to her, other than that you made use of the information in your analysis that you did for her. It cannot be left out because you do not want to imply that you made up numbers (data) or picked them out of thin air. But those numbers are "behind the scenes," to say the least, and they, on their own, say little about the vehicle. The calibration values are something like x volts equals y Newtons, and you have a function for the linear relationship of volts to Newtons, for example, $y = 2.8x + 3.55$.

Here are two additional report topics with some quick assessment of primary information for each:

(1) A procedure/instruction document: primary information is the procedure itself, neither your opinion of the procedure, nor variations of the procedure, nor a lengthy analysis of the benefits of implementing the procedure.

(2) A socio-economic estimate of the effects of doing X: primary information consists of your predicted results of X; secondary information would be your methods for making those predictions, alternatives to X you might wish to propose and their predicted effects, a recommended data collection scheme to assess effects once X is implemented, and a procedure for implementing X.

A rule of thumb for identifying your primary information, therefore, and segregating such from material of a less-important nature, is this: whatever information, if given to your immediate recipient, would make that recipient believe you have both (1) finished the project and (2) identified the crucial information required by the recipient, which would preclude a client or supervisor from asking, "When are you going to supply me with the information I need?"

> Primary information, if given to your immediate recipient, would make that recipient believe you have both (1) finished the project and (2) identified the crucial information required.

In a few genres, the bulk of the document is primary, and very little is lower-level information. This includes procedure/instructions and proposals. In the latter, each item has been specifically requested by a sponsor or funding organization. You must include nothing not requested, so it is ALL primary. With a proposal, furthermore, you must not forget to provide ALL items requested; it is one of the few documents that consists of purely primary information.

To end this chapter, we can look at some typical report genres with an eye toward prioritizing the required information and, thereby, sketching the overall structure of the document.

Journal article: Brief analytical conclusions are primary; background, data, and results are secondary; methods are tertiary. (Methods may be primary if presenting a novel method is the article's singular focus.)

Contract or business agreement: The essential and specific terms linked to the purpose of the agreement are primary; all the rest is considered a mix of "boilerplate" and "fine print" that might be standard to all similar contracts: obligations, assurances, signatures, and introductory matter. An example of essential and specific terms linked to the purpose of the agreement might be something like "install a test well at location x, at a cost of $200,000, by August of the following year, using the patented and proprietary 'bore and sheath' method." Additional obligations and assurances are secondary, while clauses on cancellation, delays, breach, and so on, are tertiary.

Technical memorandum: When such a memorandum contains an answer to a question, a solution to a problem, or a request with justification, the main generalization—answer, solution, or request—is the primary information. Conveying this single item is the point of the document. This single message is primary, followed by secondary information that proves the author has a correct, effective, and rational answer, solution, or request.

Table 3.4.1 summarizes the main points of this chapter. It shows, for a handful of technical report genres, a simplified categorization of information across primary, secondary, and tertiary levels.

GENRE	PRIMARY	SECONDARY	TERTIARY
Proposal	All		
Procedure	Nearly all	Alternatives and options	
Design Presentation	Description of design	Proof design is optimal	Construction of prototype
Scientific Article	Conclusions drawn from research	Motivation, data, and results	Methods of research or testing
Request	Item(s) requested	Justification	Alternatives

Table 3.4.1 Distinction of Primary, Secondary, and Tertiary Content for Five Genres.

Chapter 3.5

Organize the Information

In Part 3 so far, you have been instructed to have a point and stay on message with each and every report. Furthermore, you must be comprehensive, gathering all the required information needed to fully convey your message. In addition, clarity is served through normalization, and structure is served through prioritization. Thus, only a single purpose-oriented technique remains: enable reading navigation by organizing the report's Discussion into a hierarchy of sections.

This last lesson of Part 3 is an easy one. The bulk of the conceptual learning required to master organization has already been introduced in the preceding four chapters. All that is left is to understand the importance and benefits of organizing your Discussion into a hierarchy of sections. The concept is fairly simple. You do not want to have one big, undifferentiated, monolithic body of a report for a Discussion, labeled or not. Such a large block of text will intimidate readers and completely eliminate any possibility of their navigating through the Discussion in a way that allows skimming, selective reading, or skipping, all of which should be permitted by any reader if her position, responsibilities, and informational needs allow it. Instead of employing a single heading, Discussion, or absolutely zero headings, you must break up your supportive text into suitable sections, each representing a cohesive sub-topic of the assembled information.

The technique is not just to add a few section headings here and there. Maybe throw in a "method" or an "analysis," to stretch the whole thing out and seemingly add some professorial panache to the clump. Quite the opposite. You must add as many sections as logically fit the material, and you want them to occur on more than one level. That is, some sections, as the material warrants, will have subsections. For example, if you have a Section 3, you might have a 3.1 and a 3.2. Within Section 3.1, you might have 3.1.1, 3.1.2, and 3.1.3. If and when the material allows for hierarchical organization, you must do it. You owe this to your readers, and you owe it to yourself. (As you revisit your report on future occasions, you will thank yourself for creating a multi-section hierarchy.) Now, producing a hierarchy may seem simple, but the lack of such is a common problem in ABERST documents. Simple or not, a hierarchy is imperative. If it seems rudimentary and comes natural to you, all the better. If not, you are hereby notified to create a hierarchy, and you should find that your ability improves over time. Practice and experience will make it a familiar task. Once you master it, it becomes second nature, and you can use the technique in every technical document you write.

Let's look at a few examples, and then you will be ready to try your own hand at creating section subordination.

Recall the scanning electron microscope (SEM) report where the author requests funding support for purchase of a new SEM. We looked at this when discussing normalization. From an organizational perspective, we can identify a number of obvious sub-topics:

- Background of research goal
- Focus of research on polycarbonate diamond (PCD) coating for stents
- Need for SEM
- Problems with existing SEM
- Grant # specifics, with limitations
- Need for separate funding
- Research of new SEMs to purchase
- Payment and financing options for purchase

After walking through the first "sweep" of organization, the author might come to see that two of the sections above warrant further subdivisions. While discussing the problems with the existing SEM, the author wants to make the point that these are not trivial flaws. (See subparts a and b in revised outline below.) They are serious deficiencies, and they fall into two broad groups. Similarly, when proposing purchase of a new SEM, the author is requesting some significant infusion of funding. The author would be wise to show that she looked at the different models of SEMs and analyzed the relative advantages and disadvantages of each one, so that the funding request is necessary to buy the optimal model, hopefully the cheapest but, if not, at least the one representing the best overall value. This analysis is reflected in the further subordination of third-level subsections under each SEM model to present "features" and "costs."

- Background of research goal
- Focus of research on PCD coating for stents
- Need for SEM
- Problems with existing SEM, with subparts:
  a. Not enough magnification and
  b. Tedious controls for rotating for a full-spherical view
- Grant # specifics, with limitations
- Need for separate funding
- Research of new SEMs to purchase, with subparts:
  a. SEM model 1
    i. Features
    ii. Costs
  b. SEM model 2
    i. Features
    ii. Costs
  c. SEM model 3
    i. Features
    ii. Costs
- Payment and financing options for purchase

Perhaps the author then realizes that the first three sections could be regrouped into a hierarchy, so this adds another level of subordination as well.

- Background of research goal
  a. Focus of research on PCD coating for stents
  b. Need for SEM
- Problems with existing SEM, with subparts:
  a. Not enough magnification and
  b. Tedious controls for rotating for a full-spherical view
- Grant # specifics, with limitations
- Need for separate funding

- Research of new SEMs to purchase, with subparts:
    a. SEM model 1
        i. Features
        ii. Costs
    b. SEM model 2
        i. Features
        ii. Costs
    c. SEM model 3
        i. Features
        ii. Costs
- Payment and financing options for purchase

Lastly, the author ponders the idea of subordinating "Need for separate funding" under the preceding heading, "grant # specifics, with limitations," but remembers that a section cannot have only one subsection. Any breakdown into subsections necessitates at least two subsections. One reorganization that could work is to move up "Payment and financing options for purchase" and put it as the first-level heading, then subordinate the other two under it.

- Background of research goal
    a. Focus of research on PCD coating for stents
    b. Need for SEM
- Problems with existing SEM, with subparts:
    a. Not enough magnification and
    b. Tedious controls for rotating for a full-spherical view
- Payment and financing options for purchase
    a. Grant # specifics, with limitations
    b. Need for separate funding
- Research of new SEMs to purchase, with subparts:
    a. SEM model 1
        i. Features
        ii. Costs
    b. SEM model 2
        i. Features
        ii. Costs

c. SEM model 3
   i. Features
   ii. Costs

The above refinement of structure demonstrates a step-by-step process of reorganizing the material into a sectional hierarchy that depicts a logical grouping of information. Instead of the original 8 sections, all at one level, the final version has 19 sections at three levels.

This organizing process can be applied to all technical genres. Another example worth some study is a procedure/instruction document, also called a "user manual." This type of document (more on this genre in Part 5, Chapter 5.1) should be organized into three main segments. Although structural variations exist for this genre, you should use a consistent overall structure for all such documents within one organization. I recommend a three-segment structure, to perpetuate a single approach advocated by my colleagues, in particular J.C. Mathes and Dwight Stevenson.

The overall document is broken down into Objective, Background, and Procedure. Within Background, certain helpful information needed before a user starts the procedure is introduced. This includes tools, equipment, materials, and parts needed to do the procedure. Background also encompasses concepts that must be understood, including equations, formulas, theories, assumptions, and prerequisites. Lastly, Background includes warnings to be given to all users prior to starting. With this understanding, the Background section is very easily organized into a hierarchy of sections, such as the following:

Background

- ▶ Materials needed
  a. Reused material
  b. Newly purchased
- ▶ Equipment
  a. Self operated
  b. Technician operated
- ▶ Tools
- ▶ Concepts
  a. Materials science
  b. Chemistry

- Warnings
  a. Temperature-based concerns
  b. Chemical-based concerns
  c. Noise-based concerns

As more and more necessary information is assembled to complete the document, the natural groupings and subordination should be identified. Each author must become skilled at organizing the bulk (Body or Discussion) of every report into a hierarchy of helpful sections.

I mentioned at the top of this chapter that readers might approach the Discussion with different expectations and responsibilities. They may skim, read selectively, or skip most of it, depending on their position, responsibilities, and needs for information. Looking at the two report examples above from the perspective of different readers should reinforce this basic principle of hierarchical organization.

Imagine these different persons looking at the SEM report after the supervisor has forwarded it to them: contract lawyer, laboratory manager, vice president for finance and procurement, and new research assistant. The contract lawyer spends most of her time looking at the "payment" portion (excerpted below) because she wants to make sure that the plan for funding and expenditure of research funds from the grant is correct and consistent with federal contract regulations:

- Payment and financing options for purchase
  a. Grant # specifics, with limitations
  b. Need for separate funding

The laboratory manager looks carefully at the "problem" portion, to see if the author understands the current SEM and has made a convincing case of its deficiencies:

- Problems with existing SEM, with subparts:
  a. Not enough magnification and
  b. Tedious controls for rotating for a full-spherical view

The vice president for finance and procurement pays closest attention to the portions on the competing SEM models so she can affirm the choice made, or voice a contrary opinion:

- Research of new SEMs to purchase, with subparts:
  a. SEM model 1
    i. Features
    ii. Costs
  b. SEM model 2
    i. Features
    ii. Costs
  c. SEM model 3
    i. Features
    ii. Costs

The new research assistant, a bit overwhelmed with issues about old and new SEMs of which he cannot speak with any meaningful contribution, reads the "background" portion, which explains the research context, so he can get up to speed. He will sit out the debate about purchasing a new SEM, and use whichever SEM he is told to use. Of critical concern to him is learning about the project he is going to start working on. He knows that he will be using one SEM sooner or later, so he also reads through the portion discussing the problems with the existing SEM to familiarize himself with the technology and its perceived drawbacks:

- Background of research goal
  a. Focus of research on PCD coating for stents
  b. Need for SEM
- Problems with existing SEM, with subparts:
  a. Not enough magnification and
  b. Tedious controls for rotating for a full-spherical view

Certainly any of the readers can choose to read the whole Discussion, and all of them will (should) read the full Overview, as discussed in Part 1. Most will read the whole Discussion, but they might break it up and read some parts before others, and over several different days. It does not matter to an author. The reader's prerogative is to choose to either take apart the report into pieces or read the whole document from start to finish. The hierarchy enables a reader to navigate through with intelligence and personal preference. If the whole Discussion is a single, undivided block, readers can do little but work all the way through it. Many will not. They will quit the effort out of frustration and exhaustion. An organized hierarchy gives readers a document they can enjoy rather than dread.

Part 4

Persuading Your Reader

As stated in the introduction of this book, purposeful writing in an organizational context is imbued with an overall goal to get something done that may be any of the following: helpful, innovative, profitable, exciting, challenging, necessary, urgent, imperative, protective, or strategic. A purposeful document is instrumental and moves matters along in some way. A purposeful document cradles within it something of intellectual value to people at the author's organization or an affiliated organization. This thing of value might be a design, decision, explanation, theory, hypothesis, justification, request, proposal, plan, procedure, and so forth. This valuable intellectual "property" needs to be described at the very least. In addition, it must be presented favorably, with its advantages, benefits, and overall justification spelled out, or the outcome may be that no one within the organization is motivated to pursue and implement it. It may not materialize. Again, it is not enough to merely describe a new idea. To actuate implementation of the new idea, the author must defend it, or, you might prefer to say, "sell it." This latter portion of the presentation is an author's argument.

Readers may or may not concur with the message of the document; indeed, they may disagree and reject, or partially agree and suggest refinement. The author's preferred outcome, certainly, is that they concur with the main idea, seeing the facts of the situation as the author has presented them, and becoming persuaded to approve, support, and implement the author's proposal. In a sense, in the preferred outcome, the readers "buy" the "property," that is, the intended message.

Whichever the outcome and response from readers, the author must make a sincere effort to present the message of the document in its best light, clearly, cogently, and reasonably; in other words, the author must present a case, or, an argument. Persuasive arguments, therefore, are inherent to a vast majority of ABERST documents, so are cautious and rightfully skeptical readers. This must be recognized as a starting point for this part of the book; thus Part 4 begins with an explanation of argument's prevalent and necessary role in ABERST, found in Chapter 4.1.

After that preliminary chapter, Part 4 continues with a proven and straightforward approach to writing an argumentative section of a document. As much as Part 3 espoused the notion that an ABERST report's main contents are determined and selected to concisely and effectively fulfill the author's writing purpose, Part 4 is focused on how to present the argument necessitated by that purpose. By following Part 3's instructions, you have identified the essential information to include in the report's discussion, using the Purpose Principle. Part 3 furthermore provides a go-to method for normalizing, prioritizing, and organizing the required information, and using a hierarchy of sections to facilitate easy reading. Furthermore, when a report has a persuasive purpose—and most do—I have intimated that additional organizational skills are required geared to persuasion. That is the focus of Chapters 2 to 5 of Part 4: a fail-safe method for writing persuasively, that is, presenting information in an effective argumentative structure. Thus, in Part 4, you will learn how to organize the essential pieces of information that constitute the argument within the report.

Part 4 is organized into these chapters and sections:

- 4.3.3. Strategic Construction: Provide support and make it readily apparent, through persuasive structure
- 4.3.4. Audience Alignment: Align argument with readers' needs, as best as you know them
- 4.4. Maintain and Enhance Credibility
  - 4.4.1. Leverage Corroboration's Power
  - 4.4.2. Admit Flaws of Any Type
  - 4.4.3. Consider Alternatives
- 4.5. Reinforce Argument with Flow and Transitions
  - 4.5.1. Create Flow Between Sections
  - 4.5.2. Remind Readers of Main Points with Transitions

This five-chapter outline took much thoughtful consideration on my part because argument is a difficult topic to cover succinctly. Five chapters? Why not five hundred? Presenting a quick primer on argument is a task tantamount to summarizing the history of Europe in one page or less. The topic is deep and wide. Rhetoric, debate, persuasion, argument, litigation . . . these related topics are covered by whole books, hundreds of them, probably thousands. (Certainly you can do well, if you have the time, by looking at a few other books on this topic; they all offer insights, theory, and skills to draw on when writing in one's field.)

As the outline suggests, I want to accomplish five objectives in this part of the book, each linked to a major chapter: introduce the essential role argument plays in ABERST reports, review argument's noble lineage, concisely summarize a four-step approach for creating an argumentative structure, offer some suggestions for enhancing one's credibility, and lastly present an approach for reinforcing your argument with flow and transitional sentences.

Throughout all of the discussion in Part 4, I am connecting argument to the work done in ABERST: research, experimentation, data collection, simulation, calculation, and, generally, all the common methods for acquiring data and testing ideas and hypotheses. This work is the source of the support one uses in an argument, and a thesis is the result of a cumulative inductive analysis of the data discovered. In this way, you will see that your work and your writing go hand in hand, permeated by argument.

The prior three parts of this book have prepared you to incorporate argumentative writing into your documents. Argument complements the concepts already covered, such as starting strongly, fulfilling purpose, and normalizing information. Adding an argumentative component adds completeness. With an argument, the document's essential message, as promised from the start, can be presented effectively and persuasively.

On a final note for this introduction, I want to recall that this book is intended as a guide to technical writing in both college and career. One genre that fittingly links the two realms is the resume. A resume, accompanied with a cover letter, is a technical document that must be both personal and persuasive. The principles of persuasion covered below apply to it, as do many of the concepts connected to genre-specific writing covered in Part 5. It is, therefore, a bridge between college and career and between persuasion and genre. I have chosen to devote a full appendix (B) to resumes, cover letters, and thank-you notes to give this threefold topic the extra attention it deserves. Feel free to consult Appendix B before, during, or after your study of Parts 4 and 5, to help you create or upgrade these vital, career-building documents.

Chapter 4.1

Acknowledge Argument's Crucial Role

Arguments do not come naturally to many ABERST professionals, and I know the type quite well. I have worked in ABERST my entire career after graduating from college. I know my colleagues, as well as the students who are aspiring to pursue ABERST careers. These are objective, inquisitive, and assiduous people; they are not the confrontational types. They are not usually argumentative, combative, or contentious. They are scientists, engineers, and researchers. They are pursuing careers to learn things, help people, and solve problems. They entered engineering and science because they wanted to avoid "mind games" and, instead, wanted to study empirical data, the natural world, and interesting phenomena. They wanted to work with machines and technology. They are not focusing on arguments. They look askance at argument and books on persuasion. The books on rhetoric are not typically the textbooks these professionals choose first. They think theories and techniques of rhetoric are useful for people in fields that involve arguing: law, politics, public policy, philosophy, and so on.

Granted, ABERST encompasses a huge variety of career paths, disciplines, and personalities. In some areas, such as business, moreover, individuals and organizations embrace argument, and even thrive on competition and market domination. If this describes you, no further persuasion is needed on my part. You are already open to mastering the argumentative arts. My singular approach of combining numerous specialties into ABERST, even if some are superficially opposite, stems from seeing the essential role argument plays in all of them. No doubt some professionals in business and technology take as naturally to rhetoric as do lawyers, judges, and politicians. But the truth is that all the professions require persuasion. You cannot avoid it. Presenting any point of view (answer to a question, decision of policy, proposed design, a plan of action, etc.) requires the ability to persuade.

This is where I begin this part of the book: to disabuse you of the view that argument is not inherent to a field of objective inquiry and rational analysis. On the contrary, it is foundational. Closing your eyes to it leads to weak reports. Acknowledging the role argument plays in ABERST is a critical step in completing an effective purposeful report. So, in brief, this chapter explains argument's role in ABERST documents.

At the most basic level, messages in ABERST take two forms: informative and persuasive. That is, the documents you write in a professional setting should be, at minimum, informative, and, more often than not, persuasive as well. Persuasion plays a role in a significant proportion of reports written in ABERST. Aside from a small selection of genres that are merely informative (instructions, press releases, and progress reports, to name a few), most reports in the workplace that pertain to necessary activities and completion of those activities possess an argumentative element.

As mentioned earlier in this book, a large proportion of the ABERST documents are written to inspire, direct, or actuate a course of action that benefits the author, the author's organization, or the author's clients and customers. To inspire, direct, or actuate, an author provides technical information that has a message and a point of view (covered in Part 3). Importantly, this message may be a new idea or one not yet widely accepted. Reports address unique, novel, and unresolved issues. Reports journey into new territory, setting a course for unchartered waters. Solutions, designs, conclusions, analyses, theories, explanations, cost estimates, predictions, plans, recommendations, interpretations, judgments, and the like are products of an author's mind. Mental products are not established facts; they go beyond facts, putting many of them together with some interpretation and analysis, to formulate a viewpoint, or, an opinion. And, to offer and defend an opinion, an author must persuade. Persuasion entails argument.

Many writers do not see the fundamental role that argument plays in their reports. These writers expect to simply write up the work they did. They see reports as chronicles of research activities or analytical efforts. The draft reports they produce often have a journal style, with a play-by-play account, like the following: "First I did this, then this; next I did this. That didn't work, so I switched to this."

When pushed to confess that their work might have produced something of value if they could do a little further applied interpretation, authors may grudgingly say they also found data during their research, so they will be willing to report their data and results. At this next stage, they write: "Here are the results of the study, with variances, errors, and confidence intervals duly recorded at the end." They are diligent to admit the limitations of their data (much more readily than asserting the applied value of the data). With great objectivity and integrity, per scientific ethical standards, no one will falsely proclaim the data to be better than they are. ABERST authors display lots of hesitancy and caution, in other words, timidity. This timidity will not get a good document written. To be sure, I do not demand the sheer opposite of timidity: I do not want to see temerity. No one needs to exude excessive bragging and audacity. Something in the middle is perfect: reports should be built upon reasonable arguments.

Argument is roughly speaking defined as follows: a set of reasons (support) given to persuade another that an idea or proposal (thesis) is correct or worthy. To prove ABERST authors are dabbling in argumentation in most of their reports, we can prove the existence of argument in technical reports from two verifiable premises. We start with the assumption that you know that your purpose is to present a decision, judgment, recommendation, point of view, policy, proposal, etc. Any of those terms equals a thesis, and this is Axiom 1: You have a thesis. We also assume that you have some good reasons for putting forth this thesis, and the existence of good reasons is Axiom 2. So by combining Axioms 1 and 2, we move to Theorem 1: you are presenting a thesis combined with your good reasons behind that thesis. Theorem 1 is undeniably true, you must admit, yes? And Theorem 1 matches the definition of argument impeccably. Therefore, it is proven: many ABERST reports are inherently argumentative.

If you are wondering how it is that I have missed the people who toot their own horns so loudly and so often that no one can hear the coffee percolating, and how I have missed all the chest-pumping and self promotion, I can say that, yes, I see some of that in some people. But, the vast majority of ABERST professionals tend to downplay their accomplishments and their work. They say things such as, "I collected some interesting data; I'm not sure if I can discern any significant trends." They will tell you the work they did; that's fine, but they are reluctant to say it's truly useful or important. Report drafts are often good records of work done and data gathered, without having an overarching message of utility and application. ABERST professionals take some pride in not trying to play with anyone's head or push their ideas too vociferously.

The good news is that technical argumentation is not about head trips. Be assured that you are not required to be arrogant or egomaniacal. Rather, you are cool and dispassionate with arguments in ABERST. Tone is important, and ABERST arguments have no place for threats, intimidation, guilt, conceit, falsehood, or distortions. Still, you must be an expert. When you do your work, you offer suggestions, improvements, strategies, new ideas, and solutions to problems. And when you offer those helpful items to people running an organization, you are presenting an argument. You are presenting your expert opinion (from expertise, not ego) based on some logical (measurable, objective, scientific, etc.) reasoning. This is how things change, improve, develop, and evolve in ABERST. Step by step, each working person adds her contribution, usually documented in a report. This is argument.

Argument permeates academic, business, and technical settings. The work done in those disparate settings is vastly different, but the common element to professionals writing technical reports is the recurring use of expertise to present a point of view that represents the best solution, interpretation, or way forward. Argument is the conceptual framework for sharing that expertise with necessary parties to explain, clarify, convince, and reach consensus so that decisions—either unilateral or multilateral—are facilitated and documented.

Take one example from the chemical engineering business. Perhaps a team working on new projects is putting together a design for a new acetic acid production facility. Senior executives at their company have identified the possibly lucrative endeavor of making acetic acid. They have asked some mid-level engineers (the "team") to do a "paper" study, design a manufacturing plant, and generally move forward with the project. The team doing the "paper" study has finished their work, and they want to present the plant design to senior management. This team implicitly asserts that their design works, that is, it produces acetic acid. It does this, specifically, from butane, by converting the feedstock with the right equipment, reactants, and catalysts. The team might be tempted to stop at their design: here it is, it works, and it could be built with our plans and specifications.

But very likely, the team has much more to say because management expects more expertise from them. Much more needs to be assessed before the company approves the new factory. In addition to describing how they intend to convert butane to acetic acid, the team has a further assessment of the merits of the initiative. This is in response to the full instructions from management, which no doubt requested a full analysis. Thus, the team has produced an argument that the plant will (or will not) be a wise investment, and only their timidity keeps them from acknowledging this. Quite probably they have produced an economic analysis of potential profit, as well as a health and safety review. So, the argument is taking shape, thusly: a chemical manufacturing facility, in order to be a good investment, should work effectively, make money, and operate safely. Are other criteria applicable? If we probe members of the team, they will reveal many additional criteria. The factory should be environmentally friendly, feasible, adaptable, automated, locally favored, politically correct, lawful, reliable, and maintainable. So, the analysis is much deeper than the modest engineers and scientists wanted to initially take credit for. As this example illustrates, very few ABERST projects involve merely facts and information. They almost always, to achieve something within an organization, have an argument.

Recall my friend from Part 3 who researches diarrhea. Earlier we looked at many possible writing scenarios that she may encounter. She might propose collaboration with a fellow researcher in a long letter, justify her use of laboratory space via an internal memorandum, instruct her staff to follow procedures in an email with a formal attachment, and present her research findings in a journal article. All of those documents must contain an argument, each one geared to the specific persuasive objective required in the unique scenario. They are all different reports with different purposes, but each one has an argument at its core.

An argument is at the core of most reports.

This is a simple fact: an argument is at the core of most reports. In order to achieve your purpose—help advance your work and help the organization receiving the report—you must persuade readers to adopt your perspective, agree with your conclusions, and approve and implement your recommended actions. This necessitates argument.

Chapter 4.2

Revisit Argument's Heritage: The Fundamentals

The awareness that argument is part of most documents prompts the next lesson: how to formulate and present an argument. This requires some understanding of rhetorical theory. You need to understand the principles of argument in order to adapt them to your purpose in a technical document. This is no simple topic. Please understand that rhetoric has a noble history; it is a long-established field and a well-regarded discipline with many treatises, famous thinkers, and whole schools devoted to one approach or another. Plato, Aristotle, Cicero, Spinoza, and Dale Carnegie have all weighed in on the subject. Other, modern thinkers on the subject include Stephen Toulmin and Carl Rogers. The subject is also covered in most writing textbooks and technical communication handbooks. No single perspective is dominant.

Instead, different outlooks exist, each with a unique and valuable contribution. You could spend a chunk of a lifetime studying it and becoming conversant in the principles, not to mention related lessons of history, lexicography, and philology. The topic has all sorts of sub-specialties also: philosophical rhetoric, scientific rhetoric, educational rhetoric, to name a few. Rhetoric of the self, rhetoric of the Internet, and rhetoric of agriculture probably exist, for all I know. Importantly, you do not have to know much of it. This is not your field. It does not even need to become a minor field of study for you. You just need to take the best of it, as applicable to your specialty: polymers, irrigation, genetics, oncology, marketing, astronomy, and so on. So, this chapter has just a quick review of some of the foundational aspects relevant to this part of technical writing.

This book is a practical guide. It leans toward "how to," rather than "why," as you have seen so far. This explains why the next chapter (my recommended approach to argumentative writing) is longer than this chapter (theoretical background). Here, I primarily synthesize and condense. Suffice it to say that you can benefit from a familiarity with the fundamentals of rhetoric and argumentation, as a foundation for mastering an approach to argumentative writing in ABERST.

One major thrust of the field concerns dispute resolution. In fact, a simple division of the rhetorical landscape separates the field into two-way and one-way argumentation. Debate, dialectics, litigation, negotiation, dispute resolution, and mediation are some of the primary variants within two-way argumentation, where two parties are in counter opposition intellectually and often direct conflict practically. Some experts focus on two-way argumentation. Plato's dialogues are some of the finest and original demonstrations of dialectics, primarily using the Socratic method. In recent times, Carl Rogers took a psychological and communication-based approach and attempted to craft a theory to help people resolve arguments through discussion, sympathy, and compromise. One of my favorite books, *Getting to Yes*, by Roger Fisher and William Ury, presents a brilliant negotiation method that leads to a consensus among parties that is mutually satisfying.

Because two-way argumentation is so prevalent in social environments, much of the heritage, fundamentals, and theory of argument has been used over the centuries in the service of two-way argumentation, when two parties must negotiate, debate, or litigate a controversy. Each party has a personal perspective, and these differ from each other. Thus, the parties marshal their ideas, opinions, facts, and interpretation (their arguments, in a word) to present to the other side, in the hopes of convincing them, or convincing a neutral party (mediator, judge, parent, council, administrator, etc.) that their view is correct. They win the argument, in such an instance. If both sides' arguments have merit, they might reach a consensus or a compromise.

Different formats exist for two-way argumentation, but they all have the common element of allowing each side to state its argument, while the other side listens, followed by opportunities to enhance one's argument, rebut the other side, or concede some points, or any combination of these three options. Back and forth they go, until the argument comes to a negotiated or adjudicated end, or the parties disburse without resolution.

When you have a two-way argument, you find an interlocutor or adversary in your face, chomping at the bit while you speak, waiting to rebut your points. You constantly have to build your argument against pressure that is tearing it down. It is not easy, and it takes listening skills, quick thinking, and lots of preparation. It can be tense and challenging. Fortunately, as common as it is in today's world, two-way argumentation is not directly relevant to argument in ABERST writing. Indeed, the aspect of argumentation that embraces disputing or negotiating with an adversary is beyond the scope of this book.

One-way argumentation, in contrast, is the type of argument found in ABERST documents. In this format, one side (the author) presents its argument in written form, with each idea added up, in sequence, to present as thorough a set of support points as possible. Thus, in one-way argument, you need all the preparation of two-way argument but without the intensity of an adversarial discourse. You simply need to arrange all your points in an organized manner and present them in clear writing. You can do this preparation in the peace and quiet of your office, home, or favorite café. The process is organizational instead of confrontational, and it is built upon a strong collection of support points for your argument, all based on the fundamentals of rhetoric.

You probably already know some rudiments of rhetoric. Most of us have some inkling about the techniques of persuading a fellow human being to do something. Your vocabulary matters, as does your tone, your body language, your expression, and your type of appeal. In planning to persuade, you may have pondered these questions: Do you threaten? Scare? Cajole? Flatter? Beg? Seek sympathy? Present your logic? Do you size up the recipient and align the rhetoric to his weaknesses (or strengths), or do you start with the general topic or specific issue and identify the most applicable flavor of rhetoric to match the topic or issue?

The great Greek philosopher, Aristotle, asked himself these questions centuries ago, and he proceeded to identify the elements of argument in his *Treatise on Rhetoric*, finished in 322 BC. As complicated as this whole field can be, Aristotle's approach may seem refreshingly simple. Conversely, other thinkers have made it intentionally nuanced and multi-faceted. To understand Aristotle is to have an excellent foundation, so the summary below draws heavily upon his theory.

Three argument categories:

- Personal
- Emotional
- Logical

In Greek, these are *ethos* (character or ethical appeal via credibility or reputation), *pathos* (emotional appeal via suffering and experience), and *logos* (reasoning and logical appeal via words), respectively. Aristotle thought a medley of all three might be ideal when applicable. In the realm of ABERST writing, you will rely primarily on the logical perspective, with the other two playing a minor role on occasion, if at all.

Until you are an expert in your field with many publications and worldwide fame, you cannot rely on your personal aura and cult following to seal your arguments, in other words, to convince your readers to follow your recommendations. Very few engineers and scientists have cult-hero status, like Michael Jordan or Leo DiCaprio. Bono may be able to sell a wind-powered drinking water well pump to a non-profit group working in Africa, but you would have to provide an effective, logical argument to do the same. Until you attain the status of Bono, your success will hinge on the inherent logic of your arguments embedded within clear, easy-to-read technical reports.

Similarly, emotion-based persuasion is a powerful yet controversial strategy. I will leave that one to religious leaders, insurance agents, and advertising professionals. Tugging on someone's heartstrings, or appealing to his fears, vanities, vices, fantasies, and so forth, works wonders for persuading and actuating a person. "You can invest in this oil-based electrical generator or you can starve to death the next time we experience a power shortage." Needless to say, appealing to a reader's reason, and rational abilities, is the foundation of ABERST. Thus, your reports should be based on arguments of logic. In this sense, I have, again, narrowed the enormous field of rhetoric to a small slice directly pertinent to your report-writing assignments: one-way argumentation based on logic.

Logic is a general term that captures the method of reasoning rationally, from one idea to the next, starting with information that is factual and culminating in an interpretation (for ABERST, a pragmatic one) that is an opinion.

The culminating opinion is, of course, not itself a fact. (If it were a fact, an opinion would not need your logical development.) On the contrary, the culminating opinion is one of the following: a perspective, point of view, prediction, suggestion, proposal, thesis, recommendation, or policy. It logically follows from the facts, but, as such, is not obvious and widely perceived and understood as true or right. Yet, it is a reasonable interpretation or development from the facts, and via logic others can be made to see the opinion as valid, sensible, reasonable, and ultimately worthy of concurrence, approval, and ratification.

- As a theorem, the mental process of decision-making looks like this:

Facts + Logic = Thesis

The thesis is exactly the knowledge ("valuable intellectual property") required in the situation. This thesis is a product of expertise, which is an opinion and previously unknown judgment or conclusion. People want facts, for sure, but they also need an interpretation, policy, or next step beyond the facts. In short, the opinion is the item sought by those who want to move forward on the project, problem, venture, or program that is the subject at hand. This is a pragmatic use of logic.

Presenting just one end of the theorem above, sadly disappoints people depending on your expertise, as follows:

- If you present only a thesis, with no facts and logic to support, you appear to be flippant with a toss-away, unsupported opinion no better than clamor.
- If you present only facts with no thesis, you come off as unintelligible with lots of numbers but no message.

You must present both ends of the theorem (facts and thesis), and use the middle portion, logic, to link the two ends together. Logic plays this crucial connecting role. It is the diplomat that brings the two other sides together, which, if they stood alone, would not be useful. Logic ties all the elements together and is unifying, pragmatic, and sensible. We often say that logic appeals to common sense, but that is not to say it relies on urban legends, wives' tales, street lore, rumors, or a layperson's hunch. Common sense demands a cool, calm mind, looking at facts and making only reasonable inferences that can be drawn from those facts. You want your argument to be strong on common sense; underneath that common sense is a foundation of logic.

If you can show your line of reasoning, you are asserting logic. If your analysis can be described as sensible, objective, dispassionate, unbiased, reasonable, rational, sane, balanced, valid, and lucid, you are likely using logic. When a thinker or problem solver is applying objective analysis to a situation to derive sensible, rational, fair, and reasonable conclusions and recommendations from the data, logic is at work.

Putting logic to work *at work*, that is, in your occupation, is a natural application of a long-established discipline. Logic for practical purposes, as the foundation of one's occupational rhetoric, is the application of concern for us working in ABERST. Practical logic can be described by various models. One such model was developed quite lucidly by Stephen Toulmin. In the Toulmin model, each element of the full logical argument must be expressed, as they are interconnected and inter-dependent. If one element is omitted, the argument is ineffective, due to a severe gap in the logical reasoning. Conversely, when all elements are explicit and logically connected, the argument achieves its full force.

Specifically, the elements are six in all, starting with a claim. Every argument has a final conclusion (claim); this is the message (idea, opinion, conclusion, recommendation, etc.) you want your readers to agree with. To convey the claim, you must put forth your supporting facts and evidence (grounds). Your evidence persuades only if another element of the reasoning connects the facts and details (grounds) to the certainty/reasonableness of the message (claim); this connection is the key principle, reasoning, or provision of the argument (warrant). The warrant may not stand on its own as self-evident, so a rationale must often be given (backing). Furthermore, a savvy author of an argument will acknowledge any counter-arguments or weaknesses in one's reasoning (rebuttal). Lastly, an argument may be tempered, conditioned, or constrained by specific expressions of limitation (qualification). Example qualifiers are "until further data can be collected" and "acknowledging the limitations identified." (In terms of ranking within Toulmin's model, claim, grounds, and warrant are the three essential elements, while backing, rebuttal, and qualifier are optional and introduced when applicable.) With six basic elements, Toulmin's model is straightforward and broadly relevant, and it alerts us to the importance of setting forth each required element in the reasoning. If a crucial element is omitted, the message is incomplete, and the argument is unlikely to persuade. Indeed, the four-step approach in the next chapter emphasizes the same need for comprehensiveness.

A common weakness in technical writing where the argumentation is (rightly so) based on logic is that the logic is left for readers to infer. It is never stated explicitly, perhaps because it is so self-evident to the writer. (We should all be so lucky.) If the logic is obvious, all the easier for your to present it. If the logic is subtle, equivocal, and controversial, all the more reason to explain your slant on it.

I provide here an example from aerospace engineering that includes an implicit logical reasoning to support a recommendation, but the elements of the logical reasoning are not fully stated. They must be teased out, as any manager would scrutinize a proposal before approving it, which I guide you through below.

To start, the recommendation is to accept a small lot of rocket engines, rather than reject them, due to their expected adequacy in fulfilling their intended purpose. Our natural skepticism, which we all must demonstrate when introduced to an argument and before making a decision, would make us ask, "Why do you recommend accepting the lot?" The author might answer with a fact, such as this: "The standard deviation was such and such in our tests." Oh, the skeptic must ask, "You tested the rockets?" "Yes," comes the answer. "We test fired these engines." "How many?" we ask. "Three." "How many were manufactured?" "1,000." "Is it OK to judge 1,000 by 3 samples?" "Yes, because the 95% confidence level is such and such," the author replies. "What is a confidence level?" the skeptic asks. The answer might be this: "It is a statistical metric that tells us something about all 1,000 engines based on testing only 3."

"Let me paraphrase your argument so far: you actually tested three rocket engines and acquired performance characteristics and you also performed a statistical analysis to apply the results from the three samples to a full lot of 1,000. Is this correct?" "Yes," the author answers. "OK, then," the skeptic further inquires. "What are the standards you were expecting these rocket motors to meet?" And so on, and so on, this conversation goes until the author can present a fully comprehensive, logical argument in support of the recommendation to accept the lot of rocket engines. In time, with mental effort, the elements of the logical reasoning take shape, as follows:

- **Claim**: The recommendation to accept the lot of rocket engines is offered.
- **Grounds**: A statistical analysis of sampled data suggests high precision (better than 5% significance) in thrust provided by engines.
- **Warrant**: An acceptable rocket engine is one that is precise to the level of 5% significance in its thrust performance.
- **Backing**: Statistics are a proven and rigorous method for predicting probabilities of an unknown future event or behavior of a large population, from a historically recorded observation or small sampling of that population.
- **Rebuttal**: Only three data points were used for this analysis.
- **Qualification**: This predicts only the approximate location of the mean of a population, not the precise result for any single specimen of that population.

4.2.1 Streamlined Argumentation

Understanding argument's foundation puts you in a favorable position to work efficiently because you do not need to use Toulmin's full model to organize an argument in ABERST. A general and streamlined approach, consistent with Toulmin's model, can be simplified this way, in two parts: (1) formulate your overall message and (2) apply premises to facts to support your message.

Importantly, this simplification refers to presentation and communication of the argument, not development of it through research, study, and many similar work efforts too numerous to list. With development, the parts occur in the opposite order, starting with Part 2. More on this in Chapter 4.3. For now, understanding the two parts of argument is best done in terms of communicating an argument because if it cannot be expressed to others it is not worth much.

Part 1: Formulate Your Message

An argument begins with an overall message. It needs to be expressed concisely, with certainty. This will present your stand on the matter at hand in relation to some organizational purpose that can be achieved. From this message, the rest of your argument will fall into place. The overall message can be expressed in different ways, but it must be explicit. It presents a proposal, recommendation, conclusion, etc., you wish to advance, and it presents the overall reasoning behind the idea. For any idea worth pursuing, you must feel, for example, it has benefits that outweigh the costs. Put another way, it has advantages that overshadow the disadvantages. Yet another way to express your reasoning might be that the idea is a good one, with few bad qualities; perhaps it is the best of the options. If you have a design, the design is an effective one. If you propose a solution to a problem, the solution is the best available. If you have a conclusion drawn from research, the conclusion must be the most reasonable, defensible, and provable one. A prototype is ready for production, you declare, because it meets all specifications and requirements. Formulating your overall message in this way, as Part 1, is critical to your argument. From it, the other half of the argument follows.

As just demonstrated, the overall message is easy to formulate; it is usually a straightforward idea. Similarly, facts are easy to present. You need tables, charts, graphs, and backup material that verify the facts were obtained correctly. Facts (data, findings, observations, etc.) are rarely controversial. (People agree on facts, or, at least they used to.) With facts, you start Part 2. With premises and logic, you complete Part 2.

Part 2: Support Your Message

Between the message and the facts lies the crux of the matter: the logical reasoning that connects facts to your particular interpretation of those facts. This is the argument at the report's core. Your report sinks or swims at this point. This middle area is the domain of logical premises, that is, basic concepts and propositions that are reasonable, coherent, and sensible to use in one's reasoning, as long as they are stated explicitly. It contains the rationale behind your message. Part 2 has two components that are coupled: facts and premises. Explicit linking of these within a logical argument produces conclusions that are valid, that is, reasonably derived from those initial ideas and pieces of information. Premises are key. Some common logical premises are definitions, principles, generalizations, causes and effects, analogies, and indicators.

These premises deserve adequate definition. Nonetheless, you will know them from your work. If you have done work that has produced a conclusion, you have inherently used premises. If you can state your conclusion, that conclusion is a good place to start your writing. Obviously, you did not start the work there; you ended it. But, for writing, it's good to start there. Similarly, you can get all your facts organized and on hand. The part that takes some planning, and outlining is the organization of the logical argument. You simply need to walk backwards: review all your facts and findings that led to your conclusion and put them into a proper order, from most significant to least significant. But before I add complexity in further discussing the crux of an argument's structure (Part 2), I want to emphasize the two-part aspect of an argumentative message, with these three examples:

1. This is the best conclusion because the bulk of the significant evidence suggests it is so.
2. This is the best design we can propose because it meets most of the qualities and characteristics of an effective design better than other, inferior designs.
3. This is our proposed policy because it offers important advantages and few disadvantages.

To support each of the messages above, you must lay out the logic and the facts. For #1, you explain the evidence found that enabled you to reach this conclusion; for #2, you must identify the characteristics an effective design demands; and for #3, you set forth the advantages a new policy must offer. These details of support are your facts and premises. In particular, the premises are often called "criteria." Criteria are critical for defining the foundation of your judgment. When criteria are fully analyzed, a judgment can be made. This approach is generic, in a positive sense. That is, it applies to nearly all decision making throughout ABERST. Look at a few more sample projects that an ABERST professional might be working on:

- A safer method for encrypting data
- A better solvent for cleaning silicon wafers
- A simplification of a distribution network

All of these are approached by defining the criteria that make a decision or choice—such as a design, solution, judgment, plan, policy, selection, or candidate—worthy and better than alternatives.

Any message you have, or conclusion you draw from findings, is, inherently, the best one given all the facts available to you and your interpretation of those facts. Any idea worth presenting to others in a report has your tacit approval. It has been vetted. This vetting is your logic. It's your case. It's the inductive method that you used, to work from facts, putting them together, then interpreting, judging, and analyzing those facts, until a sound, valid, and defensible conclusion is reached.

Thus, to apply premises to facts to support your argument, Part 2 of the general approach has these two overt steps:

1. Identify Criteria
2. Apply Logical Analysis

The two parts go hand in hand: identification (step 1) necessitates application (step 2). Applying the logical analysis is required for each and every criterion within your argument. Within each criterion, you use a logical analysis to decide if the criterion is fulfilled or not. Thus, logic for ABERST is neither abstract nor esoteric. On the contrary, it is applied and it complements the criteria identified. With step 1 done, the necessary logic (step 2) is thereafter contained within practical bounds derived from the criteria. Subsequently, your collective knowledge, experience, instincts, education, and training will guide you to carry out the available, applicable, and sufficient logical analysis appropriate to each criterion, within the restraints of the available data and information. You put forth the best analysis or analyses, for each criterion, that you can muster. The criterion-based analysis may be any of the standard, accepted logical techniques. The following are the logical analyses I have found to be most typical and effective: authority, survey results, benchmark, comparison, testing, prediction/sampling, indicators, analogy/parallelism, subjective decision, principle, definition, and mathematics. Each is defined below and presented with at least one example from a realistic ABERST project.

4.2.2 Standard Logical Analyses

Authority: The logic here is that you are citing a verifiable source widely considered an expert.

Examples:

- We checked with a leading expert on urban planning, Dr. X from the University of Y, and, based on her review of the research, we assert that adding bike paths in downtown areas reduces automobile congestion rather than adding to it.
- We checked with three suppliers of wholesale lumber, and 8 and 12 feet are the standard lengths for red cedar posts.

Survey Results: The logic is that a reliable and repeatable sampling of suitable subjects reveals trustworthy data and predictions using probability analysis.

Example:

- After surveying 280 persons, we found that 80% found Version A of these instructions easier to understand than Version B.

Benchmark: The logic is that a reliable and objective interpretation of empirical data can be compared to an established standard and appropriate conclusions thereby drawn.

Example:

- This process produces 5 parts per million (ppm) of nitrous oxides, and the legal regulation is 20 ppm or less.

Comparison: The logic is that by direct comparison to each other, one option can be determined superior in a particular aspect under consideration.

Example:

- Airfoil A58 produces a maximum lift coefficient of 26.3, while Airfoil B85 yields a maximum of 19.4.

Testing: The logic here is that the actual characteristic has been measured or analyzed through a reliable, repeatable, and relevant test, and the test results can be trusted and suggest a particular conclusion.

Examples:

- The vehicle was driven into a brick wall at 40 miles per hour, and all the safety features activated as they were intended.

- This instruction booklet was used with approximately 200 8th-grade students, and follow-up test scores increased on average by 21% when compared to test results achieved prior to use of the booklet.
- We conducted non-destructive testing of the welded piping, and all junctions were verified as adequate.

Prediction/Sampling: The logic is that a small (but adequate) set of data can be used to predict future outcomes, following established probability equations.

Examples:

- The chosen statistical method suggests 90% of future results will fall within this particular range, consisting of a mean value and a positive and negative interval around that mean.
- Using an established fatigue testing method, we determined that the part has a lifetime of no less than 10,000 hours.

Indicators: The logic is that reasonable signs and markings can be relied on to reveal phenomena, effects, and conditions.

Examples:

- The red and blue dyes did not mix until passing into the far stream, or, 8 centimeters beyond the injection orifices.
- The number of protestors exceeded 25,000, which indicates that a significant portion of the population are opposed to the administration's proposed rule.

Analogy/Parallelism: The logic is that an event or condition in one realm can be indicative of a similar condition in another realm.

Example:

- The mechanical vibrations of the metal beam reveal a similar level of potential vibration and resonance from sound waves moving through pipes.

Subjective Decision: The logic is that, for some parameters to be evaluated, a subjective and idiosyncratic judgment is expected and useful, and providing that subjective decision contributes to a reasonable judgment.

Example:

- The mayor and city council agree that Design A looks better than B.

Principle: The logic is that invoking an established principle contributes to an argument by introducing collective knowledge, precedent, or proven accuracy.

Examples:

- In sizing the vertical studs for this skyscraper, we added a safety cushion of 15 % as a commonly accepted safety margin.
- The simplest solution for eliminating agricultural runoff into the lake follows the principle of Occam's razor.
- The principle of margin of diminishing returns suggests we cap the number of heat exchangers in this process at four.

Definition: The logic here is that the accepted definition incorporates an inherent objective analysis and conclusion.

Examples:

- The policy includes a primary plan as well as a contingency, which offers flexibility; by definition, flexibility implies being able to respond to unknowns, should they arise.
- The system's waste products are non-hazardous and include only X and Y: X is potentially released to surrounding soil in the event of a storage leak, but X is classified by the state as a non-hazardous chemical if released to soil; similarly, Y may be released to surface and ground water, but Y is classified as non-hazardous when released to water.

Mathematics: The logic is that mathematics is inherently objective and logical and leads to a correct result, as long as the calculations, formulas, and inputs used are correct ones, and this can be checked and re-checked by multiple parties.

Examples:

- The proportional-integral controller produced a lower steady-state error than did the proportional-integral-derivative controller, 2.5% vs. 4.8%.
- The new, linear function developed from the data is simpler than a prior fourth-order polynomial developed before the system was modified.
- We calculate the rate of return on investment at 11.5 % averaged annually over 15 years.

- We used a known formula for noise attenuation to calculate the predicted sound intensity as dropping 85% after 10 meters.

4.2.3 Common Logical Fallacies

In contrast to the aforementioned logical analyses, a handful of initially deceiving analyses are often invoked as being logical, objective, and peremptory when used in argument. Unfortunately, they are not. They may appear to be logical, but they represent erroneous thinking. Indeed, they are well-recognized fallacies, which lack accuracy, repeatability, and reasonableness. The fallacies deserve a quick mention because alerting you to them provides (1) a note of caution so you can recognize and avoid them and (2) an opportunity to chuckle at extreme cases of human silliness. The following fallacies are explained below: false causation, sweeping assertion, unfounded prediction, unfinished conclusion, circular reasoning, *non sequitur*, assumed conclusion, false alternative, and straw man.

False Causation: Famously known by its Latin expression, *Post hoc, ergo propter hoc*, this is the mistake of assuming an initial action caused a later event because of the order, one following the other: In other words, it happened after, so it was caused by the preceding event.

Example:

- Barack Obama assumed the presidency on January 20th; the San Francisco 49ers won the Super Bowl three weeks later. Certainly, Obama inspired the team to victory.

Sweeping Assertion: This error is usually indicated by one of the following tell-tale signs or some similar phrase: on the whole, fully, all, none, never, only, and everyone.

Example:

- We propose implementing high-speed rail because all Americans want this new form of transportation.

Unfounded Prediction: This error is marked by a lack of proof, supporting data, and calculations, and it often is merely self-serving fortune telling.

Example:

- Neural disorders such as Alzheimer's Disease will never be cured, so we must put our efforts into early diagnosis.

Unfinished Conclusion: This error arises when the putative analysis produces no real outcome, yet that is presented as an outcome.

Example:

- Using the data collected so far, we conclude that more studies need to be done.

Circular Reasoning: This error involves asserting something has merit by simply restating what that something is.

Examples:

- This new ordinance is necessary for this community because a regulation of this type has been needed for decades.
- We propose creating a color-printed handout for the citizens of the community because that will put the information into written form for them.

Non sequitur: This error involves a lack of crucial connection between the premise(s) and the conclusion.

Example:

- The proposed hydraulic assembly line eliminates 65% of lifting activities for workers. Thus, we can eliminate physical therapy coverage from their health plans.

Assumed Conclusion, also known as "bootstrapping": This error assumes the conclusion as a premise for proving it.

Example:

- The vegan diet is the most effective diet for people who wish to follow the vegan principles.

False Alternative: This error assumes a false alternative in order to make the proposed option seem better. This is also known as the "either-or" fallacy. It is closely related to the "unfounded prediction" discussed above and the "straw man" fallacy described next.

Example:

- Generous government grants must be given to entering college students, or enrollment will drop severely and young adults will not be able to replace the retiring professionals over the next few decades.

Straw Man: This error uses a distorted, misrepresented, or exaggerated assertion to undermine or refute another, possibly reasonable, assertion.

Examples:

- The congressional candidate should not get your vote because he has promised that all citizens will be given unlimited years at college free of charge, including tuition, room and board, and spending money.
- The proposed dam construction is ill advised because it will produce a disaster that will wipe out the three nearby communities when the water behind the dam surges during a prolonged storm.

Before I end this important chapter and turn to the four-step approach presented in the next chapter, I wish to add one final note on the heritage of argument. One important western philosopher, Baruch Spinoza, offered a sweeping and elegant demonstration of the fundamentals of argument, though he did not set out to write a textbook on argument. Rather, he wrote on Metaphysics and happened to demonstrate a powerful technique: application of mathematical proof to discovery of the deepest truths of the universe. He wanted the conclusions of his life's metaphysical explorations to be more than just speculation; he wanted them to be actual certainties, so he followed the method of proof used in geometry to deduce and derive the full truth on the origins of the Universe, the essence of existence, the characteristics of a correctly lived life, and the nature of God. His classic, *Ethics, Demonstrated in Geometric Order*, uses logic in its very essence to deduce certain conclusions, following step-by-step derivations from initial, self-evident definitions and axioms. I mention this philosopher and his classic work because it is a pure example of logical argument used for no less of an objective than explaining the entire world and the purpose of life. If logic works for that noble objective, it can certainly work for your technical reports in ABERST.

Chapter 4.3

Use a Four-Step Approach to Argument

Using the various logical maneuvers discussed in the preceding chapter, you can develop an argument. The focus in this chapter is communicating that argument.

By giving you a prescription for communicating your argument, I may actually be helping you formulate, expand, and solidify your argument. These tasks work together, in mutual support. On the one hand, to communicate and write forces you to think in a fresh way; on the other hand, if you have been thoroughly analyzing and thinking fully and have a fleshed-out argument, the natural next step is to put that argument into words and into a document. So, this chapter guides you through the four-step approach for communicating your argument, starting with confirming it. I emphasize the objectives of *confirm* and *communicate*. I also incorporate *construction* and *customization* within the four-step approach, and these two additional objectives complement the analytical tasks of expanding and refining one's argument. In all, the four-step approach entails *confirming, communicating, constructing,* and *customizing* an argument. This is the 4-step 4C approach. (And you achieve the concurrent objectives of formulate, fix, and fortify as well, giving you a 3F bonus.) The 4C is the simplest approach I know that will be sufficient for arguing for anything, from an aortic aneurysm surgery to Zollinger-Ellison syndrome.

Before the details are presented, a quick summary of the four steps may be helpful. As stated above, the four-step approach starts with confirmation, and confirmation starts with the Opinion Check. In Step 1, you are checking to see if you have an opinion; your argument's thesis must be an opinion. This is to verify that you have a true thesis. The Opinion Check is a natural starting point in preparing your argument. Before moving on, you are wise to confirm your argument.

Next, the argument must be communicated to a diverse audience. Communication is only successful when the readers understand your message. You achieve little if your audience misses your point. You must understand and communicate to your audience, and Step 2—Communicate via Deductive Writing—is the technique of focusing on audiences, not yourself. One of the principal ploys of Step 2 is to let audiences benefit from all your work without having to slog through it themselves. You want to present your crucial support in a streamlined manner, not in a manner that matches your detailed experience with the material. Readers do not need your experiences; they need your expertise. Although your method and experiences are valid and likely derived from the scientific method itself, namely the inductive method, you do not communicate by repeating the scientific, or inductive, method. In other words, you USE the inductive method in your work, but you do NOT communicate with that method. Instead, you communicate deductively, and that is Step 2.

Step 3 places all the elements of your argument into an ideal structure, so you can think of this as the construction step. You will have a structure configured at the end of step 3 that is strategic and effective. This step is aptly named Strategic Construction. Step 4—Customize via Audience Alignment—comprises the final customization required for audiences that have a special hold on your work. Any particular audience member that has a demand on you, to which you are beholden, must be considered, and the communication must be complete and customized for this person or persons.

The details of each step are presented in the four sections below.

4.3.1 Step 1: Confirm via Opinion Check

The test for a thesis is the Opinion Check. This is important. Lots of purported theses are not actually theses, or they are not worded correctly. An Opinion Check has two parts itself. First, find your thesis. Your thesis is your point of it all; it is your main message, which is usually a recommendation or conclusion, or set of related recommendations and conclusions. See earlier portion of book for a refresher on the importance of having a point and overtly conveying that point (Chapter 3.1). Second, check that your putative thesis is in fact an opinion. It should be your opinion or the opinion of the responsible parties presenting the document you are writing. If it is an opinion, in contrast to a fact or datum point, then it passes the Opinion Check and can be used as a thesis.

As just alluded to, the principal way to understand opinions is to distinguish them from facts because these are the two predominant elements in ABERST documents. (Opinions are also distinguishable from fiction, slogans, poetry, fantasy, and other components of communication, but these others are not likely to find their way into ABERST writing to usurp opinion's rightful place at the top of the information hierarchy.) Facts, however, are so ubiquitous in ABERST that they may be mistaken for opinions and often are, and this conflation renders many documents devoid of a main point, message, and thesis.

Of course, by facts, I only consider truthful and accurate facts and not errors, lies, exaggerations, etc., which are intended to be facts (but are something worse). Facts are non-controversial. We must accept that they are actual truths and verifiable. Persons can disagree over their opinions, but we ideally should not be disagreeing over the facts. Facts must be conceded by all parties, even those parties who are in disagreement. The power of argumentative writing is that it moves beyond discussion of facts into the area where people should disagree, namely, interpretation of those facts. Any professional writing in ABERST does not have time for quibbling over facts and explaining known truths to others; the purpose of your work is to go to the next and rightful level of debate, namely, discussion of opinions.

Please be clear that I acknowledge that truthful pieces of fact have a place in ABERST reports. They are the building blocks of support to one's argument, but they are not the argument itself. The argument must always be intended to advance an opinion of some kind. Opinions are not universally held. They may be, at moments, too complicated for others to see, or unusual, or innovative, or controversial, or a bit of all of that. Other people may disagree with them. Hence, argument is required, as an author with an opinion has the need to persuade others. A thesis must be an opinion, albeit one logically derived in a reasonable manner, but still an opinion.

I can walk you through some "opinion check" examples from three scenarios so you can gain experience in separating facts from opinions. The first scenario is the recent failure of a silicon-chip-based, blood-testing device. The second involves the shape of a wing for a new airplane. The last comes from my friend's diarrhea research we have discussed before.

Fact or Opinion?

High humidity levels caused the failure of the fluid-flow controls on the chip.

Answer: Opinion. The cause may have been poor manufacturing, inferior materials, or operator neglect. So this is an opinion. It passes.

Fact or Opinion?

High humidity levels can cause failure.

Answer: Fact. Upon a first read, I am not sure if it is an opinion about the particular failure or just a generally accepted relationship. More than likely, this is a statement of a possibility, and few would dispute it. Possibilities, after all, are possible. Thus, this fails the Opinion Check. It is a fact (or a very equivocal opinion). I am going to rule equivocating opinions as NOT theses. Authors must learn to state a thesis unequivocally, unambiguously, and explicitly.

Fact or Opinion?

Humidity levels were high when the chip failed.

Answer: Fact. This is not an opinion, as it is just a general assessment of facts, although another researcher might look at the actual levels and not call them "high," per se. Opinions on the appropriate adjective to use could vary. Perhaps another researcher would see them as "moderate." This would matter if the purpose of the report is to decide on the best way to describe the levels measured. But, in context of a report devoted to identifying the reasons a small, silicon-chip-based, blood-testing device failed, this would not be a thesis. This would be merely a statement of conglomerated facts, with a descriptive term added to convert quantitative data (80-84% humidity) into qualitative expression (high).

Fact or Opinion?

Use this airfoil as the underlying shape for the wing on this new airplane; we think it is the best option.

Answer: Opinion. Nothing in the statement is proven fact. Thus, it is an opinion, derived from some facts to be disclosed in the report. It can be a thesis.

Fact or Opinion?

This airfoil has a ratio of thickness to chord length of 20.

Answer: Fact. A ratio can be measured, so it is a fact. It is not something one needs to opine about; thus, it cannot be a thesis. If your report purpose is to convey this fact, you can skip a report and instead send a quick email with the information.

Let us lastly revisit the diarrhea documents with an eye to our Opinion Check:

Fact or Opinion?

I propose we collaborate because your flu research complements my diarrhea work, and we should find some efficiencies.

Answer: Opinion. This is her opinion that this collaboration will be fruitful for both parties. It is an opinion and, therefore, a thesis.

Fact or Opinion?

I need everyone in the laboratory to begin following the new instructions, as outlined in the attached document.

Answer: Opinion. This is her opinion that laboratory staff are not currently doing tasks adequately and some new approaches will be beneficial and possibly required by prior commitments to follow certain protocols. Her desire to implement new procedures is based on an opinion that the new procedures are required or will be beneficial, or both. This statement may seem to be a fact, and it could be confusing. It is a fact that she is making a request, true. But the request itself is an opinion as to the best manner to proceed. The fact is that some commitments or contractual requirements (regulatory, advisory, internally, best practices, etc.) exist, and these must be followed. It may be that a requirement is pertinent. That is a fact. But the author's command to staff to start following new procedures is not the same as that fact. It is a different cognitive entity. She might know that, in fact, the requirement exists but tell her staff to do something different, while she seeks a waiver of the requirement. She might want to improve laboratory procedures but need to delay until a new laboratory manager is hired. In this scenario, facts and opinion are closely intertwined, but they are separate nonetheless.

Fact or Opinion?

I can justify my facility usage, in particular, freezer space, by asserting that I have many samples in the freezers and my research is approved and funded by a federal agency.

Answer: Opinion. This is her opinion that her use of freezer space is a prudent and beneficial use of space at her university. It is a thesis. Again, this may look like a fact: she needs to continue using the freezers. But, that need (to manage her samples) could be accommodated, possibly, in other ways. Thus, she might have more than one option. The underlying need for sample management is the fact. And there are other facts that concern the specific characteristics of the samples, as well as equipment operational features and space constraints and realities. Those are the facts. My friend has analyzed all the facts collectively and reached the opinion that she wants to request continued use of the freezers, which is her thesis.

Fact or Opinion?

In this journal article, I present my primary finding that chronic diarrhea is alleviated by fecal transplantation.

Answer: Opinion. Her prediction regarding the outcome of fecal transplantation is an opinion. It stems from analyzing and interpreting her results, enabling her to offer a reasoned conclusion (possibly recommendation) from those results. Very few journal articles would be published with just recounting of facts. Journals almost unanimously expect an article from a researcher to offer an interpretation of the findings, results, and data, that is, the facts discovered. The opinion, or thesis, is the culmination of research and analysis. In this instance, the researcher has collected data from various study participants, using controls and approved statistical analyses, to reach a reasonable conclusion—but by no means universally accepted—that one intervention may prove helpful in alleviating chronic diarrhea.

Because I used the diarrhea examples earlier in this book when explaining the importance of having a point when you write, you should not be surprised to see that each of these reports is built upon an opinion and are not merely presentations of facts. They marshal a set of facts in support of an argument (opinion). Thus, each report has a thesis, and each requires an argument structure. When a thesis is confirmed via Step 1, Opinion Check, you can move to Step 2.

Before we leave Step 1, however, I will point out three common grammatical mistakes that are made when presenting one's opinion. The mistakes involve incorrect past tense, vague passive voice, and a misplaced modifier. I will use a fictitious example to illustrate, namely, a recommendation to begin using the Barracuda 8g in a factory.

- **Past tense mistake:** "The Barracuda 8g had the lowest price and the best construction of the four options."

The problem lies in "had." If it had, perhaps it does not anymore? Consider this: The Barracuda 8g had the lowest price and the best construction of the four options... but the manufacturer recently raised the price and switched to a plastic housing.

- **Passive voice mistake:** "We researched reviews in various trade magazines. The Barracuda 8g is the recommended option."

The problem is the second sentence where the subject doing the action (recommending) is not identified. This is classical passive voice. Is the author saying he recommends the Barracuda 8g following the research, or the reviews consulted overwhelmingly recommend the Barracuda 8g?

- **Misplaced modifier mistake:** "In consulting the experts, the Barracuda 8g had much to recommend it."

The problem is that the introductory clause is a modifier of the noun that immediately follows it. In this sentence, the following noun is *Barracuda 8g*. Thus, the sentence says the Barracuda 8g consulted experts, when, in fact, the author did that consulting, not the item of technology.

I end this part on these three notes of caution because stating your thesis is your single most important communication task. It is your main message, and it is an opinion that will need support. So, you will not want to start out on the wrong foot by failing to express this all-important personal opinion in a direct, clear, and certain manner. Thus, pay close attention to the grammar and syntax of your thesis statement, and then you can move to Step 2.

4.3.2 Step 2: Communicate via Deductive Writing

One cause of the difficulties in communicating complex information in ABERST is that the work and the presentation of that work are best done in opposite ways. The work is done successfully following the inductive method. Communication of the work, however, is most effectively accomplished in a deductive manner. Understanding the difference between the two methods and using the latter for your reporting tasks is Step 2. In brief, here is the rule of Step 2: Use the inductive method for doing the tasks, and use the deductive method for reporting the tasks.

Inductive methods are perfect for research, design, assessment, and evaluation in ABERST. In the inductive method, the outcome is unknown at the start. Instead, perhaps a hypothesis is put forth, or lots of options are gathered. Then, through meticulous and recorded tasks, and an objective process of gathering more and more information, an ultimate conclusion is reached, inductively, that weighs and analyzes all the available information against some reasonable standards, requirements, or expectations. This is the logic of your argument. The inductive process is complex, thorough, and time consuming. Often it moves along in fits and starts and does not follow a logical or rhetorical order. Logic and rhetoric must be imposed upon it, after it is complete, when all of the results can be seen and put together to make a strong argument.

The inductive process, if it were to be presented as it occurred, is best suited for detective stories, mysteries, and memoirs. It meanders, takes detours, reaches a dead end, backs up, retraces earlier paths, finds a new branch, reaches a complete road block, turns around again, waits for a new route to be built going around the road block, then follows that route, etc.

Authors often want to provide their argument in this inductive way, very faithfully recreating the process they followed in reaching their conclusions. This is often a default approach: Here is what I did first, then I thought this, so I tried this next, then I collected this data, followed that with corroboration, followed by an unexpected variation that needed explanation, so I did this next task, and so on. Eventually, the final conclusion is reached from this process and the author has led the reader to that conclusion by retracing the steps he took to get there.

Sadly, this is the most ineffective way to present an argument. Authors who try this approach can be commended for wanting to provide details and ultimately present a thesis. But, that's all that is positive here. For the most part, these authors need help. If an author writes inductively, as he did the work, the journey will be apparent, all the twists and turns, and clues uncovered, and data points gathered, step by step, but the argument will not be apparent. The best way to fix this problem is to understand the difference between the inductive scientific process and a deductive presentation of information.

The proven technique for communicating your opinion and the support for it is to write deductively so that your argument is front and center, and not lost on readers. (Recall from the previous chapter that part 1 of an argument is to formulate your overall message, or thesis. You start there.) This means that the conclusion is presented first, right away, and all the supporting material is delivered in an intentional sequence to be cogent, organized, persuasive, and efficient, without any extraneous information. You basically eliminate the story and keep only the salient features that support the argument. Why complain about this? Most ABERST professionals did not wish to be novelists or writers, so what is so bad about finding out that your reports do not need to be presented in a chronological or suspenseful fashion? That should be welcome news.

> **Deductive Writing:** Conclusion is presented first, right away, and all the supporting material is delivered second, in an intentional sequence.

As you start outlining your report's argumentative segment, keep in mind that a critical, impatient reader is hounding you with the demand, "Cut to the chase, pal!" As this critic wants you to get to the good parts, the message is that your reports need to provide the good parts (that is, the important parts) quickly and immediately. And, you need to be sure to identify all your good parts. Thus, you start with your thesis (tell the impatient critic the bottom line right away), then you identify the principal points of support for that thesis, then you identify your source or sources for those principal support points, and so on.

If you have a probing critic asking you to "cut to the chase," you probably can imagine that person additionally asking, "Why should I believe you?" Well, this imaginary critic forces you to identify your support and put it in order. And you only have so much time that an impatient critic will give you. Thus, you put your best support points first, in order of importance and impact. Your weakest points are presented at the end. If you flip this, you might undermine your argument just as you think you are advocating it (with weak points presented early on). Even worse, you might mislead readers to reach the opposite of the conclusion you have reached. Inductive writing can lead to misleading messages. As a writer in ABERST begins to think about the document and answers a probing critic's questions, the full spectrum of the necessary content becomes clear. You are not just trying to write correctly; you must write thoroughly.

Therefore, Step 2 helps an author organize a project, possibly retroactively. The work may have been completed but its overall cohesion may not have been clear. If forced to write about work, a writer must impose a structure, often one that had not been explicitly articulated while doing the work. Remember work is inductive, but writing (and persuasion) is deductive. Many times I look at draft documents and begin to ask the author questions, and as the answers come forth, a project's structure emerges that was not there previously. The parts had been there, but not the whole. Step 2 may not change the work, although that can happen, but it may change the way the work is understood. Furthermore, once you do Step 2 enough times on various projects' reports, you can impose Step 2 during the planning stage of future projects to create a strategic work schedule with a view to the argument that may emerge after the facts have been collected and analyzed. Whether or not you are planning future projects, Step 2 is imperative for all current projects that demand an argumentative report. Deductive writing imposes a rhetorical structure on the work accomplished in support of your conclusions, and Step 3 addresses this structure in detail.

4.3.3 Step 3: Construct via Strategy

Deductive writing has one primary goal and one only: make the support for your opinion (thesis) readily apparent, front and center, in order of its rhetorical significance. This dictate implies that the support is arranged in some way. This arrangement is necessary, and it facilitates both your reader's ability to follow your argument and your ability to persuade that reader. A strategic structure eases communication and underscores argument. When you arrange your support into a strategic structure, you take your writing to a new level of excellence. Constructing a strategic, argumentative structure is Step 3.

In this step, as in discussing argument's heritage in the previous chapter, I simplify extensively. From academicians, I might take the most heat here; from my readers, I hope, the most praise. This is where I squeeze 2000-plus years of analytical thinking and scholarship into a tiny, little approach called "Strategic Construction" with the singular goal: support your thesis. After a brief introduction to strategic construction, I divide the approach into the following five sub-topics: criteria-based argument, generic criteria, syllogisms, criteria rationale structure, and supporting facts.

The approach is simple:

- First: Identify your support points.
- Second: Write them out as bullet points or full sentences.
- Third: Arrange them in order of significance, as seen by either your primary reader or yourself.
- Fourth: Outline the sub-points of each support point, arranged for greatest clarity to all readers. (Assuming a primary reader is a non-expert guarantees comprehension by all readers, non-expert and expert alike, as covered in Chapter 3.3.)
- Fifth: Write out the report to correspond to the outline created in the four preceding tasks.

Identify support points in these ways:

- The thesis meets these, X, requirements
- The thesis has these, X many, positive characteristics
- The thesis offers these, X, advantages
- The thesis fulfills these, X, objectives
- The thesis is supported by these, X, items of evidence

Whichever form X takes, these are support points. A generic, or universal, term for all of these terms—requirements, characteristics, advantages, objectives, and evidence—is *criteria*. In ABERST, we say that X many critical criteria are met or surpassed. Using the framework of criteria is a straightforward method to present an ABERST argument.

4.3.3.1 Criteria-Based Argument

Criteria are universally applicable to most arguments because the other argumentative approaches, such as process of elimination, comparison and contrast, problem and solution, and cause and effect, boil down to an analysis of criteria anyway. Just to briefly show how these other approaches collapse into an analysis of criteria, I present the following quick summaries of each:

Process of elimination: As you eliminate candidates (options from which to choose), you are assessing them against required criteria. Candidates that fail to fulfill critical criteria are removed from further consideration. The final candidate that remains, therefore, meets the critical criteria.

Comparison and contrast: This approach involves looking at the available candidates and assessing them against criteria; the relative strengths and weaknesses per these criteria compose the essence of the comparison.

Problem and solution: A problem is resolved by finding a solution that sufficiently fulfills the criteria of an effective and feasible solution.

Cause and effect: When asked to find the cause of an event, either unfortunate (breakdown or failure) or fortunate (victory or achievement), you can set forth the required evidence for any particular hypothesis as to the cause. If you ultimately determine an event to have been caused by your hypothesis (a certain variable or set of variables), you can say that such and such evidence indicates that this is the case. As you present each piece of evidence, you are giving proof that your judgment is correct. Listing the necessary evidence is tantamount to listing criteria of a plausible cause. Walking through proof of each piece of evidence is providing support per criteria. The same process works for predicting an effect.

No matter the argumentative approach, criteria will be involved, and specific criteria arise from different sources. Sometimes the customer or sponsor dictates the expected requirements or characteristics. (You will want to align to customers' needs, as best you know them—see next section, Step 4.) The customer asks you to find a solution or pick a design/candidate that meets X requirements. In such cases, your support consists of walking through each dictated requirement or characteristic. Other times, your project is open ended and you are initiating the proposal, so no third party has dictated the expectations. Instead, you have developed a set of advantages or positive qualities by which to evaluate your proposed thesis, and you assert that your thesis does in fact possess most, if not all, of those advantages or qualities. In other words, the thesis meets or surpasses the criteria for a favorable thesis, and you have the proof of such.

Criteria, perforce, are different for each project. The issues of concern are topic-specific, and each analysis is unique. Even when doing successive reports within the same field, say, lithium-ion batteries, the priorities and technological features change over time and across applications. All that can be said about specific criteria is that each writer is responsible for identifying the ones that apply. If you are working with a team, the collective perspective of all team members can be applied to identifying the criteria. As a writer, you are likely to know the obvious ones; within your responsibilities is included the need to find the less-obvious ones also.

To help in identifying all pertinent criteria for a report, one heuristic has proven helpful across all disciplines: work from a set of generic criteria that have wide applicability to almost every ABERST project. Their generic nature makes them useful as starting points. Each writer must adapt these generic criteria to her particular analysis. Below, I introduce the generic criteria, followed by explaining how to adapt them to specific projects. This approach is adapted from the work of my colleagues, Olsen, Mathes, and Stevenson, who used the ideas of their predecessors in developing their framework.

4.3.3.2 Generic Criteria

The four generic criteria are as follows: effective, feasible, affordable, and desirable. All metrics of quality can be traced to one or more of these fundamental concepts of assessment. Each is briefly defined below, but these definitions are based on the terms' common definitions.

- Effective: it gets the job done; it fulfills its intended function; it works; it solves the problem it needs to solve; it offers the functionality demanded; it explains the occurrence of the phenomenon in question; it takes into account the identified causes.
- Feasible: it can be implemented; it is realistic; no significant obstacles prevent its implementation; it is available; it is practical; it has been proven at bench-scale or in actual performance.
- Affordable: its cost is reasonable; it is the cheapest option; it represents the best value of all options; its capital costs, as well as operations and maintenance costs, are acceptable; it has acceptable short- and long-term costs.
- Desirable: it offers additional benefits; it provides extra advantages; it meets requirements on our "wish list"; it adds convenience; it improves or positively impacts some other issue, such as public opinion, morale, or the environment.

Some additional aspects that may be appropriately considered in the assessment are the following: meeting quality or safety standards, increasing output, diminishing loss, boasting simplicity, allowing versatility, maintaining consistency or precision, incorporating advanced technology, possessing inherent legality, satisfying employees' expectations, preserving environmental quality, preventing leaks and spills, promoting good resource stewardship, reducing land acquisition, limiting use of eminent domain, promoting repurposing, conserving energy, enabling quick implementation, promoting diplomacy or political harmony, presenting psychological benefits, and satisfying stakeholders.

If you have more than one candidate/option to evaluate, the one that best meets the most criteria is preferable. If you have only one candidate/option/solution to consider, you simply look to see that it meets all the essential criteria, in an acceptable manner. If it fails one criterion, you might still be able to recommend it. Even if it fails more than one, you might still present it as your choice, as long as the organization can live with a solution that is not flawless. If all criteria are critical, then you must wait to provide a recommendation until you have found one that meets all criteria.

I find that priorities within projects vary: sometimes cost matters severely, and sometimes cost is a consideration but not a deal breaker. Sometimes aspects considered only desirable can be ignored if they are not fulfilled. Other times, even desirable qualities must be present. Most of the time, the qualities that are used to determine that a recommendation is both effective and feasible must absolutely be met. It is hard to recommend a design/solution/policy/candidate that is neither effective nor feasible for its intended function.

Your approach to criteria is the crux of your argument. You win or lose here. The facts (measured data, visible physical characteristics, simulation results, calculation results, etc.) are not debatable. Assuming the facts are correctly collected and expressed in your report, no readers will dispute them. Readers will, however, disagree with your thesis. In other words, your point may be moot, but not your facts. How is that? People judge things differently. We evaluate facts differently, providing different analyses and interpretations. Your framework of criteria (which characteristics matter and how much so) is an explicit ranking that must be agreeable to readers. If it is, and you have the facts ready at hand, you will win the argument (succeed in convincing readers); if you do not have an agreeable criteria framework, you will not convince readers no matter how many facts align with your point of view. When an argument fails to persuade, the following is the explanation: the un-persuaded reader disagrees with your slant on the criteria, or, your argument structure, not the facts.

Your slant on the criteria is your rationale, and it should be developed independently of the facts that arise in support of your thesis. Ideally, you form most of your rationale as you initiate the project. A good rationale helps you plan your work and determine the data and facts that need to be collected. Sometimes the rationale evolves as you uncover more knowledge. This give and take with the criteria is normal, but one approach is strongly discouraged: developing your criteria after all your work is done, to match the findings of facts. This flaw is a variant of "bootstrapping" (recall discussion of fallacies at the end of Chapter 4.2), and it indicates sloppy work and bias. If you wait to see the facts, then squeeze out some criteria from those facts, you are forcing an argument into your data.

For example, if you find that the laptop you purchased on impulse has a nice display and a fast graphics card, you do not want to retroactively introduce (1) nice display and (2) fast graphics card as the two critical characteristics of a good laptop. Such a two-point rationale for a high-quality laptop is flawed. This is an accidental, retroactive rationale. It basically uses the actual candidate selected to formulate the argument for selecting that very final candidate. (This is "bootstrapping" pure and simple.) This methodological flaw taints your inductive process. It is a root problem, not just a communication weakness. The flawed inductive process excludes other features and possibly other candidates from competing fairly against the criteria. It is biased, limited, and subjective.

Instead, you must determine the characteristics of a high-quality laptop independently of assessing any single candidate. You determine these characteristics from a range of sources, experiences, and assessments. You develop a rationale prior to completing the project. Then, you use that rationale to evaluate the facts and make a decision (select a candidate, choose a solution, improve a design, and so on). A strong inductive process has the potential to yield an equally strong deductively structured argument, and a weak inductive process cannot be masked by even the best communication. The rationale must be solid (and not bootstrapped). To emphasize the importance of the rationale, I need to introduce the concept of a syllogism.

4.3.3.3 Syllogisms

Logical arguments, in a pure and simple form, often are referred to as syllogisms, with three parts, as in the following generic example:

- Major premise: A design that has characteristics X, Y, and Z should be used.
- Minor premise: Design Alpha100 has characteristics X, Y, and Z.
- Conclusion: Design Alpha100 should be used.

In syllogisms, no one disputes that a good deal of proof is needed to support the minor premise. This is where all the facts and support points (data, results, records, statistics, testimony, etc.) are used. A mistake is made, however, when proof is presented with the major premise. The major premise requires no proof, per se, and this may be confusing initially. Quite simply, when presenting the major premise, which encompasses the criteria and the rationale for those criteria, no support for any particular thesis is used. It would be inappropriate there. It must wait for the place in the report where the minor premise is proven, with facts and support. Nonetheless, the individual nuances of the major premise, that is, the criteria selection and ranking, must be explained. Therefore, I say the major premise requires no proof, only a rationale, which is a crucial argumentative component. An author must explain the reasons for selecting and ranking the particular criteria.

Observe one engineering example: our company wants to design and manufacture a new electric-powered automobile with outstanding acceleration, top speeds, and high safety ratings. To prove that a new model meets these criteria, the support is required that this model, in particular, boasts outstanding acceleration, top speeds, and high safety ratings. In addition, a rationale must be given for highlighting those three characteristics above all others. We do not "prove them," per se, but, rather, make a case that these three criteria should be considered crucial in an evaluation of a new car model. In terms of a syllogism, the major premise must be explained and justified in order to complete the argument for the new model.

Indeed, as mentioned above, without a rationale, the argument collapses. Fortunately, the structure for organizing your rationale and presenting a criteria-based argument follows a common, universal pattern. You can use it for every argumentative report you write in ABERST. I explain it below.

4.3.3.4 Criteria Rationale Structure

At the highest level, you determine which of the four generic criteria apply. Certainly, if you can argue that all four are applicable and your thesis meets all four, you will have a watertight argument. Otherwise, you must at least argue that your thesis is effective and feasible.

The next step is to customize the generic criteria for your specific project. This involves proceeding through three more levels of specificity. At specific level 1, you identify the characteristic(s), or metrics, that you are using to assess each generic criterion. This is the measurable characteristic, quality, feature, etc., that represents the generic criterion, as in these examples:

- Effective rocket engine: metric is total impulse
- Feasible space station repair: metric is achievable by on-board astronauts

- Affordable solar array farm: metric is eligible for tax credits to offset capital costs
- Desirable emissions control system: metric is total weight added to vehicle

For each criterion, you may have more than one metric, but you always need at least one metric. There is no such thing as an effective X (solution, design, product) because it is effective; that would be circular. A criterion requires a metric, which is within the power of the authors to identify (or a customer has identified it).

At the second level of specificity, you identify a benchmark for each metric listed at level 1. The benchmark must be selected (or identified by customer) and adhered to.

Imagine a sensor to detect diseases via a patient's breath. Here is a customization of each generic criterion through specificity levels 1 and 2.

- Effective: must accurately detect (metric) at least five diseases (benchmark) and not give any false readings (benchmark). If it currently detects only two diseases and occasionally gives false readings, it is not yet effective.
- Feasible: can be designed eventually into a suitably sized package (metric) that a medical professional can hold in front of a patient's mouth (benchmark), thus measuring roughly 2 by 3 inches (benchmark). If the sensor cannot yet be designed smaller than a large truck, it is not feasible.
- Affordable: purchase cost to customer (metric) must be no more than $1,000 dollars (benchmark). If its retail cost is estimated at $1,200, it is not affordable.
- Desirable: device can be easily recharged (metric) when battery runs down with one cable (benchmark) and within 30 minutes or less (benchmark). If it requires two cables, an adapter, and 90 minutes, it is not desirable.

The third level of specificity, and the last one, is your presentation of the facts that prove each metric in question meets its benchmark, thereby ensuring that the thesis fulfills the criteria, or as many as possible.

In a nutshell, the above discussion explains the fourfold heuristic of criteria (four generic criteria), with three levels of specificity: metric, benchmark, and facts. Furthermore, each element of the rationale must be explained if it is not self-evident. If you expect a circuit box to contain 40 circuits on an all-copper bus, you must justify both 40 and all-copper. Are these expectations determined by municipal code, durability issues, or other concerns? That is the final component of a rationale: the justification of all elements that may not be obviously reasonable to any reader. With a complete rationale, you can shift your focus to the facts to be assembled with respect to the criteria set forth in the rationale.

4.3.3.5 Supporting Facts

To present the facts, you must have additional structure. Within this fact presentation, you must explain your method of acquiring facts: tests, simulations, interviews, surveys, calculations, and so on. If a test, you must explain your setup and method, including calibration, error analysis, and hysteresis assessment. If a simulation, you must explain the input variables, the simulation software including its built-in assumptions and algorithms, and the limitations of the output. If an interview or a survey, you must explain your method and provide a copy of your questionnaire(s). If a calculation, you must introduce the equations used and the input values, then the method for determining input values (such as assumptions; required, given, and commonly accepted numbers; independent variables; constants; and measured inputs).

After this, you present the facts acquired (preferably in visual form—graphs, tables of summary values, data maps, histograms, pie charts, etc.). Then you can interpret those facts vis a vis the benchmark already introduced.

This will look familiar to those of you who have some experience with writing or reading scientific journal articles; it is nothing other than introduction, method, results, and discussion (IMRD). Here, "Introduction" is the explanation of the criterion, metric, and benchmark. "Method" is the effort made to acquire facts. "Results" is the presentation of the facts, and "Discussion" encompasses the interpretation of those facts against the benchmarks (standards, regulations, goals, status quo, baseline, or competition).

The difference is that IMRD is not the complete structure for an entirely clear, whole report in ABERST. It is merely the structure for each support point within a report's argumentative section. The whole report is a collection of

IMRD-based segments, placed together into an Argument for a Thesis, which was introduced in the Overview (Executive Summary or Summary).

Before the structure that emerges as a result of Step 3 is set in stone, you must take a further look to ensure that every essential, customer-required aspect has been addressed. This is Step 4.

4.3.4 Step 4: Customize via Audience Alignment

While the ultimate and full audience for an ABERST report is impossible to anticipate because documents last for decades and may be consulted by an infinite number of readers for reasons that the author may not realize at the time of writing, the first-tier of intended readers is possible to identify. In fact, identifying the direct recipient is an essential element of report development, as explained in Chapter 1.5. This direct recipient, or recipients, must be identified, and this action is integral to understanding and fulfilling one's writing purpose. In addition, the recipients have explicit needs and requirements, and those must be attended to completely (also covered in Part 1).

The recipient again deserves consideration when one is finalizing one's argumentative structure. An author must review the argumentative analysis, with particular attention to the criteria, metrics, and benchmarks, to ensure that every known issue of importance to the recipient has been addressed. In short, authors must align their arguments with recipients' needs. This occurs at the first level (covered in Part 3) of prioritizing by isolating and highlighting primary information and distinguishing such material from secondary and tertiary information. As a quick review of the first level of prioritization: Think in terms of the critical information the primary recipient needs from you, per your purpose and main point. This drives the emphasis on argument. Readers rarely need secondary and tertiary details. If they have to respond to your thesis (purpose for writing them is to present a recommendation or conclusion for their review), your development and support of that thesis is vital information for them.

At the second level of prioritization, you double check that your criteria, metrics, and benchmarks match any and all expressed concerns from recipients as well as the concerns that you can anticipate they would have.

Recipients are readers with a special hold on the document, and they will be expecting certain information. Get that into the argument very early. Determine your ranking of criteria based on any information the recipients require. Also, as you can anticipate other secondary readers, you can double check that you have addressed their possible concerns in your ranking and inclusion of criteria. Special readers within the larger audience might include customers, clients, executives, and regulators. If someone in this special audience group would expect to see some information, you must prioritize that.

As alluded to above, if the recipient's concerns are overt, you can easily address them. If they have not been spelled out, you must be thorough in searching for them. The generic criteria are a starting place. Generally, you know that all stakeholders and interested parties always will demand an assessment of efficacy, feasibility, affordability, and all other desirable qualities. The ranking can be based on your best judgment with acknowledgement of your recipients' and possible audience members' concerns. If safety is a priority, then all analyses that reflect on the inherent safety advantages are ranked higher than other issues. If immediate implementation is a priority, then feasibility deserves a high ranking. In all, your criteria rationale must be finalized with your audience in mind, and let the facts follow.

Chapter 4.4

Maintain and Enhance Credibility

As much as you work objectively and apply dispassionate analysis to your work and put forth a credible argument with reasonable, fact-based support points, you still always run the risk of being perceived as biased, subjective, prejudiced, overly emotional, and personally motivated. Some readers may doubt your objectivity. They may feel you have an agenda, vested interest, conflict of interest, or some other personal reason for putting forth your argument. On the one hand, your facts and logic-based argument should persuade them otherwise. On the other hand, some readers are naturally skeptical, cautious, possibly paranoid, and, in the worst case, irrational. You might not be able to get them to agree to the facts themselves, let alone consider your reasoned interpretation and judgment of those facts. Moreover, even smart and sane readers may not necessarily agree with you from the beginning, and that is a good thing. Skepticism, scrutiny, and caution lead to better decisions in the end. So, you must be prepared to persuade smart and reasonable people who do not initially share your outlook.

This chapter offers three more techniques for bolstering your argument that will impress both the supportive readers and the skeptical ones. These techniques speak to your credibility. They reinforce your commitment to objectivity, fairness, and open-mindedness. The first one, leverage corroboration's power, bolsters your argument by presenting additional sources (methods, calculations, simulations, authorities/experts, and the like) that have produced similar data, results, and conclusions as your own. By showing that the same result and conclusion was reached by different methods or persons, you help make it clear that this is not your own, idiosyncratic, personal, and subjective finding.

The second technique, admitting your argument's flaws, bolsters your argument by collecting and presenting all the weaknesses you are cognizant of and that you have identified, considered, tolerated, and, if possible, accommodated, adjusted for, or addressed in some way. (This excludes logical flaws as previously covered, which are intolerable and must be rectified right away.) You admit the understandable flaws while maintaining your overall assertion that your conclusion has merit. By admitting these flaws, you might preclude a critic from thrusting them at you in refutation. If you can address a flaw of any type and still your argument holds up (to a bright light of scrutiny), then you have a fairly watertight argument. Admitting these flaws in your report (and not waiting to present them only if you face criticism) is an important technique for showing that you have looked thoroughly at all the information, that you are objective and honest, and that you understand both (1) no prediction can be made with 100% accuracy and (2) no idea is 100% perfect. If you present the flaws and still a skeptic disputes your overall thesis, you can do nothing about that. People can disagree as to how to interpret data. Your thesis is, after all, just an opinion. Someone may disagree. Most likely the critic sees the criteria of judgment in a different way than you do. That is OK. At least the dispute does not come from some omission or limitation in your work that you should have foreseen.

Lastly, the third technique, consider alternatives, reinforces the single choice you made by proving it was not the only choice you considered. You did, indeed, look at all the available options, or candidates. You did not pick the first choice that came walking down the road.

4.4.1 Leverage Corroboration's Power

Corroboration is one of my favorite words. It sounds a little like a combination of "labor" and "cooperation," and those are the ingredients of a successful effort. I like to give my students an extra 5 points for a document in which they use this word. The idea with corroboration is to find one other source to confirm that your facts are correct or your interpretation of those facts is sound. To corroborate is to strengthen by confirmation.

The technique of corroboration is common sense itself and fits nicely into a logical analysis. It starts by playing a role in basic comprehension, but evolves into an element of persuasion. At the basic level of comprehension, corroboration is there at the start of childhood development and basic cognition: when we, as very young children, hear something more than once, or hear it from two different sources, we begin to sense its importance. A parent says with evident concern and gravity, "The fire is hot. Stay away." In addition, we can feel the heat as we crawl toward that fire. This sensory experience gives us corroboration of the oral warning and confirms our comprehension that hot fire should be avoided. Similarly, when two separate friends recommend a movie, we tend to move that film higher up the "must see" list. Getting the facts one time, from one source is helpful, but hearing the same thing another time, from another source, seals the deal. And when this intuitive sense of corroboration transfers into the professional realm of ABERST, it plays the same valuable persuasive role.

It can be seen as an extension of everyday good research practices, in particular the process of double-checking one's data-collection methods. In a lab test, one conducts multiple trials of the same test, to see if the data acquired are comparable and the first test was not an anomaly. Only after two or more trials, and collecting comparable data, can one be sure that the method is useful. The data, moreover, can be averaged, with variance and error calculated, to account for unknowns, equipment deficiencies and insensitivities, environmental fluctuations, noise, and human error. Using multiple trials to make the data more robust is a type of corroboration. Similarly, hand calculations are often done to check that computer-assisted calculations are producing the right results. This is the way to check that the code in the computer program has been written without error. Again, this is a type of corroboration.

While these examples are a start for understanding corroboration, corroboration is not strictly speaking a description of your own good practices. Corroboration, in the real sense of the word, moves beyond your own laboratory or computer, beyond your own double-checking and collection of precise values, and into another approach, outlook, and perspective. You must aspire beyond your best, first effort. You gain credibility by finding confirmation and reinforcement for your work by finding comparable data, results, and conclusions from a second, and different, effort.

Here are some examples of corroboration:

- Laboratory testing corroborated by computer simulation
- Laboratory testing corroborated by theoretical calculations
- Research done in location A by team X corroborated by research done in location B by team Y
- Research done at time A by team X corroborated by research done at time B by team Y
- Survey results corroborated by focus group results, or vice versa
- Survey results corroborated by differently worded questions in same survey
- Results reached by method A corroborated by method B
- Anecdotal evidence corroborated by forensic evidence

Within the subset of method corroboration, you might find the following as familiar or inspirational:

Corroboration with multiple methods:

- Mechanical methods corroborate acoustic methods
- Pressure-collection methods corroborate force-collection methods
- Fatigue tests corroborate pull tests
- Deep-learning methods corroborate response-control methods
- Ground data corroborate satellite data
- Photographic observations corroborate sensor outputs
- Radiography results corroborate blood-test results
- Physical examination corroborates magnetic resonance imaging

The aforementioned are just a sampling. In the large and vast realm of ABERST, the options for corroboration are always expanding with every innovation and new person to enter the profession and bring new ideas and additional resources for collecting and analyzing data. While you cannot always provide corroboration for your support points in every argument, this technique should be used each and every time you are able to strengthen your argument in this way.

4.4.2 Admit Flaws of Any Type

Flaws are unknowns, simplifications, assumptions, estimates, predictions, missing data, equipment deficiencies, neglected issues, worries, concerns, drawbacks, negatives, and any other potential weakness in your inductive work and deductive argument. You would be wise to admit and concede flaws rather than appear to be hiding anything. If you wait for a skeptical reader to confront you with a perceived flaw, you can, at that time, defend and explain the situation. But, you are working defensively at that point. You already have let a critic develop a case against your argument. Human nature is such, moreover, that if a reader finds one flaw, she may become extra watchful and inclined to find others. Indeed, the critic may find a number of points that serve to refute, undermine, disparage, or reject your thesis. Defending yourself against this reader's case, especially if it escalates, may demand time and effort.

Alternatively, some or all of these purported flaws raised by a critical reader could have been addressed in your own report, as a preventative posture. You can and should, therefore, implement a preventative strategy by admitting flaws before they become overblown by a detractor and used against you. Flaws are potential time bombs that can be diffused by open admission. Overall, a flaw that is admitted by the author has nowhere near the negative impact on a reader as does a flaw the reader discovers for herself. Admitting flaws can take various forms, depending on the work you have done and the argument put forward.

If you are collecting and using survey data, you might need to admit any of the following: small sample size, biased sampling, skewed sampling, and so on.

If you are using physical modeling, you might have imperfections in the model, motion or shifting in the model when it should have been steady, inability to collect all the desired data from the model, inability to put the model through realistic conditions, or limitations in the testing of that model in any way (not enough temperature change, vibrations, shaking, speed, gravitational effects, vacuum effects, and so on).

If you are doing computer simulation, then you must admit assumptions and limitations in the software that runs the simulation or the input of data into the simulation.

If you are doing calculations, perhaps you had to assume some input, or simplify the variables for complex, stochastic phenomena.

I offer two suggestions for presenting flaws in your argument. One, present flaws twice in a report. Two, present flaws without using the generic term *flaw* in the report and especially in headings. These are explained further below.

4.4.2.1 Presenting Flaws Twice

Perhaps the best way to present flaws is to present them twice in a report. The first time you bring up a flaw is in the section of support to which it is linked. The second time is in a final section of the report dedicated to acknowledging flaws. No reader can accuse you of trying to hide anything when you present the flaws twice. Let me give an example.

If you are testing a driverless, fully autonomous vehicle, and you are presenting data collected from driving for 10,000 hours within a private proving ground, closed to the public, but designed with various roads, intersections, and realistic activity composed of auto, bicycle, bus, and pedestrian traffic, you must be explicit that this testing was not done on actual public roads. Furthermore, in the section where you present your hopefully positive driving results, you make it clear that the data come from testing within a private proving ground. Later, near the end of the report, you present a dedicated section on omissions of the study, and there you admit that no data come from driving the vehicle on actual public roads amongst real traffic. Instead, the study was limited to driving the vehicle within a controlled, test facility.

4.4.2.2 Presenting Flaws within Specified Sections

Flaws is a general term, and it may sound unnecessarily pejorative to a typical reader. I understand that you may not want to appear as someone who presents a flawed approach. I fully agree that you are not required to undermine, disparage, or belittle your own effort and thesis. You should not be your own critic, attempting to pick apart your work by finding all its fatal weaknesses. On the contrary, I am advising you to strengthen your argument by discussing flaws.

Flaws represent a category of objective and reasonable caveats and qualifications to otherwise rock-solid work. Sections dedicated to specific flaws are your final opportunities to demonstrate your thoroughness and objectivity. You want to state with confidence that you are quite aware of the imperfections of your work, though you nonetheless stand by it and feel the need to advance it.

Thus, instead of *flaw*, you are advised to use a term or terms without unnecessary negativity. Be straightforward and clear headed about these issues, so no one can undermine you with aspects of reality that you could not change. ABERST is done in a real, imperfect world, and you do not need to pretend your work exists in a fairy-tale-like land of make-believe perfection. Terms that work well as headings are these: Drawbacks, Limitations, Omissions, and Errors. If you can come up with a different term that suits your work, by all means do so. I address each of these briefly below, with some examples also.

Drawbacks: Even your best idea may not fulfill all the expectations, requirements, and constraints. You pick the best of the options. The best option may have drawbacks. It may be heavier than you would have liked. It may require a longer development time than one of the other candidates. It may be costly. It may have emissions to air or water. It may mean some people will lose their jobs. Some people may get sick during the trial period. Some neighbors may be inconvenienced. You cannot please everyone. No idea or solution is perfect. There is no panacea. You must contend with a drawback or two, even with the best recommendation you can produce.

Limitations: As much as you have attempted to develop and assemble an airtight and comprehensive case in favor of your thesis, you can never cover every single potential issue. Time and money usually preclude a total exploration from every angle, using every tool, and all possible resources. You can rarely implement every single desired test. It is difficult to leave no stone unturned. If a solution is needed, it usually is needed before all potential situations can be thoroughly vetted. Thus, in almost all arguments, some limitation can be found in the support. Limitations may enter your work through limiting assumptions and limited data. You might have limitations with your equipment, or timeframe of study, or methods. You might have limitations with the candidates available to you. You might have limitations with space, weight, funds, facilities, and materials. Be alert to all possible shortcomings in your research or study effort, and present them in a straightforward manner.

Omissions: Similar to limitations, omissions are easily seen holes or gaps in your work, along a continuum that is mostly complete. This set of flaws is typically specific and small in scope. Omissions can be of conditions (temperature, humidity, wind, rain, and snow); populations; demographics; or settings of equipment. For example, it might be that you have studied a continuum of conditions with regard to weather, but did not study freezing rain. This is an omission. You might have studied many angles of a motor's rotations, but did not look at 105 to 135 degrees. This is an omission. You might have looked at low and medium frequencies (Hz and kHz), but not high frequencies (MHz). This is an omission. In studying people, you might have contacted elementary school children and high school children, but not middle school pupils. This is an omission. These and similar holes or gaps are omissions that must be admitted.

Errors: Only one class of errors is reported; the other class must be corrected before the report is submitted. The latter class, consisting of human error, is not a drawback, limitation, or omission, but, rather, a mistake. You must fix human error before the deadline, or ask for an extension so you can finish the work correctly. The former class of error, however, consisting of random error, must be reported but is generally accepted as realistic and inevitable with any project wherein data are collected.

Another way this distinction is expressed is by contrasting systematic error with random error. Systematic error is something that can be addressed and corrected while doing the work. For example, if an amplifier drifts, a connection comes loose, or a scale shifts from zero when it should be displaying zero, the necessary mitigation measures can be taken so the work can proceed without systematic error. Corrective actions include rebalancing the amplifier, tightening the connection, and resetting the scale to zero. At more extreme levels, often work might overtly include a significant human error that goes beyond systematic error, such as calculations done incorrectly, components set up mistakenly, or data sets neglected. As in these aforementioned examples, all errors of human or systematic origin must be corrected.

In contrast, random errors are inevitable unless one has infinite financial resources that would enable nearly perfect and comprehensive equipment and data collection methods. Otherwise, random errors are out of one's control.

They creep into the work because noise arises in a system, in one way or another. This can be a result of insufficient resolution of one's instrumentation or insufficient filtering of signals. Noise and resolution issues are often referred to as "fractional uncertainties." Consider these simple examples: you want to measure to 1/100 volt, but your equipment only shows 1/10 volt. Or, you want to measure 1/32 inch, but your ruler has a scale that stops at 1/16 inch.

Another term used for random error is "residual error." If you can calculate error bars with your data, you are addressing residual error. And sometimes a "correlation coefficient" is calculated to describe the residual error, or this error can be given in percentages. Ideally, these percentages bounce around zero as closely as possible, without going above 5%. When associated with residual error, the data have been interpreted to fit a functional or physical relation, but the fitting is imprecise by some small residual. This is a random error, and must be reported, assuming it is small. If the error involves a systematic shift away from a mean of zero, this can be reported as long as it is small. Again, if the systematic error associated with residual error is large, it should be corrected before collecting and using the data.

4.4.2.3 Two Special Situations with Error

In closing, let me discuss a few special situations that might arise when confronting error.

One, if you become aware of an error right before a deadline, you have three options. First, it is best to correct the error by revisiting the work and doing it correctly, assuming you have time. Otherwise, second, seek an extension of the deadline so that you can complete the work without error. Third, if you believe the benefits of submitting a report with an omission due to error outweigh the negatives of such an omission, you can submit with admission of the omission due to error.

Two, error in no way refers to data that are not ideal or do not fit your prediction. That is just reality. Sometimes the data are unexpected and sometimes they are not what you would wish them to be. You must neutralize that subjectivity. Indeed, you must report all findings, data, results, and observations, regardless of whether they are helpful to your argument or not. Your argument must stand up against conflicting, contrary, detrimental, antagonistic, and negative data; if it cannot stand up to this, you must change your argument.

If some data point is anomalous, it is analogous to an error. It does not fit an otherwise observed or expected trend. This may not be fatal, yet it might be puzzling. You must present it.

4.4.3 Consider Alternatives

Candidates (options) for consideration arise in one of two ways. Many times they will be presented to you for evaluation. A client or customer assigns you the task of evaluating a pool of candidates and selecting the optimal one (per criteria as discussed in Chapter 4.3). Other times, you will be responsible for identifying the candidates from which to make an optimal selection.

If you are ever rushed, you might be tempted to look at a single option only. This poses a problem whether you select or reject it. If you select that option, you run the risk that readers may think you skewed the decision toward that option because it was the only one considered. If you reject the option, readers will think that you have not solved the problem or offered a plan for moving forward. In almost every situation, excepting those where only one absolute option can be considered, you are advised to explore a reasonable amount of options (candidates) before making a decision, so readers know that you were thorough. This acknowledgement of alternatives to your thesis is important to your credibility.

Consider looking at travel between points A and B, say, Chicago and Detroit. If you decide on recommending a train trip, and you choose the Amtrak train line, you have really only that option. The Amtrak Wolverine is the only passenger train that runs between those two cities. That is a weak thesis, as you have not encompassed a look at alternatives to the thesis. Your credibility is suspect. You may be viewed as biased toward train travel, or you may be seen as superficial in your research. You may appear uninformed or disinterested. You must prevent any such doubts of your professional integrity by acknowledging and assessing alternatives.

If you expand your pool of options to include bus and air, you can claim to be comprehensive, hence credible. Selecting a choice out of this expanded pool is likely to be more convincing to your audience. If you recommend the Wolverine train line over Greyhound and Megabus services, as well as American Eagle and

Delta air lines, you certainly have done a comprehensive assessment of the various alternatives. The Amtrak Wolverine appears to be a choice with merit, not merely the only option considered or available. Your credibility is intact.

In some argumentative structures, such as process of elimination and compare and contrast, a healthy pool of candidates is explicit from the overriding structure itself. In other argumentative structures, the author may make the pool of alternatives evident from the beginning. That is, in a very early section and in each subsequent primary section, the presence of several candidates under review is explicit. Still, in other structures, the discussion of alternatives is not prominent. In such structures, you must place a discussion of alternatives at the end of the argumentative segment, to prove that you did consider alternatives to your thesis.

When addressing alternatives at the end, to enhance credibility and preserve your integrity, you have a few different ways to handle them. One, you can pick a second-best alternative to your thesis and present all its strengths and weaknesses, explaining the reasons for not placing it first while still giving it credit in many respects. Readers like to know that a viable alternative exists, in case the first choice does not work out for some reason. Two, you can reject all alternatives and rationally explain the reasons they are inferior to your thesis (the first choice). Three, you can admit that alternatives deserve more attention than you were able to provide, and one or more of these alternative may prove superior to your thesis on further study. This approach leaves the project dangling somewhat, but it leaves your credibility set in stone.

Chapter 4.5

Reinforce Argument with Flow and Transitions

When readers stumble through your report and cannot follow the train of thought or find your logic, they will say the document "does not flow." We have seen flow as a concept within paragraph organization and structure (Section 2.2.1). Specifically, within paragraphs, the sentences must be connected, and this creates cohesion. Here I address the same idea, but for the whole report, which is understood as flow. Flow between ideas, from one to the next, from section to section, as a coherent whole, is the ideal quality in an easy-to-read document. Without flow, your argument will not be easily apparent.

The solution is to create flow between and within sections and to express the connections overtly in transitions. If you are having trouble explicitly describing the connection of one section to the previous one, you have a sign that your sections are not in proper, flowing order. If you can express the connection, but you are neglecting to do so, you must revise the document so that the connections are conveyed explicitly. Readers cannot read your mind, only your document. You may know perfectly well how the next section relates to the previous one, but a reader may not... until you tell her. Implicit flow and explicit transition statements reinforce your argument.

4.5.1 Create Flow Between Sections

In first drafts of technical reports, a common problem is lack of flow from one idea to the next. The ideas presented are in no apparent order to the reader, and the sections seem to jump around, instead of flow together as a coherent whole. This is solved in various ways, some already covered. First, by following earlier parts of this book, an author will start strongly, identify the document's purpose, and provide an overview before discussing details. Second, an author will put the details into a structure that makes the argument prominent, clear, and thorough. Third, transitions, as covered in the next section, will remind readers of main points as new points are introduced. These three techniques will go a long way in building flow. Still, choppiness and randomness—when ideas bounce from one to another without seamless connections—can creep into a document.

This choppiness happens for three reasons: the sections are not in a logical and coherent order, the sections are in a good order but each section begins poorly, or the sections are in a good order but the writing within each section is muddled and does not follow the proper methods for (1) organizing paragraphs under topic sentences and (2) providing generalizations before particulars. A combination of understanding the causes of muddled writing and knowing the techniques for creating good flow (five covered below) should solve the problem of poor flow.

Creating flow in a technical document will not be done unless an author acknowledges its importance, and one misperception usually precludes such acknowledgement: being convinced that readers are fellow experts with similar education, work experience, and knowledge of one's project to be discussed in the document. This is an absolutely detrimental misperception. It is not the case (as discussed in Chapters 1.4 and 3.3).

Thus, to eliminate this detrimental misperception, you must remember when drafting the document that readers do not know as much *as you* about your project. Even the true experts in the same field do not know about YOUR portion of it, of which you want to report, unless they work side by side with you. Very few readers are people who work side by side with you. If your readers were only those people who knew your project inside and out, how many people would that be? Really, you must consider if the document is for you and the other members of the "inner circle" or people outside that circle. Recall Part 1.4, which sets forth the basic axiom that most readers of your documents are non-experts who know little about the specifics of your work. For this broad audience, you need to fill in the blanks, so to speak. You must put basic linking explanations in your document that are in your head but not as obvious to others until present on the page. And you must put the various sections into an order that makes sense to people who cannot read your mind.

In summary, you have these five steps for creating flow: First, put the sections into logical order based on the lessons of Part 1 and Part 3 of this book. Second, put the argument into its ideal structure based on the lessons of Part 4. Third, use transitions as highlighted in Subsection 4.5.2. Fourth, use topic sentences for all paragraphs and place generalizations before particulars (review Section 2.2.1). Fifth and finally, insert mapping words and phrases, grammatically considered sentence adverbs and often referred to as "metalanguage," which were introduced when discussing sentences in Subsection 2.2.2.2. These five steps will ensure flow in your document.

I provide illustrations of each of the five steps below, with initially a negative example, followed by a positive one.

4.5.1.1 Flow from Logical Order of Sections

Negative Example (two adjacent report sections):

Method:

Pristine fullerenes were prepared using established methods per Deguchi et al. (2001) and Spohn et al. (2009). In addition, functionalized fullerenes, which contain terbium in their core, were prepared. This produces an electron-dense spot in each fullerene. These functionalized fullerenes were compared to pristine (C60) fullerenes in terms of zeta potential and aggregate size. A Malvern zeta sizer was used to determine zeta potential and aggregate size. A Nanosight LM 14C was used to determine and compare the aggregate size of terbium particles and fullerenes. C60 fullerenes prepared on three separate days were analyzed for charge and aggregate size on each instrument. Due to the precious nature of the terbium endohedral preparations, only one preparation was sampled three times.

Results:

The Malvern zeta sizer determined the size/zeta potential of the C60 fullerene and the terbium-enhanced fullerenes to be 91.13 nm/-44.3 mV and 99.31 nm/-64.3 mV, respectively. The Nanosight LM 14C determined the size of the C60 fullerene and the terbium-enhanced fullerene to be 55.18 nm and 63.77 nm, respectively. The difference between the two instruments' size determinations can be attributed to the tendency of the Malvern zeta sizer to skew towards larger aggregates. Regardless, the size determinations as well as the zeta potential values suggest that the aggregates are similar in size and zeta potential; therefore the terbium-enhanced fullerene is a reasonable substitution for the C60 fullerenes for the purpose of imaging.

Positive Example (two adjacent report sections):

Need for Substitute Fullerene

With the increasing use of nanomaterials in consumer products and drugs, a full understanding of such materials' interaction with human cellular components is needed. One such nanomaterial is a C60 fullerene. To conduct laboratory studies of C60 fullerenes with the potential to enter human cells by penetrating the cell membrane, fullerene suspensions can be added to cell cultures, allowed to incubate, and subsequently observed with a transmission electron microscope (TEM). The pristine C60 fullerene, however, is difficult to find using TEM. Thus, a substitute is needed that would enable accurate imaging with TEM. Such a substitute is a terbium-enhanced C60 fullerene (TbC60). As a functionalized fullerene, the TbC60 contains terbium in the core. This produces an electron-dense spot in each fullerene, which facilitates TEM imaging. While preparing a TbC60 fullerene is somewhat difficult and introduces some concerns, the crucial factor is to determine whether it adequately substitutes for a pristine C60 fullerene, so as to fairly represent the latter molecule's interaction with human cells.

Successful Creation of Substitute

Using two different methods, we determined that the TbC60 fullerene is a reasonable substitute for the C60 fullerene for the purpose of imaging within a cell and cell materials, such as the cell membrane. The first method, a Malvern zeta sizer, was used to determine zeta potential and aggregate size. The second method, a Nanosight LM 14C, was used to determine and compare aggregate size. The Malvern zeta sizer determined the size/zeta potential of the C60 fullerene and the terbium-enhanced fullerenes to be 91.13 nm/-44.3 mV and 99.31 nm/-64.3 mV, respectively. The Nanosight LM 14C determined the size of the C60 fullerene and the terbium-enhanced fullerene to be 55.18 nm and 63.77 nm, respectively. The difference between the two instruments' size determinations can be attributed to the tendency of the Malvern zeta sizer to skew towards larger aggregates. Regardless of this small discrepancy, the size determinations as well as the zeta potential values suggest that the aggregates are similar in size and zeta potential.

4.5.1.2 Flow from a Structured Argument

Negative Example:

- The next test we completed was to change the flow speed over the cylinder and use the hot wire anemometer (HWA) to detect vortex shedding frequencies. According to theory, for each flow speed, a distinctly correlated peak shedding frequency should be identified. We did, in fact, confirm this theory, proving the HWA can detect the actual shedding phenomenon accurately. The testing was set up as follows.

Positive Example:

- As our first reason for recommending the hot wire anemometer (HWA), we believe it detects the phenomenon of vortex shedding accurately, finding each peak shedding frequency unique to different air-flow speeds. We did confirm this accuracy during our testing, so the HWA fulfills our first criterion of merit. Our test setup is discussed next.

4.5.1.3 Flow from Transitions with Reminders

Negative Example:

- Equations 5, 6, and 7 were used next to show the aircraft's performance in turning. Turning is another aspect of flight performance. Turning is an instance of accelerated flight.

Positive Example:

- While we have shown that the aircraft performs exceptionally well in static flight with no acceleration, we now move to our proof that the aircraft's performance in accelerated flight is equally impressive. We analyzed accelerated flight in the form of steady turns. Turning performance was quantified using three fundamental equations, 5, 6, and 7, shown below.

4.5.1.4 Flow from Topic Sentences and Generalizations

Negative Example:

- The first priority of our company is to ensure safety of all employees at the new plant. A safety release valve opens to reduce pressure in the event that the maximum pressure of a tank is reached. Each tank at the plant has one such valve. We also want to strictly protect the environment. In case of accidental spills, overflows, and leaks, a 400,000-gallon dike surrounds the plant to minimize contamination to the surrounding area.

Positive Example:

- The first priority of our company is to ensure safety of all employees at the new plant, followed by strict protection of the environment. To this end, the plant incorporates a regulated process with appropriate control equipment to ensure employee safety and environmental protection. Controls on the process track and maintain suitable pressure, level, temperature, time, mass, and pH balances. For pressure control, in particular, each tank at the plant has a safety release valve. In the event that the maximum pressure of a tank is reached, the safety release valve opens to reduce pressure well before it could jeopardize safety. For level control, furthermore, in case of an unexpected drop in tank level, which suggests a spill, overflow, or leak, a 400,000-gallon dike surrounds the plant to minimize contamination to the surrounding area.

4.5.1.5 Flow from Mapping Words and Phrases

Negative Example:

- Two rocket motor characteristics are of fundamental importance. Total impulse has two components. Specific impulse additionally takes into account fuel weight.

(The pitfall with this negative example is that readers might lose the flow and misinterpret these sentences in one of two ways. First, readers might think that the two rocket motor characteristics of fundamental importance are the two mysterious components of total impulse, thus misinterpreting the relationship between the first and second sentences. They might not understand that total impulse and specific impulse are, in fact, the two characteristics of fundamental importance mentioned in the first sentence, which unclearly previews the two subsequent sentences, not just one. Second, readers might think that total impulse's two components are listed in the next sentence: specific impulse and weight, thus misinterpreting the relationship between the second and third sentences.)

Positive Example:

- Two rocket motor characteristics are of fundamental importance: total impulse and specific impulse. Total impulse has two components. In contrast, specific impulse has three components: thrust, time, and fuel weight. Total impulse incorporates only thrust and time.

4.5.2 Remind Readers of Main Points with Transitions

As you write the Discussion of your report, particularly the argumentative segment, you will realize that support points build one upon another, and you ultimately create a hierarchical, multi-layered case for your thesis. This complexity is powerful but makes writing difficult. In terms of power, if you have only "one leg" of an argumentative "stool," it will fall over. If you have three, four, or more legs, the stool will stand upright and support the weight of the thesis that rests upon it. In terms of difficulty, however, a complex argument is hard for readers to absorb. Indeed, when an argument is multi-layered, readers may struggle to grasp and retain the whole presentation. You are asking a lot of readers: (1) keep track of past points while learning new ones, and (2) follow how subpoints fit within higher-level points, and so on. Perhaps a majority of readers will be able to wrap their heads around three or four legs of your argument, but each of these legs has many subparts as well.

As a result, abrupt beginnings of sections and subsections may make absolute sense to an author, but readers will likely be confused. Thus, beginning a section with "Harrison and Shankar found that deep learning could be applied to optical analysis (2016)" may be exactly the starting spot in your mind, but readers may not grasp that you want to tell them that your work with optimization of power distribution systems through voltage-control algorithms derives from earlier efforts applied to optical analysis. Indeed, you have borrowed from prior work and applied certain approaches to your specialty, power distribution. So, to improve the transition, you might want to start with a reminder, followed by the new information: "My proposed control algorithm for energy distribution management is similar to one used initially in optics, to assist in analysis of camera images for flight control of unmanned aerial vehicles. I want to explain this initial research here, before I show my adapted version of the deep-learning control algorithm."

Skilled technical writers provide readers with assistance in keeping track of the hierarchy and remembering how each piece fits within the whole. Thus, you must use transitions between sections as both connectors (flow controllers) and reminders. Both of these writing elements reinforce your argument and help readers remember your key points and follow your train of thought, as it is logical, lucid, and persuasive.

Wherever you are in your argument's presentation, reminding readers of your case up to this point is always a good way to lead into the next support point. At the very least, you can review the overall message, which is to present a particular thesis. Thus, a good writing rule for reinforcing your argument is to repeat the thesis as you start a new section. For example, if the thesis is to recommend reverse-osmosis, you might begin your third portion of the argumentative segment like this: "Our third reason for choosing the reverse-osmosis method is that it has very low long-term maintenance costs." Similarly, as you start subsections within a support segment, you can also review the support point to which the subsection is a contributing factor. For example, "Although our own testing indicates that the device is precise, we also corroborated this finding by conducting a computerized simulation, and the inputs and assumptions of our simulation are discussed in this subsection." Thus, the reminder can be at either a primary or secondary level.

At the primary level, any of the following are helpful items to review as you progress through the presentation of your argument:

Helpful primary reminders:

- Overall finding or thesis
- Overall purpose for doing work
- Overall information unknown to make known
- Overall objective to fulfill or question to answer

On a secondary level, reminders can be tied to the higher-level support point to which the new section is linked. Each transition sentence essentially begins with the idea that the new information is connected to something you have previously written, such as the following:

- The section that follows presents my proof that the aforementioned criterion X has been fulfilled.
- The section that follows describes the physical testing conducted to support the point immediately above that criterion X has been fulfilled.
- The section that follows describes the analysis of the data acquired and presented above.
- The section that follows describes an effort I made to corroborate the aforementioned three key findings.

Part 5

Mastering Important Genres

You may find yourself writing a particular genre of report over and over in your working career. That is good. We say, "No reason to reinvent the wheel." If a particular style and format works for diverse communication needs, it evidently must be a practical and adaptable genre. On occasion, you may find you need to write a document of an entirely unfamiliar nature. You will need to be flexible in such instances and use principles from other genres and any guide (samples, heuristics, or requirements) you can find to help you organize and format the document. Aside from such instances of blazing new territory, your basic, familiar technical report genres should be your first choice when preparing a formal communication in ABERST.

Interestingly, each specialty has its own common genres that are useful and necessary in the course of conducting the work and completing the job. For example, if you are in construction, civil engineering, development, or natural resources, you might write environmental impact statements or mitigation plans. If you are in law, you will write pleadings, answers, and briefs. If you work in social work or psychology, you might write individual case studies. If you are in engineering or manufacturing, you might write a document referred to as an FMEA—failure modes and effects analysis. If you are in medicine, you will produce a patient's medical chart. If doing research in sciences or engineering, you document daily work in a lab book or notebook. If you are in a legislative branch of government, you might draft a bill or a white paper. If in municipal government, you might produce a zoning plan. There are hundreds of these specialty genres, and I cannot cover them all. Thus, I have spent the first 80% of this book explaining universal and widely applicable technical writing principles that can and should be applied to all writing, even highly specialized documents. Similarly, I use this remaining 20% to cover widely applicable genres that should be common to all the diverse fields within ABERST.

In particular, I use Chapter 5.1 to provide some specific guidance for ten useful and adaptable technical report genres. These are the ones that should cover the vast majority of your writing needs.

In Chapter 5.2, Enhance with Visual Aids, I present my advice for designing and using diagrams, photographs, schemata, maps, graphs, tables, and charts. These important technical report elements constitute a sub-genre all of their own. They must be designed to fully stand alone, that is, be understandable in isolation separated from the surrounding report, so each visual aid is in a sense a little "report unto itself." Although there are many types of visual aids, some basic functional and design principles apply to them all, and those principles are covered, with numerous examples to demonstrate the good and bad qualities of visual aids.

Closely linked to the topics of genre and visual aids is document format. Formatting ensures that the overall document looks polished and appropriate for its genre and that visual aids are integrated into the document seamlessly and attractively. My experience in ABERST has convinced me that authors care about, and take pride in, their documents' formatting as much as readers notice, appreciate, and even demand a polished look. Formatting is covered in Appendix C. In the appendix, I offer some simple and easily implemented suggestions for formatting your documents to make them attractive, easy to read, and useful as subsequent references. In addition, I provide two more appendices with advice for two other genres that are not truly documents in the normal sense but are nonetheless common in ABERST: conference posters (Appendix D) and oral presentations and projection slides (Appendix E).

Chapter 5.1

Match Genre to Purpose and Audience

As intimated in the introduction above, this book cannot cover all possible types of documents. Individual companies have their own proprietary document types. Within ABERST, an author may encounter examples and instructions for documents that are unique and customized to a type of business or a particular field, responsibility, or specialty. Imagine a scenario where you have investigated a product failure or a system malfunction, looking for possible problems, and you do this routinely in your job. Your company might have a standard format for this, such as the aformentioned failure modes and effects analysis (FMEA). You will have to learn the FMEA format and other proprietary genres once you are in place in a particular organization. As a foundation, you can learn the broadly universal genres covered in this book and apply them to nearly all situations. In addition, in many cases, using a universally known genre may be more effective than attempting to communicate an important message via an esoteric genre. Returning to that FMEA report, for example, I would be confident in suggesting you use the venerable memorandum, a standard genre covered below, for that communication situation.

After all, the goal of technical communication is to be clear to as many possible future readers as possible. Starting with a familiar and known genre might foster that effort, whereas relying on an esoteric form may sabotage it. Thus, becoming familiar with the genres covered in this final part of the book has a twofold benefit: one, it reduces an open-ended field of possible report styles to ten core genres; two, it prepares authors to use a set of common genres that should produce wider comprehension and communication efficacy than would an unlimited set of proprietary report styles. This is analogous to the difference between open-source software being readily available to all versus private, proprietary computer code restricted to a small population.

Being familiar with a handful of useful, widely applicable genres is only a starting point, however. You must also understand how to select the appropriate genre for each communication purpose. If you send an email, and a letter should have been used, you could lose some business. If you produce a white paper, when a proposal would have been better, you could miss an opportunity. If you write a status report, when a memorandum is called for, you could jeopardize safe operations. Sometimes the correct genre is immediately obvious, and sometimes it is not.

For most situations, an ideal genre always suits the communication purpose, with an understanding that purpose broadly encompasses both the organizational context and the intended primary audience as well. Thus, linking your communication purpose to the correct genre is vitally important. As alluded to in this book's opening introduction, using the wrong genre for an important communication can lead to confusion: perhaps the intended recipient ignores the document, or perhaps this person reads only haphazardly and takes away an incorrect message. Communicating complex material in ABERST is a difficult proposition; you do not need the additional obstacle of using the wrong genre.

In some instances, the required genre is determined by the customer, publisher, client, funding agency, course instructor, and so on. When the genre is dictated, your task is simply to satisfy the request as emphasized in Chapter 1.3. In a college course, you may be asked to prepare a document within a certain genre, for example, a business start-up proposal or an internal memorandum. By a funding agency, you might be asked to deliver a research progress report. A publisher might ask for a technology review paper. The genre is determined for you, but you can use the guidelines in this chapter to help you structure and format your document. When the genre is not determined for you, however, you select the best genre to support your communication purpose.

For example, if you want to ask for $10 million dollars in funding, an email may not be the best genre. You would likely need a formal proposal. But if you want to let a supervisor know that the laboratory is nearing the end of its supply of latex gloves, you can certainly use an email for that purpose. If you work in the laboratory and your team wants to prototype a propeller guard with a carbon composite layup, you may need to seek approval from facilities management. They might ask you to write up your layup procedure, so it can be reviewed and approved. For this purpose, you want to use a procedural instruction, or training manual, genre.

What if you have discovered something wrong with equipment, and scheduled tests cannot be conducted until this equipment is repaired, replaced, or modified? As you understand the situation, crucial equipment must be improved in some way, with some design effort, planning, and cost. What type of report do you send to your superiors to notify them of this situation and obtain their approval for the retrofit? An email does not convey enough severity, a business letter seems wrong for internal readers, and a formal format may be unnecessary; people are too busy to write and read formal reports for every internal business issue. Applying a process of elimination, you can resolve that an internal memorandum is best used for this purpose.

The existence of various genres is consistent with one of my primary maxims: "I cannot teach you how to write a technical report. I can only teach you the fundamentals of technical communication, and you can apply those fundamentals to many different types of technical documents." No single "technical report" archetype exists. Full competency of technical writing in ABERST requires familiarity with the common genres, as presented in the remaining portion of this chapter. I list them below and elaborate on each one in the sections that follow.

Ten Common Genres

- Email
- Progress and status reports
- Business letter
- White paper
- Proposal
- Pre-proposal
- Memorandum
- Procedural instruction
- Formal report
- Journal article and conference paper

I have omitted genres that are primarily used for personal and private writing, such as journal entries, diaries, and field or laboratory notebooks. Although important, these genres are used for notes meant for the author's later reference, rather than to provide information to another reader. Because you must do your best writing when you intend to distribute, publish, or deliver the document to an audience beyond yourself, this book focuses on documents to be *submitted* (recall Part 1 and, specifically, Chapter 1.5).

For each genre covered, I provide a brief explanation of its typical uses, linked to purpose (including context and audience). In addition, for each genre, I suggest an ideal structure, inclusive of necessary content and format. For example, a business letter as a genre necessitates a particular structure of information as follows:

Business Letter Structure

- Heading with basic information
- Opening paragraph with purpose and (possibly) a personal greeting
- Middle paragraphs with content
- Closing paragraph with courtesy and next steps requested (of reader) or promised (by author)
- Signature block

These guidelines for structure, content, and format should help you compose and arrange your message into an effective genre-based deliverable. Overall, this chapter should give you a solid understanding of ten useful genres for writing in ABERST. (Additional genres are covered in appendices: resumes, job cover letters, and thank-you notes are covered in Appendix B; conference posters, Appendix D; oral presentations and projection slides, Appendix E.)

Except for a few genres where I provide full examples, my suggested genre-based structures are provided in outline form in the figures of this chapter. You can use them as references. Again, they are your structural skeletons. You must, in turn, fill in the outline and turn the skeleton into a living, breathing document.

5.1.1 Email

Email is an excellent method of communication for day-to-day operations. It can be used to do any of the following: confirm plans, request information, ask questions, announce news, send brief notices, provide reminders, present prices, add follow up, propose (small) changes, transmit attachments, offer answers, give feedback, check in, re-establish contact, dispense compliments, share information, and similar communication tasks.

Within the same organization, emails may be sent up and down the organizational hierarchy. If desiring to communicate outside one's organization, emails may or may not be appropriate. If permission has been given to communicate by email from the intended recipient, it is perfectly fine to do so. If no permission has been granted, and you are trying to write "cold" to someone, consider who the recipient is in the grand scheme of things and your purpose. If you are trying to sell something to someone who does not know you, email will likely not be suitable. We all have a word for such an email: spam. If you are writing to give something to someone, most recipients will not complain of an email. If you are writing someone who is busy, important, and unfamiliar with you, an email may be filtered as junk/spam or deleted without a moment's thought. Generally, people prefer emails from known senders. You take a big chance if you are unknown to the audience. Thus, the relationship (including relative levels of authority) between sender and receiver and your purpose of communication will determine whether email is appropriate and effective.

This section refers only to formal emails that are related to projects and accomplishing actual work, with an extra restriction that these would be preserved as part of the written record on a project. Many emails are temporary, trivial, and unimportant to the actual work, and those can be written as you wish. For example, emails to check if a co-worker wants to head out to lunch with you or an email to ask someone when the new order of copy paper will be delivered are not of project significance. In contrast, emails when you need to request some project-related action from a peer or supervisor, when you wish to communicate to a client or customer, and when you want to disseminate information related to a project to involved stakeholders are permanent project documents. They are legally binding when of a contractual nature. Thus, the distinction between trivial and formal emails is important.

One issue applies to all emails, both trivial and formal, however, and this is that they can be discovered in the event of litigation. Thus, you should not put anything in an email that is either inappropriate to maintaining a professional workplace or suggests disrespect of others, carelessness, thoughtlessness, or ill intent. In short, avoid anything that compromises the good reputation of your organization, with respect to ethics, fair practices, civil rights, and abidance with all laws, regulations, and ordinances. As an example, do not put in an email: "Hey buddy, let's go to lunch and share stories of some of the silliest things we've been asked by goofy customers."

The basic requirements of formal email are to be courteous, clear, and cogent. Each of these qualities is covered below.

Courtesy is Evident by These Four Elements:

1. Formal greeting term, which is a direct salutation to recipient, such as these: Hello, Good Morning, Good Afternoon, Dear, or Waddup (just kidding on the last one)

2. Recipient's name, which directly follows the greeting term, in one of two ways:
   a) If you are on first-name basis with the recipient, you may use that person's first name followed by a comma. An example is below:
      - Dear Kumar,
   b) If you are either not on first-name basis or unsure whether you are, or if the recipient is certainly a superior, then the salutation must include that person's title:
      - Ms., Mr., Dr., President, Vice-President, Director, Dean, Professor, Counselor, and Ambassador.

After the title, you use the recipient's last name followed by a colon, as in the example below:

- Dear Vice-President Shankar:

3. Concluding kindness statement, where you should express some appreciation and gratitude, especially if something is asked of the recipient. You can never go wrong with conclusions such as these:

- "Thanks in advance for your attention to this email and modest request."

- "I appreciate a reply at your earliest convenience."

This type of closing "thank you" can be customized to fit the context. Use your good judgment and follow good practices.

4. Complimentary closing phrase (for example, *Respectfully*, *Sincerely*, and *Kind regards*) and your signature (typed format) placed at the bottom are both very helpful to recipients. If you feel your job title would add further benefit, you can place it under your signature, as follows:

Sincerely,
First name (possibly last name if not familiar with recipient)
Job title

Without this complimentary closing phrase and signature, recipients may not know with 100% certainty that you are in fact the sender of the email. They cannot always tell from your email address (ohiobatboy@gmail.com) or from the content of the email ("Will you provide your feedback on the proposal presented last month?") Even if your email address indicates your true and full name and company (tom.peppers@statebank.com), you should use proper form and end with a complimentary closing phrase and your signature. If the closing phrase comprises two or more words, only the first word is capitalized. An example is below:

Warm regards,
Tom Peppers
Branch Manager, State Bank

Clarity is Achieved in Two Ways:

1. The subject line can be specific to the content of your email. There is no need to leave the recipient in the dark. If you are sending the email, it is apparently important enough to bother the recipient. Let them know in the subject line your topic and purpose for writing. Compare weak with strong:

Subject: meeting (This is weak and vague. It should be revised.)
Subject: Request meeting to discuss clinical trials budget—Action Required

2. The first sentence must succinctly reveal the purpose of the email. Optionally, this can be shifted to the end of the first paragraph if you wish to start the email with necessary background information that ideally prepares the recipient to understand the purpose.

- Sample first sentence: I need to schedule a meeting with you to discuss the clinical trials budget; next Thursday and Friday are good for me. Would one of those days work for you?
- Optional - Sample first paragraph: Yesterday, I received a letter from my counterpart at the National Institutes of Health (NIH), stating that the NIH might be interested in expanding our scope of work to include study of a potential ebola virus vaccine. This is a positive development, and my team has the laboratory knowledge to accept the assignment. However, an additional laboratory facility must be dedicated to this research. I want to meet with you to discuss our options for dedicating such a laboratory to this work.

None of the above examples would actually be clear if not for the adherence to the 4th guideline for courtesy, namely, provide your name at the end of the email. If you do not do that and your email is a nickname of some sort, and the recipient is a busy person with lots of projects to attend to, the short email will be ambiguous. Below is a sample ambiguous email from a possibly unknown sender:

- Unclear email: "I want to ask for an extension for the report due this Friday. I have been busy with other work. Will this be OK?"

Cogency Requires Sufficiency:

Lastly, cogency is achieved by providing sufficient information so that the recipient will consider your purpose with appropriate attention and give the email a fair reading. If you are requesting some response from the recipient, the email is cogent if it motivates that response as best as you can do so. If the email is intended to provide information to the recipient, it is cogent if the information appears complete and reasonable. You will not be able to provide a full-blown argument of fact or policy in an email, but you can use email to initiate and facilitate small actions, exchange information, and lay a foundation for future work and communication of a detailed nature.

Sample full email:

From: **Brandon Knight**< bknight@nwmu.edu>
Date: Fri, May 4, 2018 at 9:05 AM
Subject: Requesting Meeting to Discuss Clinical Trials Budget
To: Hans Peterson <hanspete@ucberkeley.edu >

Dear Dr. Peterson:
I need to schedule a meeting with you to discuss the clinical trials budget; next Thursday and Friday are good for me. Would one of those days work for you? Please get back to me by the end of day, tomorrow, so I can make arrangements for a room and invite the other attendees. Thanks so much for your cooperation on this important matter.

Sincerely,
Brandon Knight
Department Chief
. . .

5.1.2 Progress and Status Reports

These are routine documents done by request of clients, funding agencies, sponsors, or customers. When not explicitly requested, a progress/status report might be appreciated if a project is taking longer than 3 months. Those stakeholders uninvolved with the day-to-day implementation would benefit from a written update, and this is a good way to impress them with your diligence and overall sense of the project's status.

Progress/status reports might be done monthly, quarterly, or annually, or a combination of those, depending on the project's full duration. A standard format used again and again for each new report is best, so authors and recipients alike can review the project at a glance and find specific and desired information easily. This is not the genre to use to communicate a major, unexpected event (such as a setback or injury); nor is it ideal for reporting exciting results or significant conclusions and recommendations. You will want letters, memoranda, and formal reports for those. Instead, the progress/status report provides routine, expected updates on a project, with allowance for some variances, problems, concerns, and issues to be reported, but nothing extraordinary. Recipients would very likely miss the unusual due to expectations. They would not be expecting anything unusual or extraordinary to be reported, and the briefest mention in a short, highly formatted document might not be noticed with proper concern. The genre is designed for neither detailed argumentative segments, as are both a memo and formal report, nor major announcements, as is a business letter.

Typical section headings within a progress/status report are the following:

- Schedule status (ahead, behind, or on time)
- Significant tasks behind (project risks) or ahead of schedule
- Budget status
- Contracts, legal, and agreements—updates
- Particular budget line items completed, under way, or requiring monitoring
- Major achievements/milestones this reporting period
- Major upcoming tasks
- Action items required of client/customer
- Upcoming deadlines, meetings, milestones, and document due dates
- Outstanding questions

Progress and status reports might be used interchangeably across organizations and fields. I do not know of an absolute rule by which to distinguish them. From my own experience, I would use progress reports for less-frequent updates with contents that might require response from recipients, whereas status reports would be used for regularly scheduled reports with an emphasis on reassurance of ongoing work and documentation of status without demanding response from recipients. In other words, progress reports might be classified as "for review and response" (FRR), while status reports might be primarily "for your information" (FYI). An example status report is provided in Figure 5.1.1.

Web Security 2.0 Bi-Weekly Activities & Status Report Monday, April 30 – Friday, May 11

Area	Milestone	Due Date	Responsibility	Status	Notes (Risks, Questions, Outstanding Items)
Programming					
REGULATORY & COMPLIANCE	Legal Review	ASAP	Team 1/Team 3	No final due date has been published to have the code complete. Each team must report back weekly to Lead as to the status of their IRB compliance. The following sites have had compliance approved: M – 10/27/17 A – xx/xx/18 E– 12/28/17 O – 2/16/18 I – 1/31/18 U – 12/18/17	Notes: Site submissions will not be ceding applications.
	Data Transfer Agreement Executed	1/3/18	Team 1	Complete U received the fully executed DTA; it was uploaded into System on 1/3	Notes: This is an additional requirement Keynet ver 2.0 DSA. A separate DTA is being used for some of the transfer.
RESEARCH (e.g. Protocol, Research Plan)	Final Protocol Received	11/20/17	Team 2	Complete Team 1 to discuss standards	
					Notes: Changes to the protocol are underway to reduce overlap. Sites are asked to go forward with protocol version 2. It will be switched out later.

Figure 5.1.1 Sample Status Report

A status report is more limited in variation than a progress report, so this generic sample might be quite close to many of the actual status report styles used across organizations. A progress report, in contrast, might be presented in various ways, depending on the organization. For example, a progress report could be formatted as a memo, presentation slide deck, business letter, or email. A recipient might dictate the content and format of the progress report (see Chapters 1.1 and 1.3). If no specifications are given to you, you can always adapt the sample status report shown here or build upon one of the other genres mentioned and discussed elsewhere in this chapter.

An example progress report is provided below (Figure 5.1.2), based on the format required by the National Institutes of Health for funded projects.

Program Director/Principal Investigator (Last, First, Middle):

PROGRESS REPORT SUMMARY	GRANT NUMBER	
	PERIOD COVERED BY THIS REPORT	
PROGRAM DIRECTOR / PRINCIPAL INVESTIGATOR	FROM	THROUGH
APPLICANT ORGANIZATION		
TITLE OF PROJECT (Repeat title shown in Item 1 on first page)		

A. Human Subjects (Complete Item 6 on the Face Page)

Involvement of Human Subjects ☐ No Change Since Previous Submission ☐ Change

B. Vertebrate Animals (Complete Item 7 on the Face Page)

Use of Vertebrate Animals ☐ No Change Since Previous Submission ☐ Change

C. Select Agent Research ☐ No Change Since Previous Submission ☐ Change

D. Multiple PD/PI Leadership Plan ☐ No Change Since Previous Submission ☐ Change

E. Human Embryonic Stem Cell Line(s) Used ☐ No Change Since Previous Submission ☐ Change

SEE PHS 2590 INSTRUCTIONS.

WOMEN AND MINORITY INCLUSION: See PHS 398 Instructions. Use Inclusion Enrollment Report Format Page and, if necessary, Targeted/Planned Enrollment Format Page.

Figure 5.1.2 Sample Format of a Progress Report (Page 1)

Program Director/Principal Investigator (Last, first, middle):

GRANT NUMBER

CHECKLIST

1. PROGRAM INCOME ***(See instructions.)***
All applications must indicate whether program income is anticipated during the period(s) for which grant support is requested. If program income is anticipated, use the format below to reflect the amount and source(s).

Budget Period	Anticipated Amount	Source(s)

2. ASSURANCES/CERTIFICATIONS ***(See instructions.)***
In signing the application Face Page, the authorized organizational representative agrees to comply with the policies, assurances and/or certifications listed in the application instructions when applicable. Descriptions of individual assurances/certifications are provided in Part III of the PHS 398, and listed in Part I, 4.1 under Item 14. If unable to certify compliance, where applicable, provide an explanation and place it after the Progress Report (Form Page 5).

3. FACILITIES AND ADMINSTRATIVE (F&A) COSTS
Indicate the applicant organization's most recent F&A cost rate established with the appropriate DHHS Regional Office, or, in the case of for-profit organizations, the rate established with the appropriate PHS Agency Cost Advisory Office.

F&A costs will ***not*** be paid on construction grants, grants to Federal organizations, grants to individuals, and conference grants. Follow any additional instructions provided for Research Career Awards, Institutional National Research Service Awards, Small Business Innovation Research/Small Business Technology Transfer Grants, foreign grants, and specialized grant applications.

☐ DHHS Agreement dated: ______________ ☐ No Facilities and Administrative Costs Requested.

☐ No DHHS Agreement, but rate established with ______________ Date ______________

CALCULATION*

Entire proposed budget period: Amount of base $ ________ x Rate applied ________ % = F&A costs $ ________

Add to total direct costs from Form Page 2 and enter new total on Face Page, Item 8b.

*Check appropriate box(es):

☐ Salary and wages base ☐ Modified total direct cost base ☐ Other base *(Explain)*

☐ Off-site, other special rate, or more than one rate involved *(Explain)*

Explanation *(Attach separate sheet, if necessary.):*

Figure 5.1.2 Sample Format of a Progress Report (Page 2)

5.1.3 Business Letter

Business letters can be divided into two broad categories: ones that transmit some other material and ones that are themselves complete and explicitly contain the substantive message intended for the recipient. These letters have common elements, and one big difference: in the former, the transmittal letter, the letter's middle portion may only briefly explain or summarize the enclosed material. In the latter, the complete letter, the letter has an extended middle portion that is itself a "report" of a kind; thus, a complete business letter contains a document within a document.

Both types of business letters have three essential characteristics and five necessary components. This is the 3x5 approach; it is easy to remember because you can recall a 3x5 index card.

- Three Characteristics: Polite, Persuasive, and Polished
- Five Components: Heading, Purpose, Highlights/Substance, Courtesy, and Closing

The three characteristics can be quickly explained. A business letter is polite because it must reflect that ABERST projects are conducted in a gracious manner, with each party cooperating and complementing the other party. A brief statement of politeness says to the recipient, "Before I ask you to do something or before I tell you what I want, I have time to express that I value your assistance and trust in me, and I am pleased we are working together." (This is similar to a bow in Japanese culture.) Memoranda, emails, and other reports are not required to contain any special element of politeness, but business letters should have one or more explicit expressions of good etiquette.

A business letter should be persuasive in one of two ways. If it is a transmittal letter, it must persuade the recipient that the enclosure is valuable, helpful, and additionally persuasive itself, if it contains an argument. If the letter is a complete report in itself, it must be sufficiently comprehensive that it presents the author's best argument or otherwise convinces the recipient that the information provided is complete, accurate, and beneficial.

Lastly, a business letter should be polished in terms of writing proficiency and format. Errors of any kind in the composition suggest carelessness and incompetence. Letters reflect the quality of work to be conducted by the author.

If errors appear in a letter, recipients may likely fear that errors will occur in the author's work. Format must be professional and normal. If a business letter does not look like one, its effect will be diminished, and recipients will, again, doubt the abilities of the author to complete the associated work.

The five components are covered separately below, in two sweeps: first, with a transmittal business letter and second with a complete business letter.

5.1.3.1 Transmittal Business Letter

The heading for a transmittal business letter is conventional and serves as a template for a complete business letter as well.

Heading

The heading provides the basic document identification data: sender, date, recipient, subject line, and salutation. These are in a special order that must be followed to avoid confusion. Most organizations have letterhead paper, which contains the organization's name, address, phone, and other locator data in a pre-printed format. If this is not available, the same information that would be pre-printed must be typed at the top of the letter to indicate the sender's basic information. Examples of both letterhead (Figure 5.1.3) and non-letterhead (Figure 5.1.4) styles demonstrate these features best.

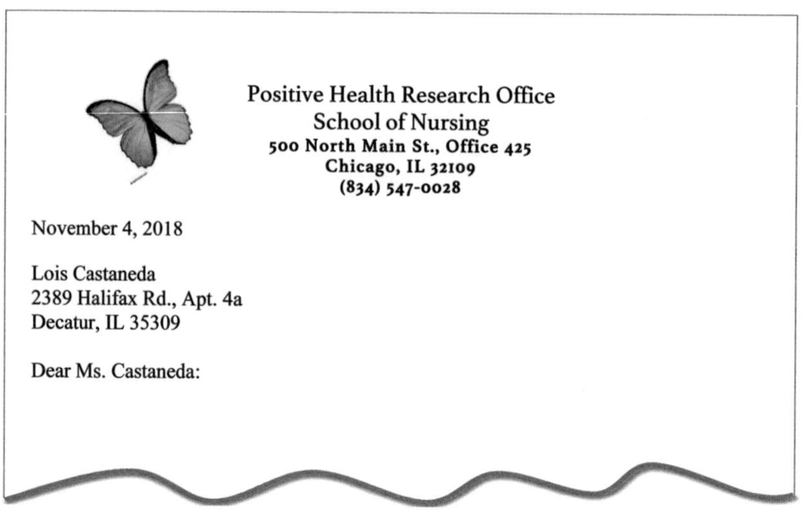

Positive Health Research Office
School of Nursing
500 North Main St., Office 425
Chicago, IL 32109
(834) 547-0028

November 4, 2018

Lois Castaneda
2389 Halifax Rd., Apt. 4a
Decatur, IL 35309

Dear Ms. Castaneda:

Figure 5.1.3 Letterhead Style for Letter Heading

777 E Madison St., Apt #4
Ann Arbor, MI 48107
(824) 248-8421
chrispat@umich.edu
April 9, 2017

Rene Tampico
Principal Investigator
Advanced Genomics
185 Spear St
San Francisco, CA 94105

Subject: Final Specifications for Gene Splicing Project

Dear Ms. Tampico:

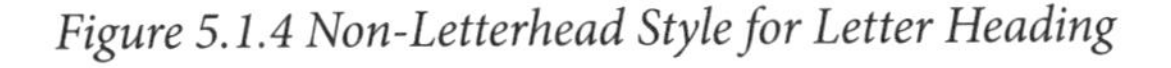

Figure 5.1.4 Non-Letterhead Style for Letter Heading

Purpose

The letter's first sentence can be straightforward and explicit. It is the place to indicate the purpose of communication. For a transmittal letter, the purpose is to transmit and enclose some material, as in this example, which gives options to choose from: "With this letter, I/we submit the named document (enclosed) for your information/approval/review."

Highlights

The contents of the middle section depend on whether the letter is for transmittal or is complete in itself. If the former, the middle section provides only highlights of the enclosure(s). If the latter, the middle section is fully substantive. A substantive middle section is covered in Section 5.1.3.2.

For transmittal letters, the enclosure highlights can be a brief paragraph or two at maximum. In this succinct paragraph, the author should set forth the gist of the enclosed document(s). This can always be done in one or two sentences, even if the document is long and complex. Imagine a complex, 500-page document presenting a detailed design of an off-shore wind turbine farm. The gist of the document can be presented briefly: "As you will see in the enclosed plan, the off-shore farm will comprise 85 turbines, covering 15 acres, and potentially generating 350 megawatts of electricity. Construction on the project could begin as soon June 2022, with completion within 2 years. Estimated persons employed during construction is 1,350, with 50 employees during normal operation for the expected 80-year lifetime of the farm."

Courtesy

As discussed above, a principal characteristic of a business letter is politeness. Thus, one of the necessary five components is courtesy. Some courtesy is mandatory. It does not have to be overly effusive, nor should it be informal.

- Too informal: When we do our site visit next month, we will definitely go out for drinks with you and the team.
- Too effusive: This is the greatest project we have ever been involved with, and we thank you from the bottom of our hearts for giving us this opportunity to bid on providing you excavation services for your new development's septic lines.

A sweet spot of courtesy exists somewhere right between flaky and fulsome. Look at these two very plain, yet acceptable, examples:

- "We look forward to working with you on this important project."
- "The information and data files you sent are much appreciated. This should expedite the work and help us during stages 1 and 2."

When thinking internationally, moreover, different countries and cultures have differing views on business letters, in particular the extent of courtesy and the extent of substantive content. Providing any substantive content may seem discourteous to some cultures, where a "light touch" in letters is normal, followed by "getting down to business" in face-to-face meetings and exchange of other documents. But this book is focused on ABERST in the U.S.A. You are advised to consult cultural guidance when doing business and communicating with customers and associates in different countries. Expertise on cross-cultural communication should be valued and utilized.

Closing

The closing consists of (1) a final paragraph, (2) a complimentary closing phrase, and (3) a signature block. These three elements allow you to end with additional courtesy and some indication of the next steps. This must be customized for each unique situation.

Options for the final paragraph range from loosely generic ("I hope to work with you in the future") to tightly focused ("Our on-site sampling team will arrive next Tuesday at 9:00 a.m."). Additional extensions of polite appreciation or offers to assist are always appropriate at the end of a business letter:

- "Once again, we appreciate your confidence in our team, and we anticipate providing the information you have requested within the next few months."
- "Please let me know if I can provide further information or any clarification of the enclosed report."

Following the final paragraph, you insert a complimentary closing phrase and a signature block, which consists of your typed name and job title, your signature above the name, and an indicator that the letter contains an enclosure (if applicable) as in this example:

Sincerely,

Jesse Davidson

Jesse Davidson
Vice President for North America
enc.

Complimentary closing phrases change with the times. They are linked to specific eras, so what is appropriate today may not be useful tomorrow. In the past, flowery and affectionate phrases were typical, but these would seem peculiar today. "Yours truly" and "Very truly yours" were normal for business letters back in the day. Today, those would be lumped together with "Yours affectionately" and "Faithfully yours" as only appropriate for intimate correspondence. Who knows who will be the next person to pen a novel phrase that strikes just the right note for business correspondence? (Members of the military often use "V/R," which stands for "Very respectfully.") I encourage you to be creative and keep your ears open, so your complimentary closing phrase can reflect professionalism, current trends, and your own style. In the meantime, if you want to have a reliable list of always-acceptable phrases, I give you the following:

- Respectfully, Sincerely, Cordially, Warm regards, Highest regards, Gratefully,

Keep in mind that only the first word is capitalized when the closing phrase consists of two or more words.

Transmittal business letters should be restricted to one page only. This may seem arbitrary, and I am not claiming to be the rule maker on this. Yet, based on my personal experience and observations, I see the one-page limit as very helpful and justified, for several reasons. One, out of respect for a reader's time, one page is best. You do not want to force a reader to take the time to flip to a second page simply to see a few final remarks and your signature. Two, a signature is ideally placed at the bottom of a letter. If you run over to a second page, your signature is likely to end up someplace other than the page bottom. Three, this letter is intended merely to transmit another document, which has the substantive material for the reader. The letter should not itself replace the enclosure, so two or more pages are excessive for the objectives of a transmittal letter. Final remarks should be edited for conciseness, as should the preceding body text of the letter, so it all fits on one page.

In addition to following the 3x5 rule, business letters must be in a specialized format, in terms of alignment. Two alignment options are available, and picking one is simply a matter of preference; Left-margin-aligned and tab-aligned options are shown in Figures 5.1.5 and 5.1.6, respectively.

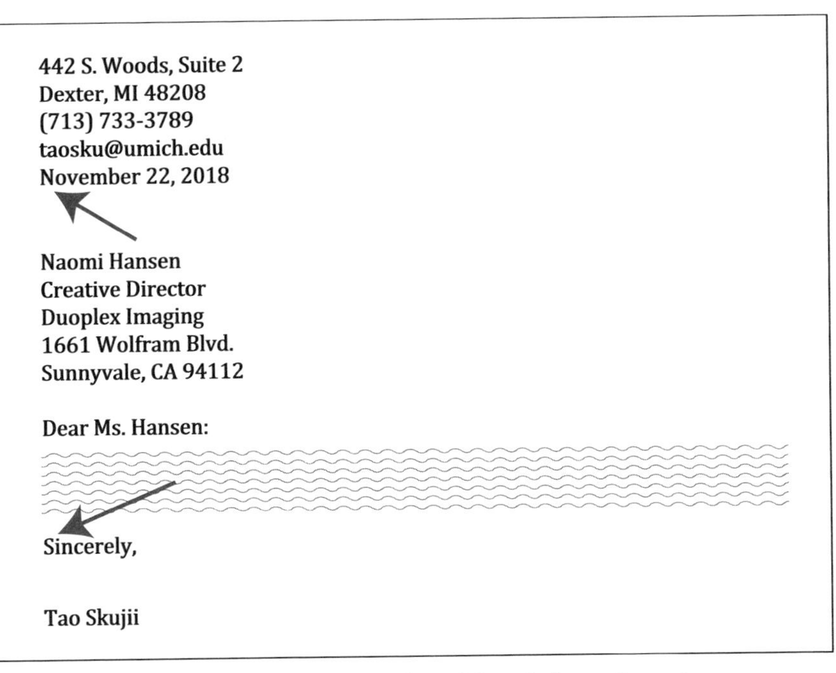
442 S. Woods, Suite 2
Dexter, MI 48208
(713) 733-3789
taosku@umich.edu
November 22, 2018

Naomi Hansen
Creative Director
Duoplex Imaging
1661 Wolfram Blvd.
Sunnyvale, CA 94112

Dear Ms. Hansen:

Sincerely,

Tao Skujii

Figure 5.1.5 Left-margin-aligned Sample Letter Format

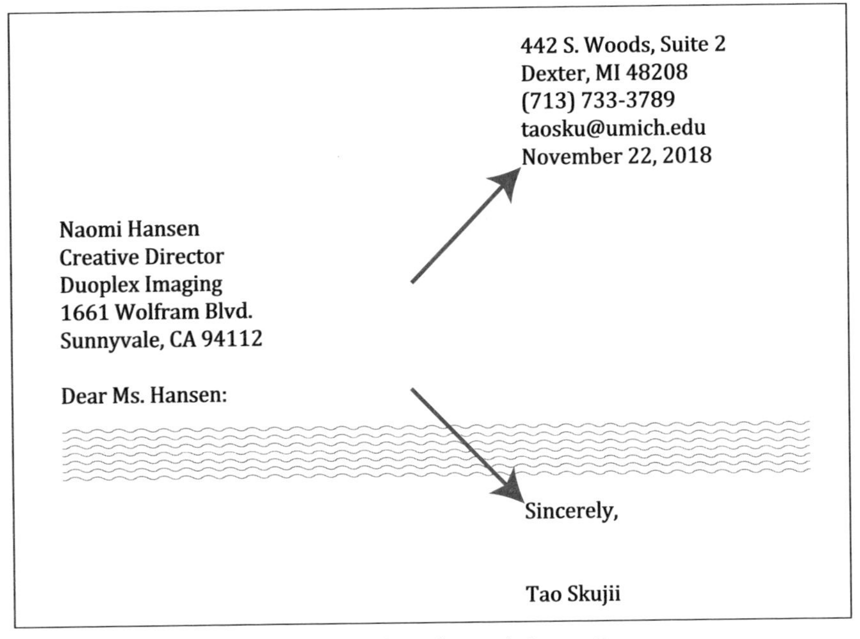
442 S. Woods, Suite 2
Dexter, MI 48208
(713) 733-3789
taosku@umich.edu
November 22, 2018

Naomi Hansen
Creative Director
Duoplex Imaging
1661 Wolfram Blvd.
Sunnyvale, CA 94112

Dear Ms. Hansen:

Sincerely,

Tao Skujii

Figure 5.1.6 Tab-aligned Sample Letter Format

5.1.3.2 Complete Business Letter

A complete letter is nearly identical to a business transmittal letter with the one exception that it contains the full communication, or message, from the author. In this way it fulfills the 3x5 rule, with the one variation that the "highlights" portion is extended to be the complete substantive message, not merely a summary of the message. Business letters like this are used as a formal medium for communication outside one's organization when an email would be too informal and a report, white paper, or proposal would be too formal or lengthy. Various uses might be to (1) provide a summary of a service with a price (repair or small effort), (2) introduce your organization, or (3) explain an issue that needs the recipient's assistance and request a meeting or collaboration. Substantive content suits the project and the author's purpose of communication.

The complete nature of the letter is revealed immediately in both the subject line and in the opening purpose statement. The subject line can easily reflect the substantive content, for example, "presentation of three services available." Similarly, the purpose statement will briefly forecast the full content of the letter, such as the following: "The purpose of this letter is to introduce our company's three core services, namely, customer personality analysis, call routing, and personnel streamlining. In addition, for each of these services, we provide some background information that explains its theoretical underpinning, practice-based origins, and current focus, as well as identify team leadership and contact information." A complete business letter fulfills the characteristic of persuasiveness by including a full hierarchy of information, from general statements to particular details. An example of a complete business letter is provided in Figure 5.1.7.

Positive Health Research Office
School of Nursing
500 North Main St., Office 425
Chicago, IL 32109
(834) 547-0028

January 10, 2019

Lois Castaneda
2389 Halifax Rd., Apt. 4a
Decatur, IL 35309

Dear Ms. Castaneda:

Thank you again for all the time and effort you offered the Student Lifestyle and Choices Study (SLACS) when you were enrolled.

This letter is to inform you that Dr. Cahn is considering a follow-up study to learn more about long-term patterns of health behaviors in Mexican-American women. We would like to know if we may have your permission to contact you in the future to tell you about our new study and talk about your interest in participating. If so, we would like to identify the best way to contact you, in approximately one year. Also, we ask that you provide your anticipated mailing address in one year, if you feel comfortable providing this.

Thank you in advance for indicating on the enclosed post card whether or not we have permission to contact you in the future and the requested locator information. The postage is pre-paid; simply drop the card in the mail. In the meantime, we wish you the best of health and continued success in your endeavors.

Sincerely,

Sarah Coletti

Sarah Coletti
Project Manager

Figure 5.1.7 Sample Complete Business Letter

5.1.4 White Paper

White paper is a fairly generic and empty genre name that can mean different things to different organizations. The name itself reveals nothing about the content and format, unlike "proposal" or "poster." In its original sense, it evolved from *white book* and referred to a policy statement developed by an official or party of a government in Britain. Two famous white papers are these: Churchill White Paper of 1922 and the British White Paper of 1939. Within such a white paper, the full policy or position on a particular matter of domestic or international concern would be expounded. The various white papers of the twentieth century advanced controversial, yet decisive, approaches to moving forward on an issue of grave public concern. Whole populations might be affected by the ideas in a white paper. The U.S. government produced a white paper of 1998 with far-reaching ramifications, as it concerned management of Internet names and addresses. From my own research, original, government-developed white papers tend to be summary statements (often to foster and allow discussion and questions) and not fully argumentative, multi-faceted, reasoned expositions of the proposed policies or positions. Today, white papers are more varied and longer than they were originally.

Outside of government, a white paper evolved in recent times (1990s) into a document used by experts in ABERST to present a policy, position, opinion, concern, or conclusion. White papers usually emanate from for-profit companies, but an expert author of a white paper may be an individual, a university research team, a laboratory, a government agency, or a non-profit organization. While the topics and objectives of white papers are widely dispersed, the vast majority of them argue for adoption or support of a new technology, product, scientific approach, standard, method, or plan. White papers of this type present a solution to a problem, introduce a technological innovation or product, or enumerate key items of importance within a specialty area. They are technically detailed but have complementary objectives aligned with both marketing and public relations. Intended recipients might be potential funders and sponsors of the project described. Other recipients include potential collaborators or beneficiaries. Basically, stakeholders and people from whom some type of "buy in" is desired compose the audience for white papers. (A note of caution: The term "white paper" is also associated with spam and email junk; an email announcing a white paper with a price tag that arrives "out of the blue" should be deleted. Legitimate white papers will be prepared by known colleagues or for them, or be requested of you by organizations of high reputation. Those are the signs of a true white paper, which may be a preliminary document prior to writing a proposal or a journal article. Beware of anything less.)

I do not know of a universally accepted format for these market-driven white papers, so my advice is to follow a style that has been used successfully in your specialty. If you are writing the first white paper for your organization, you should be able to find an example that seems appropriate to your purposes, due to some identified overlap of fields, intentions, or audiences. Whole books exist to guide writers through development of white papers, so I will leave you to consult one of those if this is a genre that will be hugely important to you.

If you are starting here as your initial guide to writing a white paper, I can offer a general suggestion for the major themes to be covered in the document, as follows:

Background:

- This section provides justification for the white paper's vision and purpose.
- This is the context for the innovation, idea, or program envisioned.
- This will likely include a problem description (recall Chapter 1.2).
- The problem might be a snag, obstacle, illness, or hole that needs to be addressed, overcome, treated, or filled.

Description:

- This is a straightforward presentation of the vision and purpose, with details of the innovation, idea, or program envisioned for implementation.
- This section encompasses the objective and goals of the project.
- Example projects include developing or commercializing new technology, implementing an intervention, creating a new treatment, working up a solution, conceiving and executing a program, or generally wishing to carry out a beneficial plan.

Argument:

- This presents the author's persuasive case that the vision is an appropriate response to the context, particularly the problem or need presented in the Background (recall Part 4).

Impact:

- This highlights the project's positive impact on people or society.
- This information may be implied in the Background section, but here it is emphasized that the project will help people and some portion of a community, if not more.
- Details of the population addressed are crucial: demographics, industries, special interests, and an estimate of the number of people if possible.
- This section includes the ways in which people will benefit.

Biography:

- This has background information on the parties involved.
- This section identifies those who will do the work or create the innovation and those who will support, fund, approve, or oversee the project. (The beneficiaries are delineated in the Impact section).

Literature:

- This section offers a review of pertinent published literature to accomplish two objectives: (1) the summary of previous work proves that the new project extends an unfinished effort or fills a gap in some way; (2) the summary reassures readers, and potential supporters, that all prior related efforts will be leveraged and utilized in some sense, so that existing resources are being acknowledged and valued, not duplicated or ignored.

Typically, a white paper is formal, akin to a proposal, but without the details of schedule, cost, management, and qualifications (see the next section for details on proposals). Despite the formality, they are not as long as final reports because of the lack of detailed argumentative support and data analysis, and they are not as long as proposals, again, due to the absence of specific terms and commitments. As a fallback option, a memorandum format can always be adapted to a white paper (see Memorandum, Section 5.1.7, below). To demonstrate a white paper's content, format, and style, Figure 5.1.8 is an acceptable example that can be a starting point for finding an approach that works for your organization and purposes.

UNIVERSITY OF MICHIGAN

CYBERSECURITY

Assessing Risk:
Identifying and Analyzing Cybersecurity Threats to Automated Vehicles

ANDRÉ WEIMERSKIRCH
Lead, Mcity Cybersecurity Working Group
Vice President, Cybersecurity, Lear Corporation

DERRICK DOMINIC
Graduate Student Research Assistant, Robotics, University of Michigan

Contents

INTRODUCTION

It's no secret that developers of automated vehicles face a host of complex issues to be solved before self-driving cars can hit the road en masse, from building the necessary infrastructure and defining legal issues to safety testing and coping with the vagaries of weather and urban environments. In addition, developers face huge risks if they neglect the vital issue of cybersecurity in automated vehicles.

Driverless vehicles will be at least as vulnerable to all the existing security threats that regularly disrupt our computer networks. That could include data thieves who want to glean personal and finance information, spoofers who present incorrect information to a vehicle, and denial-of-service attacks that move from shutting down computers to shutting down cars.

Cybersecurity is an overlooked area of research in the development of driverless vehicles, even though many threats and vulnerabilities exist, and more are likely to emerge as the technology progresses to higher levels of automated mobility. Although no over-arching solutions are obvious at this point, Mcity researchers have developed the first tool and methodology for assessing cybersecurity risks in automated vehicles. This marks not only

Figure 5.1.8 Sample White Paper (Page 1)

an important step in solving these problems, but also presents a blueprint to effectively identify and analyze cybersecurity threats and create effective approaches to make automated vehicle systems safe and secure.

There are the new cybersecurity threats unique to automated vehicles, including hackers who would try to take control over or shut-down a vehicle, criminals who could try to ransom a vehicle or its passengers and thieves who would direct a self-driving car to relocate itself to the local chop-shop.

Also, there are security threats to the wide-ranging networks that will connect with automated vehicles, from financial networks that process tolls and parking payments to roadway sensors, cameras and traffic signals to the electricity grid and our personal home networks. Consider the seemingly nonthreatening convenience of an automated car that gets within 15 minutes of your home and automatically turns on your furnace or air conditioner, opens the garage and unlocks your front door. Any hacker who can breach that vehicle system would be able to walk right in and burglarize your home.

Researchers affiliated with the University of Michigan's Mcity connected and automated vehicle center are finding that the complex and wide-ranging issue of cybersecurity specific to automated vehicles and the infrastructure that will support them is just beginning to be recognized, and will become more important as the development of these vehicles progresses. Without robust, sophisticated, bullet-proof cybersecurity for automated vehicles, systems and infrastructure, a viable, mass market for these vehicles simply won't come into being.

UNDERSTANDING THE VULNERABILITIES

The threats to automated vehicles can come through any of the systems that connect to the vehicle's sensors, communications applications, processors, and control systems, as well as external inputs from other vehicles, roadways, infrastructure and mapping and GPS data systems. In addition, the control systems of each vehicle for speed, steering and braking are exposed to attacks.

Each individual automated application will require its own unique threat analysis that maps its vulnerabilities and assesses the level of risk presented. New work by researchers working with Mcity on adapting existing automotive threat models demonstrates how

Figure 5.1.8 Sample White Paper (Page 2)

5.1.5 Proposal

Often white papers inspire and evolve into proposals. They can be companion documents, sharing some of the same information. They both entail a forward-thinking vision and an intention to pursue a venture or project. Differences between them are important. While a white paper may be exploratory and conceptual, a proposal presents a concrete plan, with definite terms, promises, and commitments. A single white paper may spawn numerous proposals, yet a proposal may be prepared without a preceding white paper. Certainly, the lack of a white paper must not prevent an author from preparing and submitting a proposal. A proposal is valuable in and of itself, and this genre should be used as often as necessary. Indeed, this may be the single most important genre in ABERST.

Proposals hold the key to funding, revenue, activity, hiring, growth, discovery, innovation, and financial success. Proposals enable dreams to come true, whether dreams of undertaking a particular type of work or receiving financial support, or dreams to bring a product, idea, or plan to fruition. When an author wants to offer a service or product to a customer or community, or seeks funding, support, or sponsorship from an organization, a proposal is the means to achieving these objectives.

Unlike many of the other genres, proposals are often competitive. Even if you write, edit, and polish your proposal, and submit it on time to the correct person(s), you may still not achieve your objective of securing the funding or contract. The proposal may succeed in conveying your message, but you have to do that better than other organizations also submitting proposals. Journal articles are similarly competitive, and you measure their success in two phases: (1) completion and (2) publication. Proposals need to be completed as well. Completing a proposal is covered here. They also need to beat the competition, and that is beyond your control for the most part. All you can control is the quality, clarity, and cogency of your written proposal.

Proposals can be either unsolicited or solicited. Proposals that are unsolicited are basically up to the author to design and develop. Proposals that are solicited are usually strictly prescribed. When solicited, submitted proposals must follow the requirements and constraints enumerated by the soliciting organization. You should not deviate from those. They might include information required, formatting, and ultimate page length. Deviations usually lead to the proposal's rejection. This is the genre that inspired the fundamental instructions that begin this book (Chapter 1.1).

As with white papers, whole books have been written on proposal writing, including books that specialize in a single type of proposal, say, for medical research. I want to give you some background to the genre overall and some general advice applicable to both types of proposals. If your proposal is a product of your own initiative and the recipient(s) have not solicited it with specific guidelines, you can use the advice below to develop the content of your proposal. It should be sufficient. I suggest that you submit it with a transmittal letter and in a format that reflects your organization's most polished and formal style for documents. If the proposals you write are usually solicited and in response to request for proposals (RFP), you can use the advice below as a starting point, after which you will need to follow the expectations and demands in the RFP to fully prepare and submit your proposal to the soliciting organization.

Overall, proposals are formal, organized, detailed, and persuasive. Each is covered below.

For *formality*, the proposal, at a minimum, has these segments: business transmittal letter, cover, and table of contents.

An *organized* proposal requires two elements: an Introduction and a hierarchy. The whole proposal starts with an Introduction, which provides a brief overview of the proposal and forecasts the remaining sections. The remaining sections are, in turn, organized into a logical hierarchy of nested subsections.

Detailed proposals cover every aspect of an organization's qualifications to do the work, deliver the solution, or provide the product proposed. Qualifications include everything from understanding the problem and need that is being addressed to having all the right personnel and experience. Resumes of key personnel are often included, as is a management plan with an organizational structure, a proposed timeline, and a budget. As a counterpart to a proposal's description of the problem and need (justification and background), the organization's proposed response to that need (the implementation plan) is crucial to a proposal. If this response involves planned work, that work must be described. If it involves development of a solution, the general approach to finding that solution must be described. If it involves creation of a product, the intended characteristics of that product must be described.

Other elements in proposals typically include proven experience, team or coalition membership, outreach and public relations plans, satisfaction of societal or demographic objectives, sample or data management plans, confidentially and conflict of interest controls, intended publications, and intellectual property rights.

Proven experience should be relevant to the planned effort, with highlights of the few most applicable past successes. Team or coalition membership is an opportunity to indicate the diversity and complementary strengths contributed from the various members of the proposed intra- or inter-organizational team, as is the case. Outreach and public relations is often required because large projects may involve or impact communities, neighborhoods, regions, or concerned citizens. Ensuring the project is seen positively by all stakeholders and is responsive to any and all concerns raised by affected persons is important if the project has any public impact, including but not limited to receiving funding from local, state, or federal governments. Satisfaction of societal or demographic objectives may be critical to an award of the work to the organization with the best proposal. Your proposal might highlight any additional social or demographic benefit that your organization offers (minority-owned business participation, small business participation, involvement of economically stressed or economic-recovery zones, youth empowerment, veterans support, etc.). If the proposal is solicited, fulfillment of such objectives may be explicitly mandated. If such is the case, your proposal must discuss such fulfillment.

Sample or data management is crucial to research proposals, so presentation of your organization's plan to accomplish strict and lawful data handling, recordkeeping, and secure storage are mandatory. Similarly, confidentiality and conflict of interest controls may be necessary, and your proposal must explain your intended control methods. Publication intentions address a commitment to communicate and disseminate project findings and conclusions. Lastly, intellectual property (IP) rights set forth a proposed agreement on assigning and handling IP claims and responsibilities.

Two additional topics might contribute to your proposal. If your proposed work is academic or otherwise has a theoretical or research-oriented aspect, a brief review of the relevant prior work, via published articles and reports, may often be either required (if solicited) or helpful to your argument (if unsolicited).

You are advised, furthermore, to cover any and all benefits and impacts that will accrue to the sponsor and all other parties that stand to gain from execution of the work proposed. This discussion of the impact (usually overall positive but possibly temporarily disruptive) is necessary and enhances credibility (recall Chapter 4.4), and many sponsors expect a thorough accounting of all anticipated disruptions, inconveniences, delays, risks, and harms. An impact assessment is necessary, warts and all, and you can offset the negative aspects by an equally thorough accounting of all the anticipated benefits.

The point of the aforementioned details is to ultimately *persuade* the reviewers that your organization, more than any other team submitting a proposal, is ideally qualified to receive and implement the contract, funding, grant, or business. A skeleton of a proposal is provided in Figure 5.1.9.

Project Title
Principal Investigator and Project Manager
Partner Organization(s)
Vision and Goals, including Innovation and Novelty
Societal Impact, including Significance, Progress, and Change
Effort Overview, including Strategy, Methods, and Analyses
Estimated Cost
Cost Sharing
Data Management
Outreach
Intellectual Property
Conflict of Interest

Figure 5.1.9 Proposal Skeleton

5.1.6 Pre-Proposal

Just as a proposal may be a companion document to a white paper, it may also be a companion to a pre-proposal, another affiliated genre. As you have just learned, proposals are detailed and lengthy documents. They require a serious commitment of time and resources to prepare. Fortunately, in many instances before a proposal is written, a preliminary document is produced. Such a preliminary document is known by several names: application essay, letter of intent, prospectus, and pre-proposal are the most common. For the remainder of this section, "prospectus" will be used as a singular term to represent the pre-proposal genre.

A prospectus may be required by a funding agency or it may be self-initiated by an individual or organization prior to preparing a full proposal. It is used to "test the waters," regarding a research or program idea, to see if it warrants further efforts, before moving to the next step of developing a full and detailed proposal. This can save time for both the funding agency and the implementing organization. If an agency can only fund so many projects, they may not want excessive submissions, especially if some of those submissions (created at considerable time and expense by the authors) are outside the scope of the funding agency's mission. A prospectus can reassure both parties that the project idea and the funding agency are a good match prior to further efforts (writing and reading) by both parties.

If self-initiated, a prospectus is an excellent vehicle for submitting a nascent concept to others who must approve, support, and fund the effort. Again, before submitting a full proposal, an individual or organization can gauge the level of support attainable from the recipient for a self-initiated project through a brief, highly summarized prospectus.

A prospectus is by definition short and general (possibly limited to 1 page or a specific word count—500 to 1,000 words is not unusual). It presents the basic outline of an idea by which someone can review it (possibly compare it to other similar options described in other prospectuses) and issue either an approval to proceed or a rejection.

A prospectus should have six basic elements and not much else (unless specific items have been imposed by the recipient). For each of the following, I provide some guidelines for preparing the ideal information.

1. Title—The title must be descriptive and specific. More often than not, the best title is NOT the title used up to the point of writing by the author and her team. Sometimes scientists, engineers, researchers, and others in business use jargon, nicknames, short codes, internal terms, and so on to discuss projects. These internal phrases are often unclear to outsiders who may be the recipients of the prospectus. (Recall Chapter 3.3.) Also, they may be inaccurate, misleading, vague, or ambiguous to anyone not involved with the work up to that point.

Below are bad and good example titles to illustrate this:

- Bad Example: Apple Driver Project
- Good Example: Development of Artificial Intelligence Code Compatible with Apple's Operating System for Autonomous Driving via iPhones

2. Project Description—The project description must outline the intended program, design, research plan, or other similar venture that requires support and funding. This description should be concise and general. Details are unnecessary and burdensome at this point. A general description should entail the project's objective (what the author intends to accomplish) along with some method (how the author intends to do it). Additional details of where (location[s] of intended effort), when (ideal project timeframe), and who (principal persons responsible for the effort as far as they have been identified) would be suitable. If key team members are known, mention of their intended participation may be helpful; otherwise, if the effort involves collaboration among organizations, the full team must be introduced, including universities, nonprofit organizations, public and private companies, government agencies, collaboratives, and consortiums.

3. Project Justification—The justification explains the reason this program, design, research, etc., is needed or desirable now. In particular, if the author can point to a specific problem, obstacle, snag, difficulty, setback, or hurdle that the recipient or society faces, and to which this project is addressed, this is fundamental information that must be provided, and it makes the project valuable, and possibly imperative. Other issues that might contribute to project justification include (1) the time is particularly opportune (and the window might slip by) and (2) the team on hand is precisely the right one to do this effort.

If the need for the project is somewhere short of dire, the justification section must nonetheless reveal the author's best explanation for believing this project has merit and warrants implementation. The author's enthusiasm for the project should be apparent; If not, perhaps this is not the right time to write a prospectus. With enthusiasm, an author explains a dream, wish, or desire to plan and implement some effort that is important to her. When the project is desired, not to solve a problem or overcome an obstacle, but to advance technology, help society, create innovation, or make life more enjoyable, the justification might be primarily expressed as project benefits. Thus, for projects that are dream makers as opposed to problem solvers, much of the justification might blend into the next section, Benefits.

4. Project Benefits—The anticipated benefits derived from completion of the project must be delineated. These may be directly applicable to the recipient, or the recipient may be committed to supporting projects that help others. Sometimes the support or funding source is involved because the project contributes to its needs and mission. Other times, a support or funding source, such as a board of directors, foundation, charity, company management, or government agency, is interested in projects that help the community or society in some way. Either way, benefits are important to introduce.

Benefits can be approached in a threefold manner: 1. Who: This should discuss the beneficiaries at both micro and macro levels. For micro, this is the people most closely associated with implementing the effort. Example, if the intended project is approved, the author will be able to create simulation software for urban planners to evaluate the effects of design ideas for community upgrades that add sidewalks and bike paths to urban areas. The immediate beneficiaries will be urban planners and their clients, such as city managers. The macro, or, long-term, beneficiaries may be all residents and visitors to cities that have been upgraded with the application of this new simulation software. When considering the beneficiaries, an author should try to be as comprehensive as possible, using the following list as a guide to possible groups to consider: demographics, populations and sub-populations, geographic regions, industries, special interest groups, vocations, and professional associations. When an approximate number of persons to benefit can be identified, this tally can be included as well. 2. When: This is the timeframe for benefits to accrue. In the near term, some individuals or parties may benefit. In the long term, a different population, perhaps much larger than the immediate beneficiaries, may benefit. Using the example above, perhaps the author anticipates the near-term benefits to accrue within 5 years, and the long-term benefits to be evident in the 10- to 20-year timeframe. 3. How: This part explains the mechanisms by which benefits accrue. It may be very simple or possibly somewhat complex, but the process of reaping benefits should be clear. As a simple example, the benefits might emanate from physicians who use the project results to change their care of patients. In turn, the patients benefit by following the improved behavioral guidelines conveyed to them.

5. Sponsor Involvement—Explain your reasons for believing this project is suitable to the recipient's needs or mission, if this has not been explicated in the preceding sections. If the connection to the sponsor is already clear, this section can focus on introducing the relationship you propose. Will you do all the work, while the sponsor provides the funding and oversight? Will you work in cooperation with the sponsor, dividing up the effort efficiently? Will you be independent, using your own locations and facilities, or will you be blended with the sponsor, doing work on site and with the sponsor's facilities, equipment, and personnel?

6. Resources Available—In this section, the author provides a quick survey of resources (facilities, personnel, published documents, internal reports, computers, equipment, etc.) that can be leveraged in support of the project to make success likely and implementation efficient. If some of this material has already been introduced in the preceding sections, repetition is not necessary. This brief, final section of the prospectus should be used for providing new information that indicates the author is cognizant of and has access to necessary and helpful resources.

In terms of formatting, the prospectus should adhere to all expectations and guidelines provided by the recipient if such have been prescribed. If the document is self-initiated, some simple formatting should suffice. Depending on the recipient and the context, it could be formatted either as a letter (letter of intent), essay (application essay), memorandum (prospectus), or formal document (pre-proposal). Format for those genres are covered elsewhere in this chapter, with the exception of essay. No matter which overarching format you use, the sections outlined above should be written in paragraphs under specific headings. A skeleton application essay is shown in Figure 5.1.10, as it is the only aforementioned prospectus-supporting genre not covered elsewhere in this chapter. Its overall structure is quite simple, comprising two parts: heading and body. The heading should be centered, at top of page 1, with project title, author's name, recipient's name, and date. The body should be divided into these sections, each announced with a highlighted heading: Project Description, Project Justification, Project Benefits, Sponsor Involvement, and Resources Available.

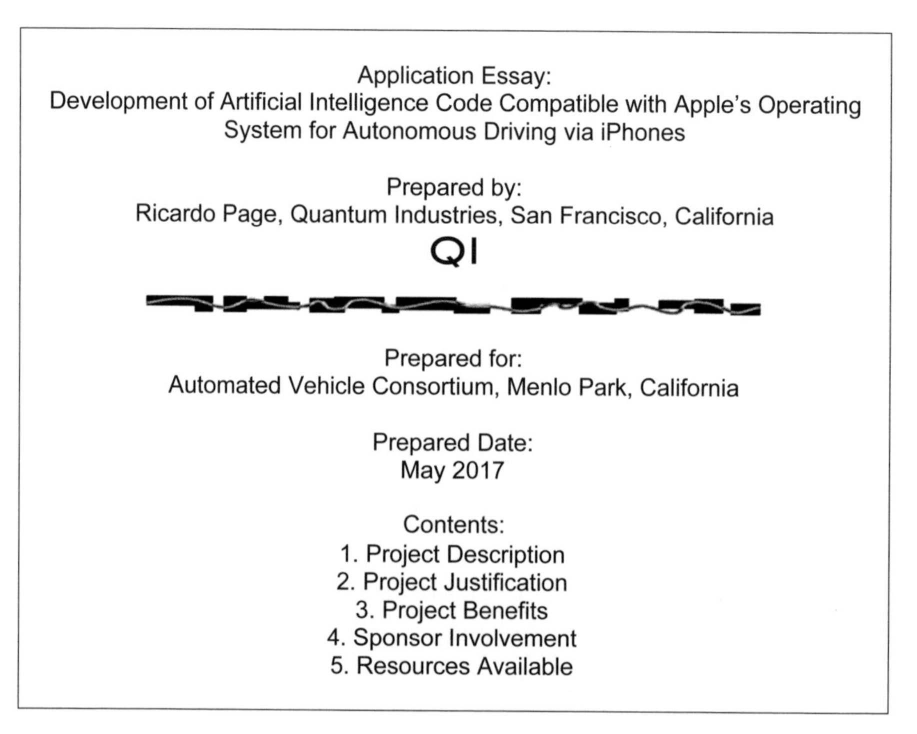

Application Essay:
Development of Artificial Intelligence Code Compatible with Apple's Operating System for Autonomous Driving via iPhones

Prepared by:
Ricardo Page, Quantum Industries, San Francisco, California

QI

Prepared for:
Automated Vehicle Consortium, Menlo Park, California

Prepared Date:
May 2017

Contents:
1. Project Description
2. Project Justification
3. Project Benefits
4. Sponsor Involvement
5. Resources Available

Figure 5.1.10 Application Essay Skeleton

5.1.7 Memorandum

If I had to teach just one single genre for ABERST communication, I would select the memorandum. Although the proposal is perhaps the most financially crucial, and the journal article perhaps of the highest profile, the memorandum has the most utility. It is the workhorse of all documents. It is versatile, and it is ubiquitous. You can use it for many communication purposes; it can be short or long. It can be distributed both internally and externally. Importantly, it requires mastery of all the core principles of technical communication, so it can showcase excellent communication skills and burnish one's reputation in the workplace. If you can write a good technical memorandum, you can very likely master the other genres. And equally, if you have difficulty with a memo, you will struggle with the others. A memo is the rightful place to demonstrate your technical writing prowess and make a reputation for yourself so your managers will assign you to work on other genres also.

The reason the memo is common and widely used is its versatility. Like a chameleon, it can be adapted to many different communication purposes and scenarios. Thus, it should be a well-understood format for anyone wanting to be highly capable in professional, non-fiction communication. Nonetheless, after mastering the general format of a memo, a writer must also understand many other patterns and genres that can be integrated into a memo to create an informal document in memo format suitable to the purpose at hand. This is the reason you must know the memo format as well as you know the back of your hand.

An example of a memo, condensed, with its five basic parts is shown below (Figure 5.1.11). Each part is explained briefly as well. Importantly, each part is essential, so the genre involves a well-established set and order of parts. You do not have to reinvent the wheel. Use the parts and adapt them to your situation. For example, some organizations add more lines to the heading, to present useful and recurring information.

AMTI

Analytical Mechanic Technology, Inc.

Intra-Company Memorandum

Date: September 19, 2018
To: R.O. Tate, Chief Design Engineer
cc: Manuel Kwatrone, Head of Research
From: Jill Brilliant and Shirley Smart, Laboratory Managers
Subject: Recommendation of Propeller Diameter based on Tunnel Testing
Ref: Project 418S-L, Division PDG

Foreword
SkyTrek, Inc., a small drone manufacturer, has hired us to conduct analysis in our wind tunnel that they lack the capability to do themselves. They have a new quad-rotor drone design and need to identify the ideal propeller size for the expected vehicle weight. We tested two diameters and extrapolated to additional sizes. In this report, we present our intended recommendation for your review before delivery to the client.

Summary
We recommend a 7-inch-diameter propeller. With this size, four propellers can lift 660 grams, which is more than twice the weight of the drone (300 grams), yielding a weight margin of 2.1. The thrust of each propeller is 165 grams, and the climb velocity in hover is 30 feet per second. Both of these values exceed the benchmarks present by the client. In addition, these thrust and velocity values are accomplished at 6,000 rotations per minute (RPM), which is produced by the motor comfortably at 80% capacity. If necessary, the RPMs could be increased to produce greater thrust and velocity for temporary maneuvers.

Discussion: Detailed Propeller Analysis
SkyTrek, Inc., a small drone manufacturer, has hired us to test a proposed propeller in our wind tunnel. They can manufacture the propeller, in the particular geometry, very easily in any diameter from 4 to 10 inches. They lack, however, the ability to conduct detailed dimensional and non-dimensional analyses. In particular, they want to know the advance ratio and the thrust coefficient. We tested two diameters and extrapolated to additional sizes. We recommend a 7-inch-diameter propeller, for the reasons detailed below.

Selection Factors
Selection of the ideal size for the propeller must take into account several important design factors and practical considerations. First, the chosen size must produce enough thrust (when multiplied by four for a quad-rotor) to lift the drone and provide some weight margin. The weight margin allows for both (1) load-carrying capacity beyond the drone's base weight itself and (2) additional thrust so the vehicle can accelerate horizontally and vertically. Second, the size must not be too large so as to place new demands on the drone's design and other ancillary components, such as the motor and battery. In that sense, the smallest effective size is ideal. Third, the client has mentioned they prefer to use a 6-inch-diameter propeller if possible. Also, sizes are best chosen at half-inch intervals, to avoid the trouble of using non-standard sizes.

Support for Recommendation
We selected the 7-inch-diameter propeller in large part because it is the smallest size that can produce a weight margin of 2 or larger. Sizes smaller than this produce a weight margin of only 1.6 or less. For example, a 6.6-inch propeller gives 85 grams of thrust, leading to 340 total grams, and a weight margin of 1.13. Our laboratory testing provided the measured values upon which we made our decision, as discussed next.

Thrust and Velocity Measurements to Support 7-inch Propeller Selection
Our testing consisted of running the two sample propellers inside our small wind tunnel at both zero wind speed (static testing) and a series of increasing wind speeds (dynamic testing). The setup is shown in Figure 1. We first tested each propellers at various RPMs, from 1,000 to 10,000.

Figure 5.1.11 Sample Memorandum (Page 1)

Next we held the RPMs constant at 6,000 and increased the tunnel wind speed to 20 feet per second (fps) in 2-fps increments.

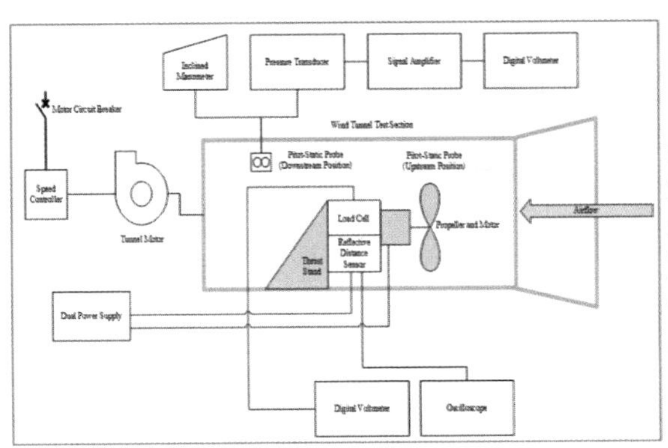

Figure 1: Propeller Thrust Performance Experimental Set-up

Test Specimens

We tested both a 4- and 6-inch propeller because those were available, and we hoped one of them would prove sufficient. If not, we knew we could extrapolate our empirical results to a suitable size. As it turned out, neither produces enough thrust for this application. Hence, the selection of the 7-inch propeller.

Static Testing

By increasing the RPMs (with added power from the motor), we observed the thrust achieved from the two propellers in still air. The data for the 6-inch propeller are graphed in Figure 2. (Collected data for both test specimens are tabulated in Appendix A.) Importantly, at the expected RPM value of 6,000, we could see the maximum thrust for the propeller. This value would be multiplied by four when implemented on a quad-rotor drone.

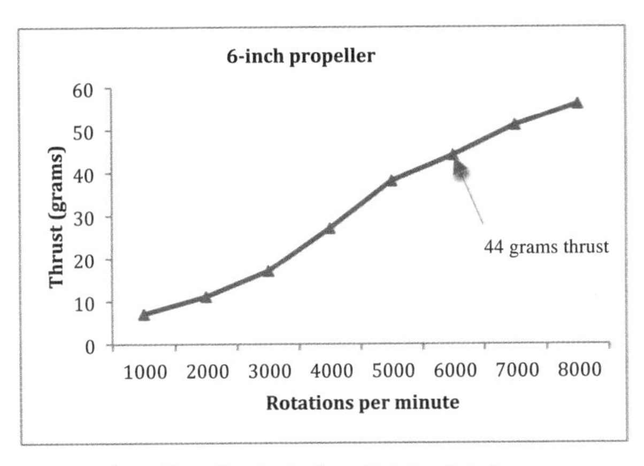

Figure 2: Thrust Rises Quadratically as Rotation Rate Increases

Before any testing was done, the load transducer was calibrated to determine the relationship between output voltage and force applied. This gave us a calibration function that we could use to relate measured voltage values into a force coming from the spinning propeller. In addition to the load transducer, we calibrated a pressure transducer to relate its voltage values to a pressure

• • •

Documentation -- Appendix A: Raw data tables

Figure 5.1.11 Sample Memorandum (Page 2)

1. Heading: Appearing at the very top of the first page, the Heading provides basic document identification information: author, primary recipient(s), date, and subject. Additional items that could be added are additional recipients (cc) or routing, contract/funding numbers, product/part/invoice/charge numbers, locations, confidentiality level, key words, and affiliations. (Organizations that use fully digital/electronic document archiving will almost certainly expect key words and other additions to the standard Heading because these items allow for filtered searches, sorts, and groupings.)

2. Foreword: As discussed in Chapter 1.2, a memo, like any technical document, should start with context before launching into the specific message. To create context, an author should identify the problem at issue and the work the author did prior to writing. These two statements, Problem Statement and Work Statement, reveal the context and may either convey as-yet-unknown information or provide a helpful review to the reader. The third and final element of a Foreword is the Purpose Statement, which conveys the author's reason for writing. The communication purpose is not always self evident, so it must be stated explicitly, and its ideal location is at the end of the Foreword. Thus, a Foreword has three elements, and only three. Any additional information adds clutter and obfuscates the context and purpose of the memo. If you are tempted to include a summary of the memo's main point or points, your instincts are good, for those come immediately next, in the Summary. In formal reports, moreover, the Foreword and Summary are merged into one section, the Executive Summary, but that is a different approach within a different genre. The two sections are best left distinct for a memo.

You can call the memo's opening section Introduction, Purpose, Context, Highlights, Background, or Preface, but whichever term is used, it serves as a "word before," that is, a foreword. One last piece of advice: use an obvious descriptive title, or heading, to announce this opening section. Do not merely start a memo with a sentence after the Heading. Readers, upon seeing an untitled, floating sentence or small paragraph, may not realize its importance. It may look like boilerplate, author notes, or something similarly unimportant.

3. Summary: As also discussed in Chapter 1.2, the Summary presents your point, main message, or thesis, as well as a selected handful of the crucial support points for that message. These are the highlights of the document, and they will be elaborated upon in the remaining portion of the memo. For now, they are summarized and are substantive though brief. If the Purpose Statement promises a particular message, the Summary is the initial fulfillment of that promise, with details to come later. In other words, the Summary contains the heart of your message, while the foundation and detailed establishment of the message is understood to be reserved for the Discussion that follows. Recall that I explained in Chapter 1.2 that a Summary enables readers to look over this brief section and become familiar with your main ideas and confident that you can defend those main ideas.

4. Discussion: This is the body of the memo. This section expands to the size required, depending on your context, purpose, and message. One, you elaborate on the Problem Statement from the Foreword, as necessary. Two, you put in the details that were summarized in the Summary. Three, you further establish and present your supporting points for your main message. All the lessons of Parts 3 and 4 of this book come to bear on the Discussion. Importantly, this part should contain specific section titles germane to the memo's content, rather than the single, generic heading "Discussion." Conceptually, the fourth part of a memo is the Discussion, but it requires a hierarchy of sections as necessary to reveal the organization of ideas presented and guide the reader through the various, related topics.

5. Documentation: The memo's Documentation—appendices and attachments—contains bulky items that support your Discussion but would be disruptive of the flow or format, or even unnecessary for all readers to review; these are items that are provided for the convenience of some readers (rather than making them ask you for the material at a later date) yet placed at the back of the document for, again, the convenience of all readers.

5.1.8 Procedural Instruction

The guidelines below are drawn from the work of my colleagues, J.C. Mathes and Dwight Stevenson, from their textbook, *Designing Technical Reports*.

A procedural instruction can be presented simply and fully in a three-part structure: Objective, Background, and Procedure. This universal structure applies neatly to almost any instruction that involves specific steps the user must follow to complete a process. In ABERST, many procedures are documented and followed for standardization, efficiency, and repetition. This is the focus of this genre. If, in contrast, the instructions are broadly educational (e.g., guide to proper nutrition), strategic (how to make friends and keep them), or philosophical (how to find happiness), this structure is less applicable. You might want to consult different examples to assist you in writing such sweeping guidebooks and instructional materials.

Some definite indicators that you are writing a procedural instruction that would benefit from the following advice are these: the procedure has a definite end point where the steps are finished and the necessary objective is accomplished; the procedure involves a set of equipment, tools, and materials; the procedure may be done over and over again by the same person at different times with slightly different materials or by different persons in different locations but with similar equipment, tools, and materials; the procedure is based in some way on either academic efforts, business processes, engineering research, scientific study, or technological development. These indicators point to a procedural instruction even if your organization uses a different term, such as "standard operating procedure," (SOP), "training manual," or "protocol."

The three parts of a procedural instruction have required elements, and each element is essential for successful implementation, as explained below.

Objective

The objective section answers some basic questions any reader may ask (what, who, why, when, and how) and has the following necessary elements, associated with each basic question:

What:

Provide both the purpose of the procedure (what will be accomplished) and the purpose of the document (to guide reader through the procedure). Example purpose statements for a procedural instruction are these:

- The purpose of this document is to provide the instructions for accomplishing specimen preservation.

- The purpose of this instruction manual is to guide the user through the process of doing specimen preservation.

Who:
Describe the intended or anticipated reader/user. Example of a statement of user identification is this:

- This manual is written for all laboratory technicians who are responsible for specimen preservation.

Why:
Explain the reason this procedure is necessary, helpful, mandatory, or desirable. This is the context for the procedure and gives its justification. An example of procedure justification is the following:

- Without adherence to this procedure, specimens that have been extracted at considerable time and expense could degrade beyond the point of usability. This procedure is followed to ensure specimens are adequately preserved so that control tests can be conducted at a later date to verify consistent findings. Without preservation, consistency, and repeatability, the data may be random and meaningless.

When:
If not already provided in answer to "why," the context of the procedure may also involve some temporal constraints or guidelines, including when the procedure should be done and how long it is expected to take.

This discussion of time may also involve any of the following as applicable: a triggering event that prompts initiation of the procedure; a time of day or night that is appropriate for the procedure; or a particular day, month, or season of the year appropriate for the procedure. An example of temporal constraint is the following:

- This procedure is best done in mid-summer, after flowering and before the buds of next year's flowers begin to set.

How:
Two elements explain how a procedure is accomplished. One, provide a quick introduction to the basic principle or method undergirding the entire procedure. Two, provide a forecast of the major divisions—stages—of the procedural steps.

The basic principle or method refers to a single core concept underlying the whole endeavor. Most processes have one underlying concept that makes the whole thing work. It is either the key method, key theory, or key principle (or some combination of these) that serves as the "heart" of the process. In brief, it is "how" this process is carried out. No doubt this method/principle/theory will be explained further in the Background, but an initial mention is ideal in the Objective. This alerts the user to the essence of the process that they are about to implement. It is, like other items in the Objective, a general introductory explanation to help provide a "big picture" overview before launching into the details.

Below are three examples of brief statements of method/principle/theory:

- These instructions for building a dining chair without any nails and screws depend instead on the mortise and tenon method.
- This is a procedure for improving physical health using the principle of resistance training.
- The process explained in this instructional manual relies on the theory of dynamic similarity to produce meaningful results for vehicle design.

Forecast of stages is a quick enumeration of the stages, named and numbered. As the Procedure component, explained below, will be a grouping of steps into stages, this is the place to forecast those stages. The Objective appropriately ends with this forecast of the main procedural divisions, with details to come later. Below is an example of a stage forecast:

- Home brewing is broken into five stages: preparing yeast, preparing wort, fermenting wort, carbonating brew, and bottling beer.

Background

The document's second part, Background, provides valuable preparatory information so the user will be fully ready to undertake the procedure before the actual steps of the process are started. The items to cover in the background are these:

Required Items: This includes parts, materials, objects, tools, and equipment that will be necessary to complete the procedure.

Relevant Concepts: This includes a wide spectrum of fundamental knowledge that users must understand while completing the procedure. In addition to reiterating the core method/principle/theory as introduced in the Objective, this section explains all the additional concepts that are relevant to accomplishing the procedure. This includes principles, theorems, special functions of equipment or tools, math and statistics needed, principles of science and engineering, explanations of innovative technology, laws that are relevant, and assumptions.

Cautions and Warnings: All cautions, alerts, reminders, and warnings must be clearly presented. Any statement that can be made on behalf of human health and safety, as well as preservation of equipment, must be provided in the Background (and repeated again in the applicable stage within the third and last part, Procedure).

Here are a few examples of a category with infinite possibilities:

- This process may take 4 hours or longer.
- Safety eyewear should be worn at all times following the brief stage of preparation as discussed below.
- Once the amplifier is turned on and warmed up (30 minutes), do not touch or move it.
- While the car is hoisted on the jack, do not let anyone sit in or on the car, or push it, or lean against it, or open any door, or hood, or trunk lid.

Procedure

An easy-to-follow instruction manual synthesizes and groups the numerous steps of the procedure into a handful of logical stages. As with any hierarchy, the benefit of logical divisions helps readers to see an overall structure and grasp the essence of the work to be done, without becoming overwhelmed by dozens, possibly hundreds, of small steps that run together into an undifferentiated long list. The stages make the procedure seem manageable and organized.

The number of stages is always tailored to the specific process, with the following guidelines worth considering. If the procedure is a task that can be accomplished in less than one day, the stages should be three to six. If the procedure is multi-day and involves numerous employees, you may need more than six stages. Each stage represents a logical and practical grouping of steps that, together, achieve a particular sub-objective of the whole procedure. Thus, each stage has its own objective. Something very specific must be accomplished (and articulated) at the end of a stage, and that specific sub-objective is obviously a contributing factor in finishing the whole procedure. The synthesis and grouping of steps is crucial. An author should know the process well in order to break it up into ideal stages. Aside from having a distinct output accomplished, a stage may be a natural division based on duration or complexity (trying to divide the whole process into roughly equally demanding portions), location (trying to group steps together in space), or skills used (relying on groupings of either tools, equipment, or expertise). After the steps are placed into useful stages, the stages can be presented in your document in the following, repetitive structure:

- Stage Structure: Objective, Background, and Procedure

This structure, for each stage, mirrors the entire structure for the whole procedure. So, each stage is a microcosm of the full procedure. The repetition of the structure at both levels is beneficial for user comprehension and retention. The mirroring of the pattern fosters concentration of the content upon predictability. The content appropriate to each of the three overall parts at the stage level is discussed below.

Objective:
This is a statement of the intended objective of the stage. In other words, all the steps put together produce this particular outcome. This is brief and simple, and you should be happy to see that no other information from the overall Objective (first part) needs to be repeated; you need only provide the "what" for the stage; however, if some additional explanation for this stage is necessary for added context, e.g. why it is done or why it is done in this particular order in the procedure, it can be provided in the next part of the stage, Background.

Background:
All the information from the Overall Background that is relevant to this particular stage must be provided here, before you delineate the steps of the stage and expect the user to begin doing tasks. Each of the items here should be receiving their second mention in the document. That is, if you need something (e.g. tool,

warning, concept, etc.) for this stage, you already mentioned it when you covered *all* the items needed for the whole procedure. Every item of background is mentioned at least twice in the document: once in the Overall Background and once in the stage (or, stages) where it is applicable.

We can study a simple example to reinforce the documentation method I am suggesting:

- The procedure for home brewing beer has elements that match processes used in medical care, chemical engineering, and the food industry. Home brewing, as mentioned earlier, can be broken into five stages: preparing yeast, preparing wort, fermenting wort, carbonating brew, and bottling beer. To complete the whole, five-stage process, at least 14 items (parts, tools, and equipment) are needed (in addition to the numerous ingredients). All of these 14 items are listed initially in the Overall Background. In Stage 5, specifically, a brewer needs to use only 5 of those 14 items: spigot, hose, bottle, bottle caps, and bottle capper. These items must be listed in the Background section for Stage 5. They do not need to be listed in the Background sections for stages 1 through 4 because they are not needed in those stages. Similarly, the warning to "sterilize all equipment right before use" is applicable to Stages 2 though 4, so it must be listed in the Background sections for those stages, but not for Stages 1 and 5, where it is inapplicable. Lastly, to reiterate, the Overall Background section contains the warning "sterilize all equipment right before use," along with four other important warnings, and these warnings individually apply to some of the stages, where they are listed in each stage's Background section.

Procedure:
In this final stage segment, you provide the to-do list of steps, or, tasks, for the stage. All of these steps should be related in the sense that, put together, they accomplish the stage's objective. Also, they all can be done using the stage-based items, and nothing else, listed in the stage's Background. If you have created practical and useful stages, one or two persons working together in a finite period of time at one location can likely accomplish the steps of each stage.

If the steps require many workers, a large time delay, or several locations, you might consider modifying your stage groupings. A stage is feasibly finished at a single location, in a single working session, with a manageable basket of items.

The *style* of writing is important for this segment of the document. Given that you have provided ample context and explanation leading up to this point in the document, additional and lengthy prose is unnecessary. For the steps within the Procedure, a terse and succinct style is ideal. In addition to using a numbering system for the steps, you can follow these three simple composition rules for writing your steps, with both poor and ideal samples provided for each rule:

1. Use terse, verb-based, imperative sentences:

   Poor Style:

   - Taking the rubber mats and placing them over the opening in the wind tunnel is suggested at this point to make sure that the tunnel is sealed.

   Ideal Style:

   - Seal the wind tunnel's openings with rubber mats.

2. Start with the familiar, then add the new information at the end:

   Poor Style:

   - We can find the Z statistic now because we simply needed to know the given specification, the sample mean, and the variance of our samples.

   Ideal Style:

   - Using the sample mean and standard deviation, determine a probability value, which is the Z statistic.

3. Provide positive, rather than negative, instructions. (Negative instructions are best included as warnings/cautions.)

   Poor Style:

   - Rotate the motor control knob but do not exceed 100 Hz.

   Ideal Style:

   - Increase the motor frequency by 10-Hz increments to a maximum of 90 Hz.

The very last point I want to make about this genre is that a procedural instruction document should begin with an identifying Heading, just as most documents do. This heading should include the document's title, the author's name, and the date, as in the example below:

Instructions for Creating Fullerenes for Experimental Study
Prepared by: Phil Buckminster, Chemist, Division II
Last Revised: February 2018

Let me emphasize one point regarding the title. It must do more than reveal the topic: "beer making" is not a sufficient title. That could be the title of a novel, poem, or short story. The title must indicate both the topic and the purpose of the document, namely, to provide a "how to" guide or set of instructions: "How to Home Brew Beer" or "Instructions for Brewing Beer at Home."

This title will reveal the purpose of the document, which is a crucial component of all technical documents. Revealing it as soon as the title is an excellent tactic. And the purpose will be reinforced and explicitly stated, once again, in the opening Objective part, so that purpose appropriately permeates this informative genre, just as it does in persuasive ones.

5.1.9 Formal Report

Formal reports complement all the other genres covered in this chapter by providing a means to communicate the final results, conclusions, recommendations, judgments, and similar high-level decisions at the end of a long-term project. The other genres serve well for communication objectives at the beginning and during the ongoing stages of a project, while a formal report is suitable when the project is complete and an author wishes to convey (and support) her solution, policy or program proposal, design, determination of cause, prediction of effects, answers to questions, or decisions. For this reason, formal reports are also described as final, formal reports.

The types of projects in ABERST that may require a culminating final, formal report are vast, but they have in common two essential features: (1) the various projects seek to achieve some type of significant goal in support of an organization or field of interest, and (2) the projects demand a final, detailed, and persuasive written report so that stakeholders can be informed of the important points that must be considered, evaluated, and acted upon. Final, formal reports are required in projects of research and development, design development, investigation into causes or effects, evaluation and comparison of optional policies and programs, expert witness studies, solution quests, product improvement, and so on. Such projects aim to solve problems, create innovation, expand knowledge, seize opportunities, reach decisions, and effect change.

Formal reports are usually inter-organizational, meaning that they are prepared by one entity for the benefit of another. Within a single organization, often a memorandum can be used to convey the results of a project. But, even within a single organization, a formal report might be suitable following a long-term effort, when the topic is significant, or when the subject is a major disruption or change at the organization. Some document examples in this category might be the following: a report from an operations vice president of a large manufacturing company desiring to present a proposal to introduce the six sigma program throughout the organization; a report from an information technology (IT) manager at a research laboratory desiring to shift to cloud-based computing away from local, server-based computing; and a report from a human resources (HR) manager desiring to introduce a new line of medical insurance and related benefits to eligible employees of a large non-profit organization.

As examples of inter-organizational formal reports, here are a few: a wetlands restoration plan, prepared by an environmental engineering firm on behalf of a site owner, to be submitted to the appropriate overseeing government agencies (local, state, and federal); a site and building development plan, prepared by a condominium developer submitted to a local city council and planning commission for site plan approval; and an investigation report of potential unlawful activities by an elected official, prepared by a specially assigned independent counsel submitted to a justice department of a government.

Formal reports are usually long and detailed, but that does not mean they have to be painful to write. When faced with the task of writing a formal report, your outlook should be optimistic for several reasons. One, by the time a project is at the stage where a long, detailed formal report is required and appropriate, many other preliminary and project-related reports have been written. The good news is that often much of the text needed in a formal report has been written and used in other project-related documents. Thus, formal reports get a boost from previously written material, which is incorporated into the document in suitable places. The key is to keep all your project-related documents neatly saved, filed, and organized for easy retrieval, so you can copy and paste necessary sections and elements into the formal report. Often you can look back at prospectuses (pre-proposals), proposals, progress reports, white papers, memoranda, and posters to find the building blocks of the formal report.

Two, the basic report rules and communication principles explained in the four previous parts of this book are fully applicable and implemented in much the same way for formal reports as they are for any document: deliver the right document (Part 1), write elegant prose (Part 2), fulfill your communication purpose (Part 3), and persuade your reader (Part 4). Importantly, if your formal report contains an argument, it is presented in the same manner as any argument in a different genre, using the same structural and rhetorical guidelines previously presented in Part 4. You do not need a different method.

Three, the particular characteristics that mark a report as formal are based, for the most part, on formatting and cosmetic issues, and not a new level of writing power or intensity. In other words, formality is mostly a matter of style and detail, not intellectual breakthroughs, transformative insights, or professional innovations.

While those three reasons should temper your trepidations, one note of caution is necessary with respect to the advantage of borrowing previously written text. You must be vigilant to adjust text from one document as you export it to another document. Some necessary changes of tense and tone are required, although the excerpted and borrowed text might be 99% usable from a substantive perspective. Without changes, the borrowed text may diminish your credibility. For example, when writing a formal report, you may reuse text from your proposal where you explained that you "will do x and y, in such a way." If you place that text verbatim into your formal report's Method section, you will be appearing to say that the work is yet to be done. But, that is incorrect. The work has been done, hence the final, formal report. The bottom line is this: be resourceful by using text sections but also be attentive to the necessary adjustments.

A review of the overall three-part structure of all technical reports is a good place to begin when planning a formal report. The three parts are start, body, and base. Formal reports adhere to this structure with an added formality and cosmetic component. The terms often used for "start, body, and base" are these: front matter, body, and back matter. (Keep in mind that a formal report is usually delivered with a business transmittal letter, which is covered in Section 5.1.3 above.)

Front matter encompasses all items that are associated with a document's start. In an email, business letter, or memo, these items are presented in condensed format, to help keep the document compact and short in length. In a formal report, short length is not of paramount importance, as readers anticipate a significant document that has been prepared at some expense, with considerable effort, attention to details, and thoughtfulness. Thus, the start is elongated into front matter, which includes the following six mandatory items:

- Hard cover, title page, table of contents, list of figures, list of tables, and executive summary.

Additional items that may be included in front matter but are not absolutely required include these:

- Distribution list; definitions of abbreviations, acronyms, and symbols; key words; glossary; notices; and acknowledgments.

These are used with some projects and under some contracts, and you will be made aware of the suitability of adding these extra items for your formal report by a client/customer, supervisor, archivist/librarian, IT representative, HR administrator, or knowledgeable peer. (As was mentioned for memoranda, organizations that use fully digital/electronic document archiving will almost certainly expect key words and other enhancements to the front matter because these items allow for filtered searches, sorts, and groupings of documents.)

Some other items are mistakenly added to the start of formal reports and these are really unnecessary, for the reasons given:

Abstract: This is covered by the executive summary, so would be redundant.

Preface: Like an abstract, this is covered by the executive summary, so would be redundant.

Index: This would be included in back matter, if at all.

Some additional explanation is helpful regarding the six absolutely necessary items of front matter listed above. Formal reports are intended to be either bound (spiral, comb, free, or stitched) or placed into a 3-ring binder, so they borrow style elements from book publishing. Like a book, a formal report has a hard cover with, usually, the title and the name of the originating organization. The cover might also have an illustrative piece of art or an organizational logo (or logos). The title page, like one in a book, is placed immediately inside the cover, and it contains the title, date of authorship, author names, recipient names, and any other notices, project numbers, funding codes, etc., required by the involved parties. The title page is followed by the table of contents, list of figures, and list of tables, which collectively offer a preview of the report's organizational hierarchy and enable quick look-up (keyed to a particular page) of all topics and sub-topics covered in the report, as well as all visual aids. These five front matter items constitute the Heading with necessary identification data as required for all ABERST documents. The elaborate and elongated Heading of a formal report exudes professionalism and credibility. The elements suggest to readers that this document has been prepared with care and, although it is a long document, the authors have made efforts to help the reader find specific information desired.

The final item of front matter is the Executive Summary, and, as covered in Chapter 1.2, it provides the report's essential beginning information. In a memo, the report beginning (following the Heading) comprises a Foreword and a Summary. In a formal report, these two segments are combined into a single Executive Summary. This is the best way to think of an Executive Summary. As I explained in Chapter 1.2, each genre has its own way of organizing the information required to effectively start the document. These are the "strategic initial sections." They all overlap with some common elements, and those elements are best explained as the items needed in a Foreword and a Summary. Thus, an Executive Summary, without any further sub-headings and ideally limited to one page (though not a universal rule), provides the following: problem statement (at all levels necessary to quickly justify the work done), work statement, communication purpose statement, and highlights of report. Recall that report highlights are the main points of your message, including substantive statements of primary support for your argument if a persuasive report.

Moving on from front matter to body, a reader reaches the substantive portion of the report, in its full form. Although the main points are highlighted in the Executive Summary located in the front matter, the body is the report itself (not just its Overview). The body can rightfully begin with an Introduction, followed by a hierarchy of sections arranged to communicate, explain, and emphasize the main message: make it clear, reiterate it, and support it. One final section that ideally ends the body is the references list. These are the end notes, and one of the standard, proper bibliographic formats should be used. Approved formats are usually associated with a publishing style (Associated Press) or a well-known professional organization (American Psychological Association). Alternatively, a formal report can use footnotes, and this is a discretionary matter. Lastly, some reports have both a general bibliography and a references (end notes) section. This, again, is a decision made on an individual, case-by-case basis, as appropriate.

Key sections within the body will be determined by the subject matter and project-specific details, but a handful of likely topics can be listed here, for your reference as a checklist.

Problem or Opportunity: This information ranges from high-level issues (macro) to low-level issues (micro). At the very lowest level, a report may be addressing the problem of something as small and specific as the inner diameter of a 10-inch-long pipe. At the highest level, a report may be about stopping an epidemic, preventing drastic climate change, or finding a way to send humans to Mars.

History or Theory: Some projects require introductory information on the relevant historical or theoretical background. Details for such sections may be as varied as a chronology, a literature review, or a set of equations. Sometimes, a legislative or regulatory history is needed, other times a chronology of developing technology provides useful contextual information. The goal is to provide the necessary information so that the main message of the report is presented in its context—politically, technologically, socially, industrially, economically, chronologically, and so on.

Method: The work performed must be explained so that support points drawn from the results of work are given a reliable and verifiable source. Although this may be a lengthy portion of your report, with a hierarchy of subsections, it is not the argument itself, only the underpinnings of the argument. Keep that in mind as you provide this information: method is neutral information (tasks and facts), while the subsequent analysis and interpretation produce an argument. Details vary with type of work performed. If the method involves computer modeling or simulation, the description may be surprisingly brief. Basic requirements for computer-based work are to explain (1) the software used, (2) assumptions and limitations in the software and your use of it, (3) inputs to the software, including fixed and changing variables, (4) software validation, and (5) software output expected (information form and type). If the method involves surveying, the basic requirements are to explain (1) development of survey tool(s), (2) proof of tool quality, and (3) implementation of tool, including pilot tests, focus groups, and full-scale distribution. If the method involves physical testing or other empirical efforts, the basic requirements are the following: (1) equipment and tools, (2) setup and calibration, (3) models and construction, (4) testing or study procedures, (5) data collection, and (6) data transformation, conversion, and normalization.

Argument: Needless to say, a final, formal report will nearly always contain an argument, which entails a full hierarchy of interrelated sections designed to persuasively present a thesis with support. The argument is so crucial an element of a report that it is the sole focus of Part 4 of this book. The method information, although highlighted directly above as if such information is included as an independent section, may be appropriately integrated into the argument and not truly a stand-alone report segment.

Before I finish discussing a formal report's body, I want to offer a bit of warning by revisiting the topic of flow from Chapter 4.5. As one result of poor flow, organizing the sections in the wrong order for a formal report may lead to slow reading, misunderstanding, reader frustration, report abandonment, and possible report rejection. Poor flow will come from, among the other causes outlined in Chapter 4.5, poor organization of the body, based on structural inefficiencies. Some approaches may seem expedient when first attempting to write and outline a document, but they result in poor flow and reduced rhetorical quality. The following are three structures not to follow: Story, Options, and Results.

1. **Story**. Do not tell the story of a project: If you are describing false starts, dead-ends, drama, or an evolution of a design, you are probably telling the story. It may be interesting to you, but readers have little time for your story. They need your message and argument for that message. Leave the story for your memoir or an essay written on your own time.

2. **Options**. Do not emphasize more than one option, alternative, or design. Indeed, if you are flipping back and forth between options that were considered, you may be failing to emphasize the one, chosen selection. Readers can become very confused when all the alternatives, versions, and options are presented in a see-saw manner, as if the author is recreating the thought processes and comparisons conducted. If you use a proper argument structure, mention of alternatives, options, and other candidates will occur in only two places: (1) as the inferior choices each time the preferred selection is supported by a point of analysis and data; (2) at the end of an argumentative segment, to add credibility to your work in selecting the preferred choice (recall Section 4.4.3).
3. **Results**. Do not recount the results of your methods in the order those results were obtained. Your thesis deserves full, persuasive coverage, from strongest support point, to next strongest, to final support point. An incorrect structure presents results in order of their acquisition, and this confuses readers and brings the argument into doubt. Results are often obtained in a mixed chaotic sequence of "good," "bad," "confusing," "good again," "mysterious," and so on. Skilled authors of formal reports reorganize this chaos into clarity.

Having skillfully prepared the document's front matter and body, you can now finish your document with back matter. Back matter comprises all documentation considered helpful to readers but disruptive to the flow of the body (see Chapters 1.4 and 3.4 for a full explanation and examples). The documentation can be divided into appendices (created by author) and attachments (created by others). If an index is prepared to help readers, it can be included as back matter.

5.1.10 Journal Article or Conference Paper

Scholars, scientists, researchers, experienced business and industry managers, theorists, empiricists, analysts, creators, inventors, and other inquisitive individuals with expertise and knowledge may desire to share their ideas, findings, and theories in a professional journal covering a specialty relevant to the material. Professionals use such journals to share and exchange information, to communicate and promote their points of view, and to engage in a long-term dialogue with similarly curious and knowledgeable people, so ideas can be critiqued, revised, and improved. Importantly, journal articles are different from most of the other genres considered in this book because they do not have specific recipients who must receive the information to keep work moving forward. That is, a journal article is not an effective genre for communicating directly to another professional or organization on some matter of vital importance to them. Reading of journal articles is optional, and you cannot expect that another person will read a journal article. (They may do so regularly, however, and they may be required or motivated to do so by a supervisor even. Still, that is not enough guarantee that a journal article will be read closely by an intended recipient.) For day-to-day operations and making things happen within an organization or across organizations, a report must be addressed to a specific, suitable person, conveying its communication purpose right at the start. Thus, for communicating a purposeful report within an organizational context and advancing activities within that organization, the other genres serve correctly, such as email, memoranda, and letters.

Although not ideal for reaching a specific recipient within an organizational context, journal articles serve a communication purpose consistent with the principle (emphasized in Chapter 3.1) that technical reports in ABERST are written for a reason that encompasses a work objective in an organizational context. Journal articles have a tightly defined reason in that a professional is finally at a point in the work that sharing some findings, theories, or new ideas is justified, without the worry of the workplace and the various documents that are needed to keep things going there. The purpose with a journal article is to present information to various persons (mostly unknown) outside one's organization to assist and advance the collective state of knowledge, as well as the author's prestige, the author's organization, and the author's professional affiliations (field, discipline, or specialty).

Before I break down the structure of a journal article, I should clarify that the genre usually refers to several similar types of documents, and *article* is commonly yet confusingly used to refer to all of them. In addition to an article, journals also include papers, research notes, letters, tutorials, and rapid communications. Here I cover the two prominent variations, articles and papers.

The lessons throughout this book should be of some assistance if and when you need to write one of the four remaining minor variations.

Papers, or, scientific papers, are used to support conference proceedings. At a conference, a speaker may give a talk, and the paper is the published version of the talk in the conference proceedings. These often match the structure of articles. Thus, this section focuses on journal articles, as well as conference and scientific papers, as the genres overlap. I use *article* to refer to both journal articles and conference papers.

Articles are written to provide either (1) a survey or review of prior research or (2) results of original research. The suitable time to write an article is of utmost importance. The following are appropriate moments:

- After completion of some high-quality science or research
- Upon realizing that one can report a significant contribution to one's field
- When one has a strong desire to communicate to specialists in one's narrow research area
- If one's approach or finding is seminal, unique, or radical

Another way to know that the work warrants an article is if any of the following describe it:

- The problem that needs redress is important
- The specialized focus is fascinating and intellectually challenging
- The method is creative and innovative
- The results will truly surprise people
- The sponsor or funding agency expects article publication

All of these motivating factors will embolden you to write and submit an article to an appropriate journal, and, if published, the article will foster the aforementioned critique, revision, and improvement of your work and ideas, instigated by comments from peers who read the article. The article contributes to continued growth, development, discovery, and evolution in your field.

As with all technical reports, journal articles have the three-part structure of start, body, and base. For the start, you must explain the context. Journal articles are similar to formal reports in that they will be based on some context—a problem or an opportunity—that motivates and justifies your work. As stated in the section just above on formal reports, problems and opportunities extend from high-level issues (macro) to low-level issues (micro), and you must explain the full picture at the start of your article (in the Introduction). This is your explanation of the significance of your work, and significance is essential if the article is to be considered worthy of publishing. Here are two "fill in the blank" options to use at the start:

1. An exciting opportunity, O, presents itself; innovation X promises to help promote O, but obstacle A stands in way. My work is directed at issue B, which is needed to overcome obstacle A, and ultimately realize opportunity O. The work's objectives, goals, and hypotheses come next...
2. A worrisome trouble, T, faces us today; frankly we must work together to solve trouble T; while we do have solution X as a hopeful option, obstacle A must be overcome first. My work is directed at issue B, which is needed to overcome obstacle A, and ultimately resolve trouble T. The work's objectives, goals, and hypotheses come next...

Expressly explaining the underlying opportunity or problem is crucial for reader understanding of the big picture and, hence, your narrow pixel-size piece of it. Sadly, most technical writers of journal articles rarely put into words these key, introductory ideas:

- The problem underlying the research
- The objectives and questions driving the investigation
- The significance of the research

My colleague Dwight Stevenson expressed it this way: professionals in ABERST rarely feel the need to state "the obvious." But it is precisely the obvious that helps all readers to understand how all the pieces fit together, when projects are complex, innovative, and unfamiliar to them.

To alleviate this common article ailment, one of the two text samples above, which provide the "obvious" context for the work, should be placed into the Introduction. To complete the Introduction, a few more standard pieces of information are normal, following your explanation of the opportunity/problem:

- Literature Review (up to the place that knowledge stops)
- Research Focus and Aims (your objective, goals, hypotheses, and questions to answer)
- Forecast of Remaining Sections (self explanatory and optional)

Literature Review is a quick recount of the important stepping stones that preceded your own effort. This may include one or more lines of research upon which your work builds. If multiple lines, they transect at some point to serve as multiple precursors of your work. You must emphasize the point that the previous research and efforts leave place for your work to go one step further or take a new direction.

Research Focus and Aims allows you to introduce both focus (your narrow topic) and aims (expected outcomes) of your research. One can imagine a variety of ways to express these ideas. One or more of the following terms should be applicable to your focus and aims:

- Objectives
- Goals
- Hypotheses
- Questions to be answered

No matter how you say it, you are covering in the Introduction "why" you did something, not "how" or "what," which come later. The final portion of an Introduction, Forecast, is optional and may be determined by the journal's guidelines. It is a simple listing of the major sections into which the article is divided, in the order they appear.

Although an article's foundational beginning, the Introduction is usually the second section after the Heading. The first section following the Heading is the Abstract. Thus the start encompasses three parts: Heading, Abstract, and Introduction. (Some publishers will ask authors to create a Key Words list also, and that list may be placed alongside the Abstract, before the Introduction.)

The Heading should contain the basic document identification data: title, authors, affiliations, date, etc. Follow the journal's expectations for precise format and style of the Heading. You can look to the journal for author guidelines or at samples in existing issues of the journal.

The Abstract is a short overview of the whole article. It is similar to a formal report's Executive Summary. It is read first, but I advise writing it last. After you write your full article, especially the body sections, you are in a place to combine the main points of each section into an Abstract. That is the simplest way to produce an Abstract, and you can also revisit the guidelines for an Executive Summary in the preceding section.

The body of the article will be organized into these standard sections:

- Method/Experimental
- Results
- Discussion
- Conclusion

I offer here some thoughts on these sections. In Method, you want to explain the intellectual process for problem-solving, not a chronology of laboratory or other activities. (Do not submit a retyped version of your laboratory or field notebooks.) This is best stated as a general approach and justified as appropriate with respect to previously explained objectives, goals, and questions. Examples of general approaches include laboratory studies, computer simulations, experiments, field research, and mathematical computations. After a general introduction, the method is elaborated upon, with details, such as equipment, experiment setup, materials, chemicals, procedures, locations, specimens, participants, statistical tools, equations, and physics/math involved. Communicating your method gives an implicit argument that results are valid. It allows repetition and verification of work. Importantly, you do not include any information related to results or analysis of those results. Such information would disrupt flow, as it would neither fit under the "Method" heading nor would be expected by experienced readers.

As a heading, "Method" is replaced with "Experimental" when your work effort comprises solely laboratory work; "Method" is more general. Many people have seen "Methodology" as a section heading. This word means "study of method," so "method" is more concise and accurate. Very few authors are studying their method. Rather, they are using their method to study something else. "Experimental Method" might be the best heading when your work comprises primarily experiments or laboratory studies.

The Results should be placed into logical order, linked to objectives, goals, hypotheses, and questions stated earlier. You must explain all results in sentences, and complement those sentences with figures and tables (see Chapter 5.2). A basic premise that helps in creating clear figures and tables is to think of them as "extras," while designing them to be self-contained and clear on their own (without need to refer to surrounding text). Importantly, you must NOT mix in any analysis of the results; Analysis is reserved for the next section, Discussion. When analysis is inserted into the Results section, a major confusion arises: facts become obscured by opinions, rendering both dubious in readers' view. I offer two more pieces of advice for presenting results:

- Present selective results (not all data you acquired)
- Present results that enable you to perform analysis and draw conclusions

The Discussion section provides your analysis of the results. Begin with an overview of your analysis, with a reminder for readers of both the original opportunity/ problem and your focus and aims. After that, you can provide details for each key result presented in the preceding section, using any of the following analytical methods:

- Interpretation
- Comparison
- Evaluation
- Process of Elimination

Sometimes, the results are not what you wished for. Nonetheless, you still feel one of the other reasons for writing an article warrants the effort. In such cases, your earnestness will impress readers. In other words, whichever way your analysis goes, it will impress readers, and your honesty is of paramount importance. (Of course, if having obtained poor or unexpected results precludes your desire to write and publish an article, by all means wait until you can conduct new and improved work.) Assuming that publishing is warranted despite the disappointing results, you should explain your thinking. Something like the following may be helpful: "We were wrong and the results were not what we expected, but . . . this is interesting because . . . ." Or you might write, "Our work is inconclusive . . . we must do more work, using this new hypothesis . . . ."

Ultimately, the analysis just presented allows you to formulate a primary conclusion and possibly secondary conclusions, which are your truly important findings. And this is the subject matter for the Conclusion section. The Conclusion section contains your main message—the principal conclusion drawn from the research along with a concise review of secondary conclusions. This main message is the reason you are writing the article. It was conveyed to the reader in the Abstract, where you both announced your purpose of communication and provided your main message. In this sense, your article is a satisfactory, purpose-based document, which started well and followed through with details to support the main message. In the final Conclusion section, the main message is conveyed as the culmination of the research and analysis that produced it, inductively. You might recall that I have explained that while investigative work in ABERST is inductive, the writing is ideally deductive (Section 4.3.2). The journal article is one of the few genres that is structured inductively, to match the process of the work. Thus, the Abstract is vital to effective communication, so the readers are not left dangling in uncertainty until the end.

In the Conclusion section, furthermore, you also want to refer to your objectives of research and explicitly assess whether those were fulfilled or not fulfilled. Optional items for this section include your plans or suggestions for further research, such as one or more of the following:

- Next stage of same research
- Parallel or complementary research
- Follow-up or corroboration research
- New objectives to pursue or questions to answer

One last optional item is to discuss the potential application of the research and your main findings. This would encompass returning to the significance (context) of the research in some sense to emphasize the real-world need for answers, solutions, knowledge, a best option, policies, standards, technology, and so on. Ending with this kind of reminder of the opportunity or problem motivating the research is an excellent technique. Other than these two optional items, no other new ideas should appear in the Conclusion. The first two pages of a conference paper are provided as a sample in Figure 5.1.12. The Abstract, Table of Contents, and Introduction are prominent on the first page, under the extensive list of co-authors. By the second page, the Methods section is well underway, along with integration of three visual aids, a fundamental topic to which we turn next.

Project Zephyrus: Developing a Rapidly Reusable High-Altitude Flight Test Platform

Hunter Hall
University of California, Berkeley
Department of Physics
847-975-3660
hall@berkeley.edu

Keenan Albee
Massachusetts Institute of Technology
Department of Aerospace Engineering
917-531-4411
albee@mit.edu

Benjamin Donitz
University of Michigan, Ann Arbor
Department of Aerospace Engineering
818-809-9407
benjidon@umich.edu

Seth Eisner
University of California, Los Angeles
Department of Computer Science
805-217-3052
seisner@ucla.edu

Leon Kim
Columbia University
Department of Mechanical Engineering
717-655-0455
lmk2194@columbia.edu

Dakota Pierce
Massachusetts Institute of Technology
Department of Aerospace Engineering
805-338-4136
dpfor3@mit.edu

Divya Srivastava
Rutgers University
Department of Mechanical Engineering
732-632-7069
divya.srivastava@rutgers.edu

Yvonne Villapudua
University of California, Irvine
Department of Mechanical Engineering
909-244-2662
yvillapu@uci.edu

Adrian Stoica
NASA Jet Propulsion Laboratory
347E: Robotics and Mobility Systems Section
818-354-2190
adrian.stoica@jpl.nasa.gov

Abstract—**Reusable, inexpensive, and rapid testing of payloads in near-space conditions is an unfilled niche that could speed development of flight-ready projects. With growing availability of commercial off-the-shelf (COTS) hardware and open source software, high-altitude ballooning (HAB) has become a viable option for rapid prototyping of ideas in near-space conditions. A product of the Jet Propulsions Laboratory's (JPL) collaborative Innovation to Flight (i2F) program, the Zephyrus system is a proposed HAB solution for high-altitude testing and atmospheric observation, a cheaper and easy-to-use alternative to traditional HAB systems. The Zephyrus system permits testing of small- to mid-sized instruments in the upper atmosphere (approximately 35,000 m maximum altitude) using a reusable common architecture. Over the summer of 2017, the i2F team demonstrated all of these key features of the Zephyrus platform via four test flights. Zephyrus I, II, III, and IV were successfully launched, tracked, and recovered over groundtracks of up to approximately 50 km and altitudes of approximately 30,000 m by a team averaging eight individuals, while accomplishing a majority of the intended science. This included a turnaround time of less than two weeks between two flight tests with differing instrumentation. Lessons from the Zephyrus I flight were successfully used in the Zephyrus II flight to speed up launch time and operations procedures (e.g. rigorous pre-launch preparation). Likewise, the Zephyrus III flight brought the cost of relaunching a mid-sized payload to 28,500 m down significantly. Many diverse instruments and experiments were tested during the Zephyrus missions (e.g. inflatable origami reflectors, flexible solar panels), demonstrating the system's versatility. From concept to first flight of Zephyrus I in just five weeks, followed by turnarounds as short as two weeks, the team demonstrated that it is entirely possible to bring to bear cost-effective, hobbyist HAB techniques in meaningful high-altitude testing using a new reusable test platform.**

978-1-5386-2014-4/18/$31.00 ©2018 IEEE

Table of Contents

1. Introduction

Currently, no infrastructure is in place at JPL for small research teams and individuals to collect data from the upper atmosphere at small operational costs and without developing an independent HAB system altogether. While quite a number of HAB systems already exist at JPL, most are for larger and more expensive missions that usually require months of preparation for ight readiness. The scope of these systems overshoots smaller research teams and individuals, who may be operating on thin budgets and perhaps with no prior HAB experience. Notably, high-altitude ballooning has undergone soaring hobbyist popularity in recent years, with groups like UKHAS[1] and others lowering the barrier

1

Figure 5.1.12 Sample Conference Paper (Page 1)

to entry with technical guides and open source software. In response, Zephyrus, a highly reusable, low-cost high-altitude balloon launch system, was developed at the JPL by summer students Benjamin Donitz (University of Michigan), Hunter Hall (University of California, Berkeley), Keenan Albee (Massachusetts Institute of Technology), Leon Kim (Columbia University), Divya Srivastava (Rutgers University), Seth Eisner (University of California, Los Angeles), Yvonne Villapudua (University of California, Irvine), and Dakota Pierce (Massachusetts Institute of Technology) under the mentorship of Dr. Adrian Stoica in the i2F program. The Zephyrus HAB system paves the way for researchers at JPL and beyond to take advantage of affordable high-altitude testing. To accomplish this, the following rough design parameters were sought for the Zephyrus system:

- Fast turnaround times: $<$ 2 weeks from payload selection to test
- Low cost: $<$ \$1,500 in recurring hardware costs per launch
- High reliability: Redundancy in all critical systems and good design intent throughout

Zephyrus has been launched four times from the Mojave Desert, each time carrying a differing instrumentation suite and an assortment of experimental payloads. Zephyrus is capable of flying payloads up to 3.62 kg in weight with integration times as low as one week and incremental material cost as low as \$800. It is also capable of carrying more massive payloads with longer lead time to obtain proper approvals from the Federal Aviation Administration (FAA) whose regulations on unmanned balloons limit system mass to 5.44 kg[2]. The Zephyrus launch system also includes operations procedures developed through the three primary phases of flight, launch, and recovery; it is a full service launch system for researchers seeking affordable and accessible high-altitude testing of instruments and experiments.

This paper details the methods and results obtained from the design and four successive flights of the Zephyrus system. It primarily focuses on the design and development of the avionics, communications, mechanical, camera, and operations subsystems. Each has been flight-proven in the relevant flight regime, ranging from 43°C on the surface to -22°C near apogee and pressure environments ranging from 101.3 kPa to 10.0 kPa. The subsequent goals and lessons learned of test demonstrations are demonstrated through the results of Zephyrus I, Zephyrus II, Zephyrus III, and Zephyrus IV.

2. Methods

Avionics

The avionics subsystem consists of two microcontrollers (Arduino MEGA and Arduino UNO), and various sensors. Two independent flight computers were used to execute a payload deployment system (via nichrome wire cutters seen in Figure 1)[1], sensory equipment, and data logging. The primary flight computer (Figure 2) consists of an Arduino MEGA, HABduino shield, and a relay shield; it is responsible for logging data to a MicroSD card and triggering the deployment system for payloads. The HABduino shield provides GPS acquistion and telemetry. The sensors incorporated into the primary flight computer are the following (as seen in Figure 2):

- Adafruit 9-DOF Inertial Movement Unit Breakout (IMU)

[1]Nichrome wire cutters were internally designed and manufactured at JPL by Michael Pauken (JPL-353K)

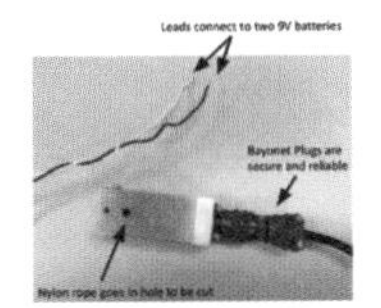

Figure 1. Nichrome wire cutter

Figure 2. Zephyrus II, III, and IV primary flight computer

- Adafruit BMP280 Barometer/Thermometer
- MicroSD Card Breakout Board with MicroSD card

The IMU measures data from its 3-axis gyroscope ($\pm$250, $\pm$500, or $\pm$2000 degree-per-second scale), 3-axis compass ($\pm$1.3 to $\pm$8.1 gauss magnetic field scale), and 3-axis accelerometer ($\pm$2g, $\pm$4g, $\pm$8g, $\pm$16g selectable scale). The BMP280 measures barometric pressure ($\pm$1 hPa) and temperature ($\pm$1.0°C). The MicroSD card records all data taken from the sensors to a comma separate values (CSV) file. The primary flight computer is powered by two double-A lithium ion batteries[2]. The secondary flight computer, an Arduino UNO, was utilized for an in-flight flexible solar cell panel experiment, demonstrating the use of multiple data-logging microcontrollers, if desired. The secondary flight computer, when used, was powered by a single 9V lithium ion battery and incorporated an independent MircoSD card shield and MicroSD card.

[2]Lithium batteries perform more effectively than alkaline batteries in colder environments

(a) IMU

(b) BMP 280

(c) MicroSD card breakout board

Figure 3. Internal avionics components

2

Figure 5.1.12 Sample Conference Paper (Page 2)

Part 5

Chapter 5.2

Enhance with Visual Aids

Visual aids are helpful and necessary in ABERST documents. By "visual aids," I refer to graphs, tables, diagrams, photographs, maps, schemata, site plans, elevations, charts, matrices, timelines, decision trees, screen shots, and any similar item. Visual aids are integrated into documents and are more common to some genres (formal report, proposal, and journal article) than other genres (business letter, email, and prospectus). They complement your text to clarify and illustrate important information, and they are often essential in support of your argumentative points. Visual aids contribute to fulfilling the communication purpose. They are purposeful, just as the whole report is purposeful.

I find it practical to think of them as their own genre, in that (1) they must be designed to be clear and informative as stand-alone items, separate from their "parent" documents, and (2) they are often used on their own in place of additional documentation to support a brief talk, poster display, speech, elevator pitch, one-to-one meeting, face-to-face inquiry, progress report, and so on. They are designed as stand-alone items because readers often simply glance at the visual aids in a report rather than read all the pages of text, and, as just stated, for some oral communication tasks, visual aids are the sole portion of documentation for the report. This acknowledgement and understanding of their stand-alone nature guides the design of visual aids.

My colleagues at the University of Michigan have covered this topic skillfully. Chapters in Olsen and Huckin's book, *Technical Writing and Professional Communication*, and Mathes and Stevenson's book, *Designing Technical Reports*, provide flawless guidelines on designing and using visual aids of all kinds, for various purposes. In addition, Edward Tufte has written entire books devoted to this single topic, and these are considered some of the definitive treatments of the art of visual display. These other scholars have informed my outlook.

My spin on this well-studied topic has three parts: advantages, objectives, and techniques. Before I teach the general objectives and the specific techniques of good visual aid design, I like to first motivate people to seek out the benefits of visual aids. Indeed, visual aids offer some very important advantages to a technical communicator. These advantages, moreover, are the reasons professionals in ABERST bother to toil over the details of designing a visual aid in the first place. One or more of the advantages discussed below (Section 5.2.1) will enable a visual aid to promote the overall communication purpose and improve your technical communication. Once motivated, you can master the design objectives (Section 5.2.2) and the techniques for 13 common types of visual aids (Section 5.2.3).

5.2.1 Advantages of Visual Aids

Built into the following four advantages are all the benefits to be realized from designing and using visual aids in a technical document.

- Deliver and Reinforce the Message
- Provide Details Better than Text
- Facilitate Comprehension and Retention
- Enable Quick Referral

Various visual aid types (covered later in this section) can be used to realize these advantages, and some visual aids achieve two or more advantages simultaneously. Each advantage is explained below, with a few examples. Every single visual aid considered for a report must provide at least one of these advantages as its underlying objective; if it does not, you can leave it out of the report. Authors occasionally sneak in a visual aid that offers no advantage and fails to support the communication purpose. Maybe it looks cool, maybe it doesn't. Either way, it ends up being distracting and useless. It is mere fluff. I have seen many that did not help readers. Sure, the author created it with data or information he acquired, but it only proves that some work was done by the author;

Remember, not all work needs to be documented in a purposeful report; only a selection of the work that is relevant to the purpose (Chapter 3.2). If a visual aid does not support the communication purpose and enhance the report in one of the four aforementioned ways, it impedes flow and offers no additional value.

To sum up this general introduction of advantages, I offer this maxim: Just as each report has a singular purpose of communication, so too must a visual aid have a pivotal purpose. You provide each visual aid for a reason. Each visual aid yields one or more underlying advantages and provides either a singular or a multivariate message that promotes the communication purpose.

Just as each report has a singular purpose of communication, so too must a visual aid have a pivotal purpose

5.2.1.1 Deliver and Reinforce the Message

As you learned previously, each report must have a main message (Chapter 3.1), the conveyance of which serves as the report's purpose. Very often a visual aid is crucial for communicating that message, that is, your main point. Alternatively, when the main message may not be captured perfectly in a single visual aid, a collection of visual aids may contribute to delivering the main message. Examples are shown in Figures 5.2.1–5.2.3, as excerpts from actual reports with preceding text to introduce the visual and the main message.

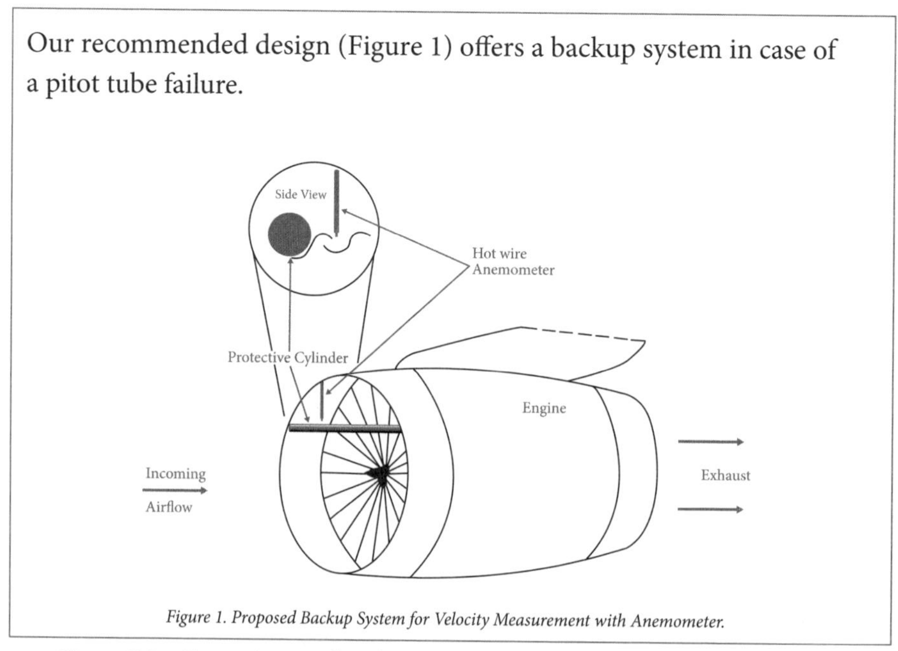

Figure 1. Proposed Backup System for Velocity Measurement with Anemometer.

Figure 5.2.1 Example Visual Delivering Main Message: Sketch with Two Views

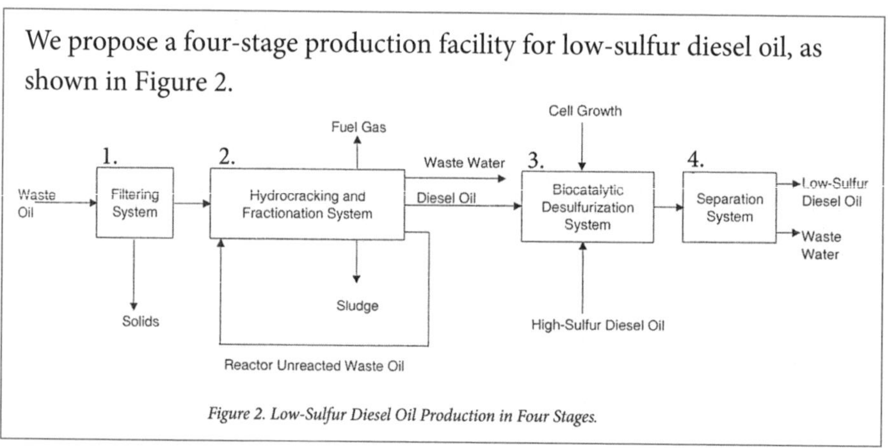

Figure 2. Low-Sulfur Diesel Oil Production in Four Stages.

Figure 5.2.2 Example Visual Delivering Main Message: Block Diagram

Our proposed methanol production facility has been optimized for the highest margin for a feedstock flow rate of 9,000 pound-mole per hour, based on calculated design efficiency in the range of 6,000 to 9,000 (see Figure 3).

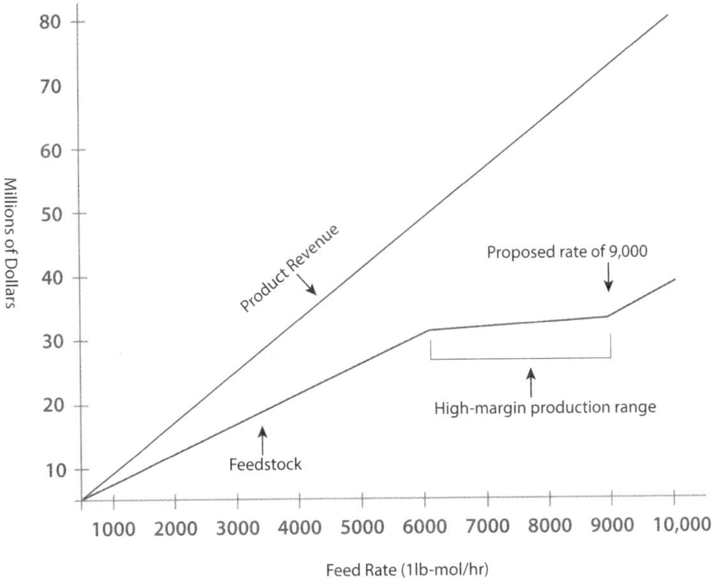

Millions of Dollars
80
70
60
50
40
30
20
10
Product Revenue
Proposed rate of 9,000
High-margin production range
Feedstock
1000 2000 3000 4000 5000 6000 7000 8000 9000 10,000
Feed Rate (1lb-mol/hr)

Figure 3. Profit analysis for proposed methanol production facility with margin increase at 6,000 pound-mole per hour (lb-mol/hr).

Figure 5.2.3 Example Visual Delivering Main Message: Plot on Graph

As in these examples, a visual aid that delivers and reinforces the report's main message is quite powerful, as it may show (1) exactly how a proposed new design or engineering solution looks, (2) the trends in data that were analyzed to yield the answer to a crucial question, or (3) the ideal choice among several candidates studied. As I discuss in Subsection 5.2.1.3 below, the advantage of delivering your report's main message both verbally (text) and visually (visual aid) is tantamount to the venerable "1-2 punch" of boxing lore. You are delivering the main message into your reader's head with two approaches. You say it and you show it. This is a communication knockout.

Only one or a few visual aids in your report are used to invoke this first advantage, so the majority of them serve the other three underlying advantages. I look at a second advantage next.

5.2.1.2 Provide Details Better than Text

Reports have numerous, important secondary and ancillary points that serve to support the main message, and visual aids are often very useful for presenting those. These rhetorical points may be either informative (for example, the engine's stator has twelve blades) or persuasive (for example, alpha is faster than beta), and visual aids follow these two options as well. As I said above, each visual aid itself must have its own clear point, which may be either informative or persuasive. It does not need to be the main message of the whole report, but it must be nonetheless a necessary point relative to one of the report's rhetorical points.

To illustrate a point in detail, a visual aid may be more efficient than text itself. It may be a better way to offer the details. In some cases, such as lots of data in a graph, using text would take too long and be excessive. In other cases, while text can describe a physical object, a diagram or photograph can show it exactly as it looks. In still other cases, complex relationships and correlations can be fully captured in detail in a decision tree or multivariate line graph rather than merely summarized by text. Some examples of visual aids fulfilling the second advantage appear below.

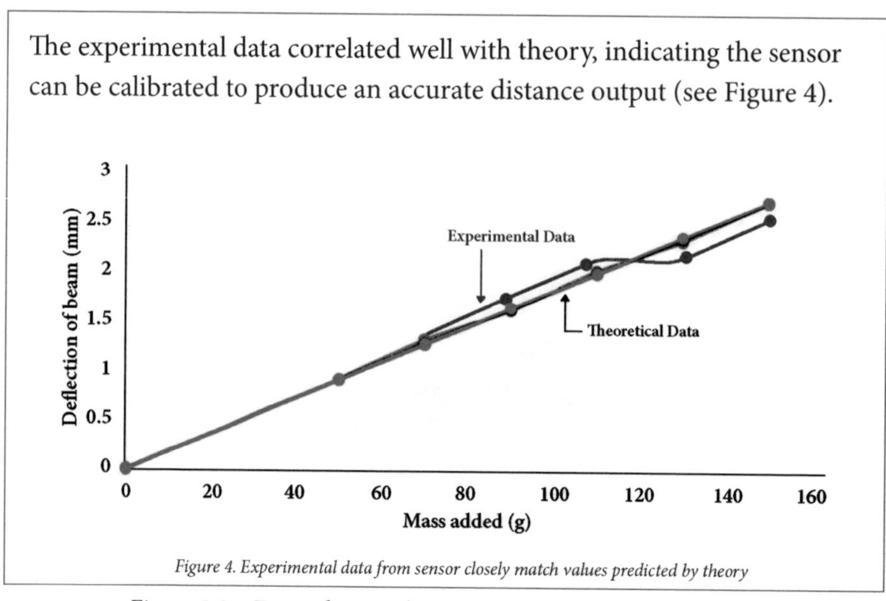

Figure 5.2.4 Example Visual Providing Details: Extensive Data

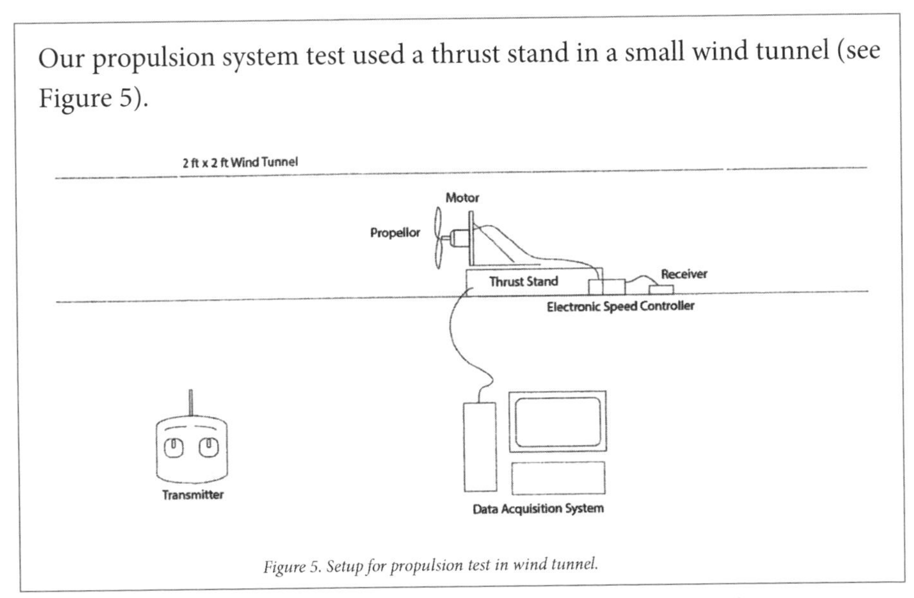

Figure 5.2.5 Example Visual Providing Details: Equipment Setup.

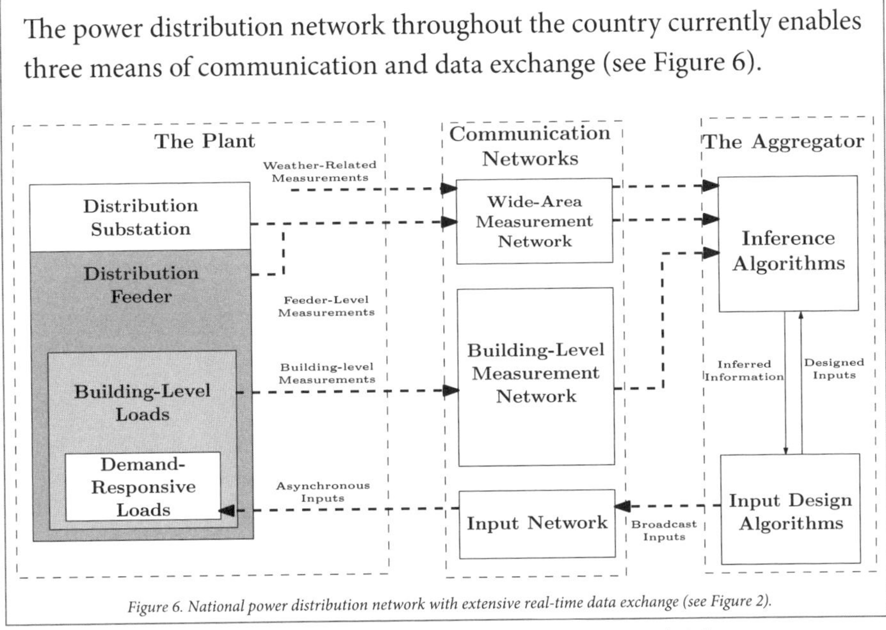

Figure 5.2.6 Example Visual Providing Details: Network Schematic

Figure 7 shows the aliasing problem produced when the sampling rate ceases to fulfill the Nyquist criterion, producing output values that no longer match the input values.

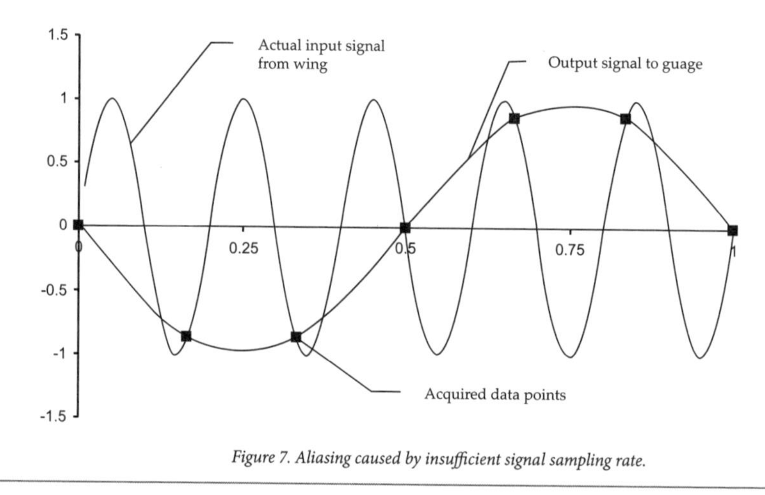

Figure 7. Aliasing caused by insufficient signal sampling rate.

Figure 5.2.7 Example Visual Providing Details: Signal Processing Standards

The experimental vertical axis wind turbine represents a radical departure from the standard electricity-producing wind turbine commonly based upon the horizontal axis design originally introduced by Vestas A/S in 1980 (see Figure 8).

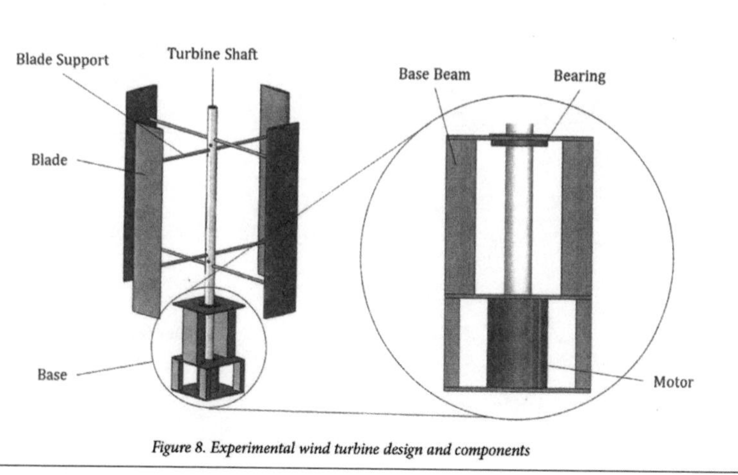

Figure 8. Experimental wind turbine design and components

Figure 5.2.8 Example Visual Providing Details: Device Components

Based on theoretical calculations, the airplane has a functional flight ceiling of 52,000 feet (Figure 9) and a maximum velocity of 880 feet per second (ft/s).

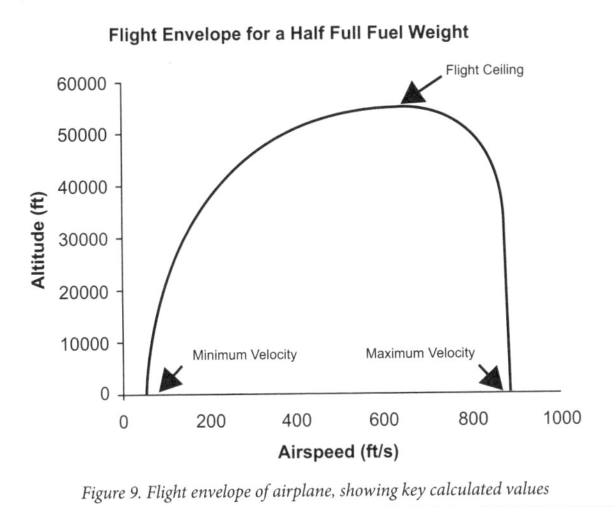

Figure 9. Flight envelope of airplane, showing key calculated values

Figure 5.2.9 Example Visual Providing Details: Calculated Data with Maxima

The timeline for the design and testing of the prototype proceeded as shown in Figure 10, with testing delayed by 2 weeks in March.

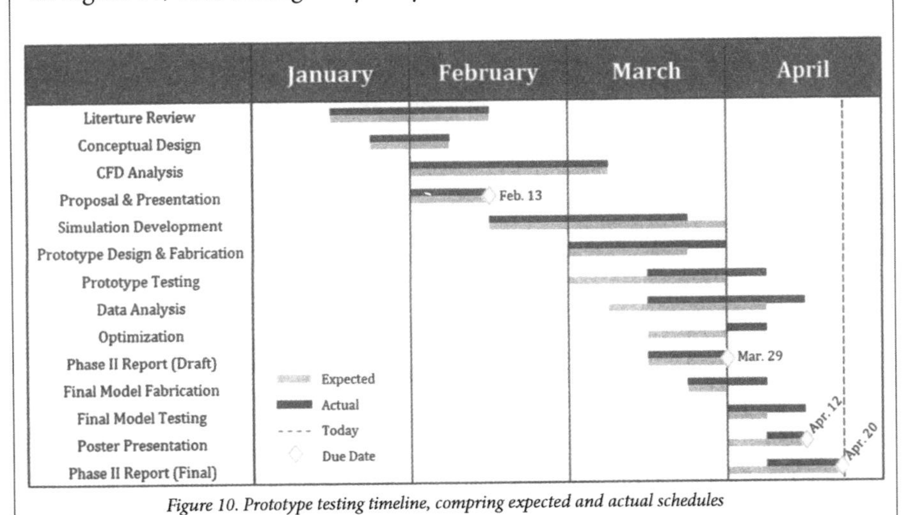

Figure 10. Prototype testing timeline, compring expected and actual schedules

Figure 5.2.10 Example Visual Providing Details: Project Timeline

As you can see in the preceding examples, when a visual aid is providing details better than text, it tends to do so in some standard ways that you are no doubt familiar with and for which scholars have established descriptive terminology:

- Iconic Illustration: show how a physical thing looks
- Quantitative Display: provide a compact summary of data, possibly several series of data
- Operational Summary: offer conceptual image of systems and relationships
- Enhanced Map: show a location of something, often combined with data
- Explanatory Summary: simplify explanation of procedures, instructions, or events

As is the case with advantages, some visual aid types transcend the boundaries and serve two or more functions as well. An organizational chart is an operational summary that is often considered an iconic illustration of the hierarchy. A flow chart, a decision tree, and a chronology all serve as explanatory summaries but can be operational summaries as well. A data matrix is both a quantitative display and an operational summary.

5.2.1.3 Facilitate Comprehension and Retention

This advantage derives more from the nature of the human brain, than it does from any inherent element of the visual aid itself. Simply put, comprehension and retention, which derive from the functions of attention, sensation, cognition, and memory, are complex mental processes that involve the intersection of external stimuli, the five sensory systems, and the brain. Although all these functions, processes, and interactions continue to pose puzzles for researchers and scientists, we have realized that learning comes from many pathways. Focusing on a single learning style may be a limiting approach. A better outlook is to be open to different methods and an overlap of those methods. All of us can learn in numerous ways, including verbally (by reading), auditorily (by hearing), visually (by seeing), and tactilely-kinesthetically (by touching, moving, and doing), as appropriate and available to each situation. We can apply this knowledge to technical writing.

Because technical documents are predominantly text, readers are being asked to put their verbal learning skills to the test. By interspersing visual aids throughout the document, you offer visual learning stimuli to supplement the verbal portions. In a very real sense, you enable one learning method to reinforce the other, and this increases your chances of success in comprehension and retention with an increased number of readers. In other words, a visual aid provides that additional stimulus trigger of visual learning (seeing), to complement verbal mastery (reading), and together they enhance comprehension and retention. Consider a few examples, below.

If you want people to understand and appreciate where the nearest streetcar stop is in relation to the main library, you can tell them to exit at Station 16 and walk 1 block to the corner of Ninth and Vine, but you can also show them on a focused map (see Figure 5.2.11). Similarly, if you want people to grasp and recall that U.S. unemployment dropped fiscal from a high of 10% in late 2009 to a low of 4.3% in 2017, you can display the data on a graph (see Figure 5.2.12). It may be hard for employees to remember all the significant steps of setting up and reading a pressure manometer, so a diagram of the device with key connections and components (Figure 5.2.13) provides the extra reinforcement and promotes memory and learning.

You can easily get around the downtown with the Connector, such as visiting the public library (Figure 11).

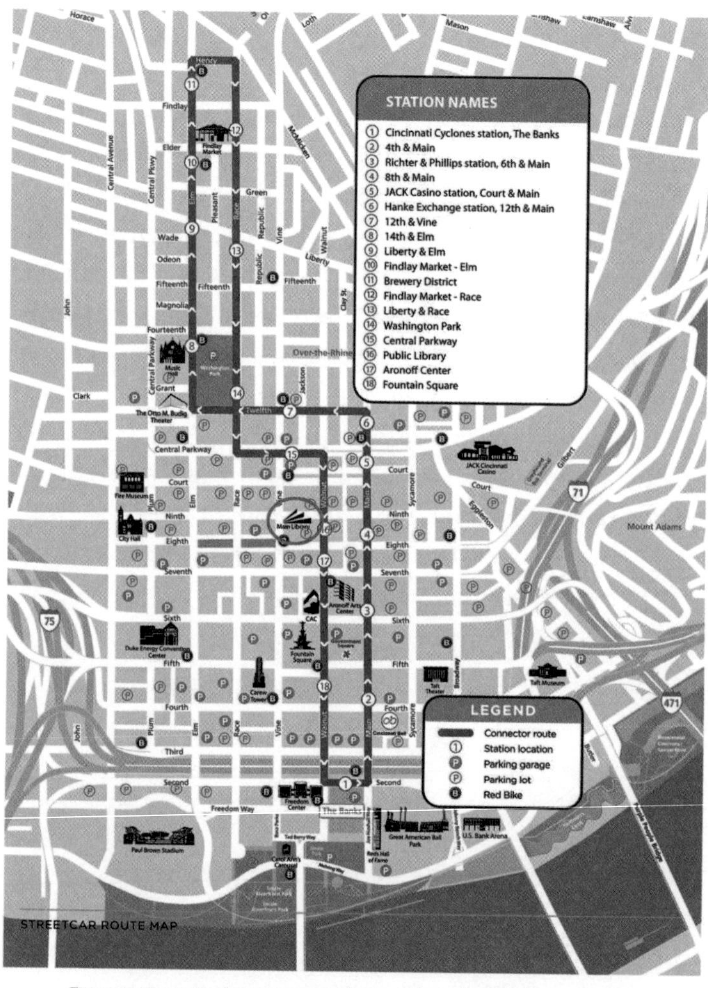

Figure 11. Connector Streetcar map, with Main Library highlighted near Station 16

Figure 5.2.11 Example Visual Facilitating Comprehension and Retention: Focused Map

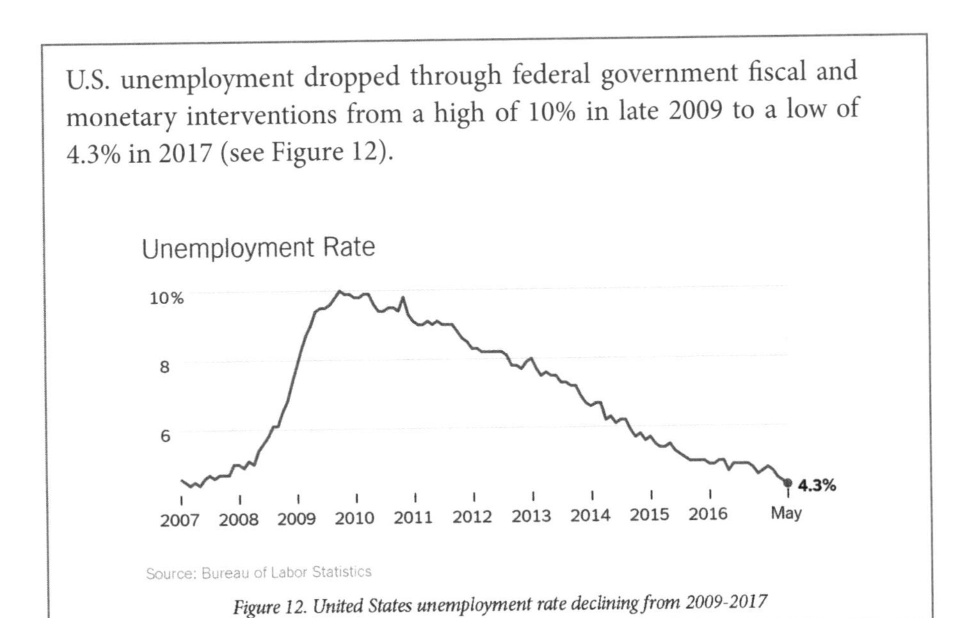

Figure 5.2.12 Example Visual Facilitating Comprehension and Retention: Data Plot

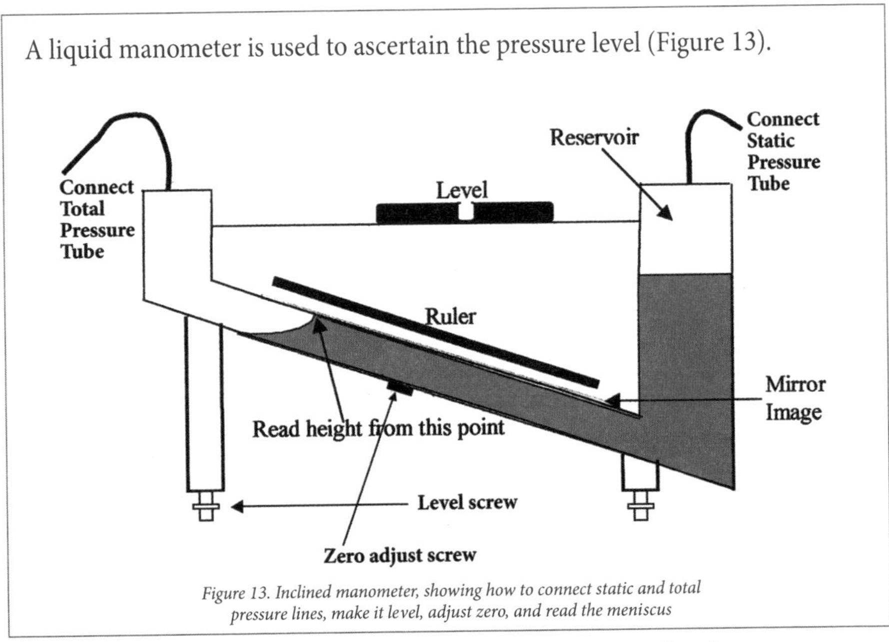

Figure 5.2.13 Example Visual Facilitating Comprehension and Retention: Device Drawing

With this third advantage, you should plainly see that almost all visual aids offer two or more advantages because comprehension and retention attach to nearly every visual aid, by the fact that the visual aid is triggering visual learning to complement the verbal learning from the text. Regardless of whether the visual aid is delivering the report's main message or providing details, it will help to facilitate a successful cognitive process in nearly all readers.

5.2.1.4 Enable Quick Referral

This final advantage is not usually foremost in authors' minds, so it arises as a bonus. You design a visual aid to gain one of the three other advantages, and this one comes along for the ride. When readers seem to recall that some information may be in a report they read previously, and they access that report to find the desired information, they are more likely to find it when it is evident in a visual aid than only in text, for several reasons. One, a visual aid is more likely to be imprinted on readers' minds than any particular passage of text. Two, a visual aid is readily apparent on the page in that it stands apart from the text via white space and can be seen easily while flipping pages. Three, a visual aid reveals itself as a particular type of visual aid upon a very quick glance, whereas all paragraphs look the same, and the visual aid type may connect with the reader's memory.

Therefore, if readers refer back to a report after having read it completely the first time, to either review some information, check something, or extract a key point, they are most likely to consult the visual aids. If readers remember something from a report yet want to double check that information, they will be able to find it faster if it is in a visual aid than if it is merely described in verbal form in the text. Here are some examples:

If you have presented an airfoil's (airplane wing's) lift force as a function of changing angle of attack (pitch up or down of the airplane), the curve of this data on a graph (Figure 5.2.14) would typically indicate the airfoil's stall point (a practical maximum for safe flight). If someone looking at your report in the future wanted to quickly refresh her memory as to the exact stall point, the visual aid you created would be the first and best place to find this information.

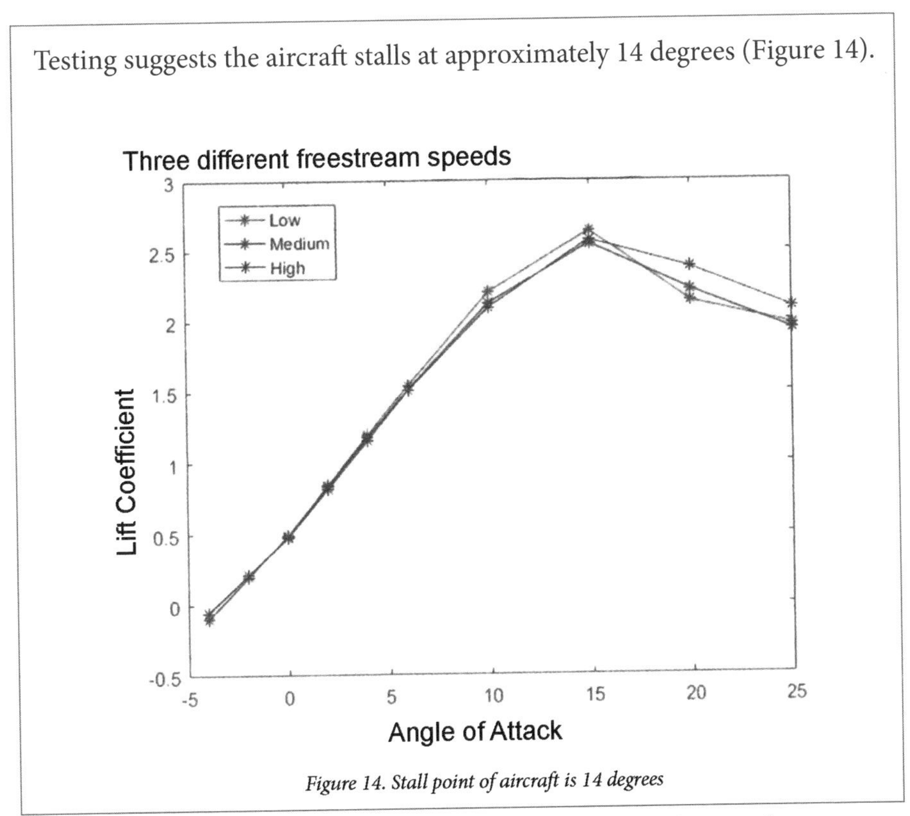

Testing suggests the aircraft stalls at approximately 14 degrees (Figure 14).

Figure 14. Stall point of aircraft is 14 degrees

Figure 5.2.14 Example Visual Enabling Quick Referral: Data Plot

If an investment advisory firm produces an annual report detailing the performance of the various investments it manages, a quick check back to a bar graph (Figure 5.2.15) created to show the four categories of investment types and their respective 1-, 3-, 5-, and 10-year returns would be the first place to look to see which type had done the best historically and which category performed best last year.

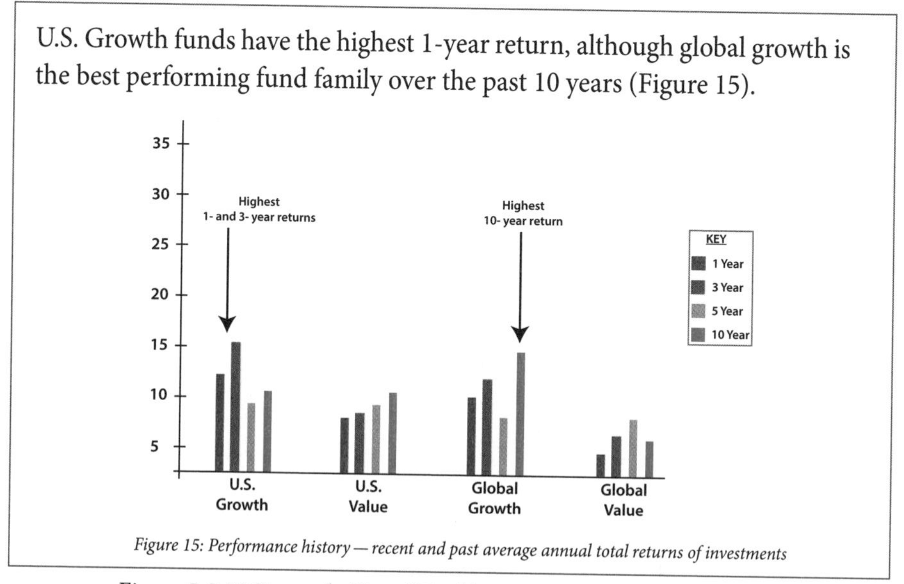

Figure 5.2.15 Example Visual Enabling Quick Referral: Bar Graph

If a nuclear engineering department designs a reactor, and a reactor core layout schematic (Figure 5.2.16) is presented in the report, this visual aid would be the ideal place someone could check to confirm the placement of Pyrex rods.

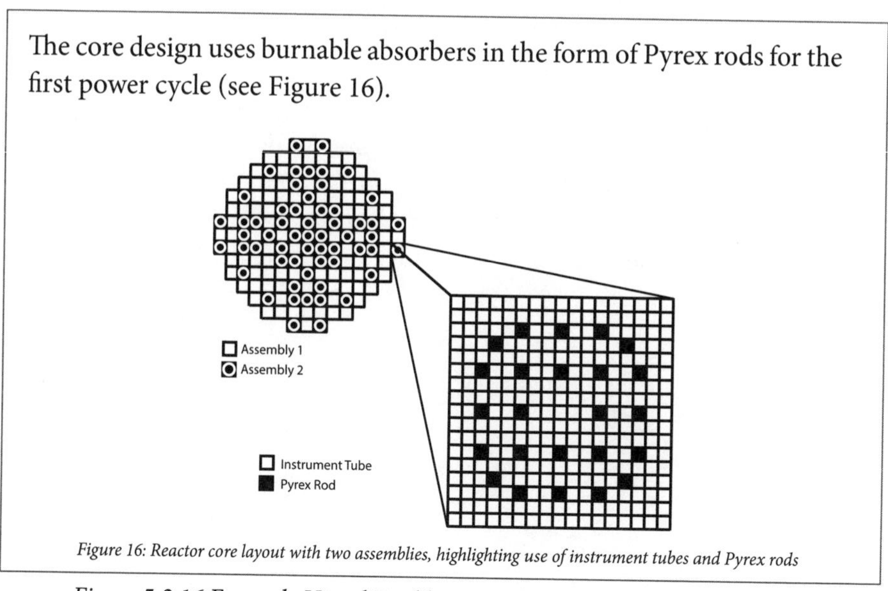

Figure 5.2.16 Example Visual Enabling Quick Referral: Design Schematic

5.2.2 General Design Objectives

With the four advantages as inspiration for creating visual aids, you are, I hope, convinced of their diverse benefits and functions and eager to incorporate beneficial visuals into your documents. For this stage of your effort, you should aim at these four design objectives:

- Efficacy
- Honesty
- Autonomy
- Simplicity

Each of these is necessary, so you can consider these the four characteristics of a well-designed visual aid. The very best visual aids accomplish all four objectives and are characterized by efficacy, autonomy, honesty, and simplicity. Each is discussed below with a few examples, while additional examples are provided in the next section (5.2.3) with an assessment of these characteristics for each example.

5.2.2.1 Efficacy

Efficacy for a visual aid is achieved when it successfully makes the point you wish it to, in that it neither obscures nor contradicts the explanation or rhetorical point in the text. To be effective, it must present what you intend for it to present, and it should either inform or persuade.

The graph below is an example of a visual aid that fails to be effective. The text explains that the new airplane design shows much promise. One of its strong features is that it is fuel efficient and can fly a long distance before refueling. In addition, to maintain its optimum fuel efficiency, it must reduce its velocity (normal for aircraft) during the flight, but not by very much. An ideal airplane should maintain its velocity throughout the flight, which is impossible, so something close is considered adequate. The visual aid created to show the airplane's performance (see Figure 5.2.17) was not effective at displaying this positive quality.

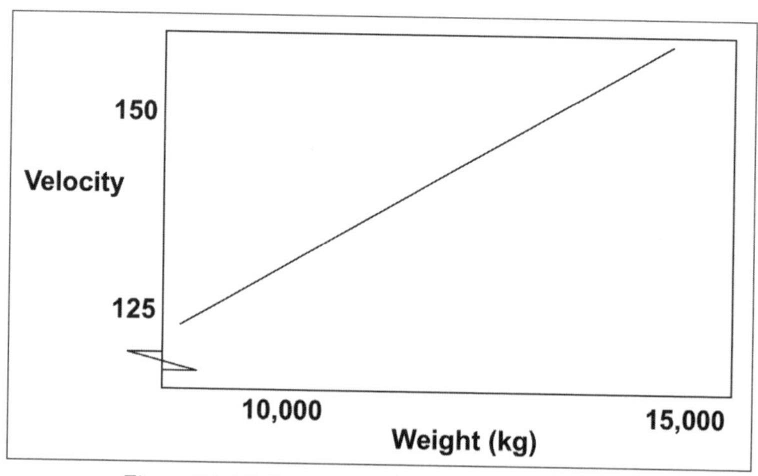

Figure 5.2.17: Optimum Velocity for Longest Range

This presents the data accurately, but it does not convey the author's message. With the single curve of data, the author's assertion of a small change in velocity is obscured and possibly contradicted. The point is to present a good design, supported by data analysis featuring a modest decline on the y-axis, not a steep one. Thus, it is an ineffective visual aid. An improved version shows a modest decline and looks like Figure 5.2.18. In addition, it includes a title that directly expresses the author's message:

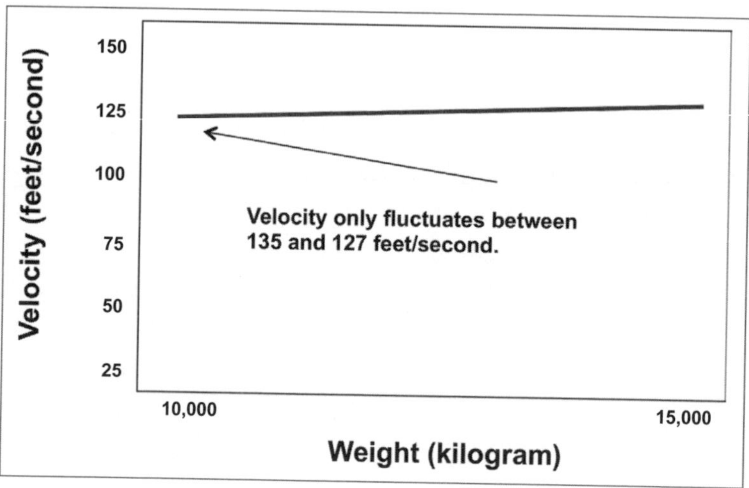

Figure 5.2.18: Airplane's Velocity Stays Nearly Level During Optimum Flight for Greatest Range

The poor version of the visual aid lacks efficacy, but it does have honesty. The data are displayed in Figure 5.2.17 accurately because of the mark indicating a suppressed y-axis. That is acceptable, but unhelpful in this scenario. Below, I discuss a version of the same visual aid that fails to fulfill the objective of honesty.

5.2.2.2 Honesty

This objective could arguably be placed first in order of importance. Above all, professionals in ABERST must do correct work and provide accurate calculations, measurements, records, assessments, and so forth. The work is not useful unless it is done correctly, and when honesty is omitted, only bad things will come eventually. Still, my experience has taught me that the vast majority of hard-working ABERST professionals are fully honest and do their work correctly, with rigor and accuracy. The problem with "dishonest" visual aids arises due to mistakes, misunderstandings, and deficiencies in design skills. Dishonest visual aids are not intentional (though cases have occurred where dishonest persons have manipulated visual aids for nefarious purposes). Thus, thinking of design as a creative process, I advise that your first design stage concern efficacy (or the result is useless), and the second stage takes into account honesty. With those two objectives fulfilled, you will have a usable (yet possibly imperfect) visual aid.

Recall the velocity graph we looked at above; a dishonest version is shown in Figure 5.2.19. This visual aid (wrongly) suggests that velocity drops to nearly zero as the flight reaches its end. The y-axis has neither a scale nor units, so readers can only see that velocity (the y-axis variable) drops precipitously. As discussed above, this is entirely ineffective in the first place and, with further analysis, dishonest as well. If a critic of the design were presenting this visual aid, it would be an intentional dishonesty. If a proponent of the design, which was the case in the original report, presented the visual aid, it is merely ineffective and dishonest, unintentionally. To remedy this poor visual aid so that it is honest, it must re-designed along the lines of the preceding two graphs about this airplane, Figures 5.2.17 and 5.2.18.

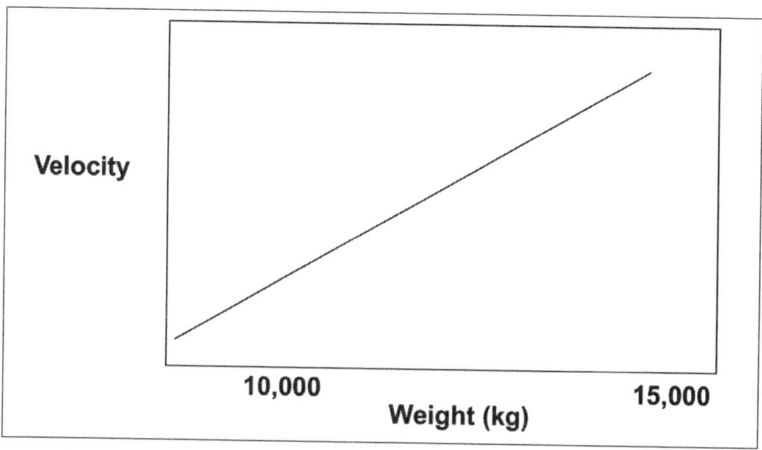

Figure 5.2.19: Optimum Velocity for Longest Range (Dishonest)

5.2.2.3 Autonomy

To transform a usable visual aid into a superior one, you must consider the third design objective. Autonomy refers to the quality of being able to stand-alone, separate from the text of the document. As discussed in the introduction to this chapter, visual aids are often looked at on their own or even distributed in an isolated format. That is the reason this is a genre of its own. Considering these facts, you must make your visual aids autonomous, which means clear and meaningful, independent of all text in the report.

You can easily ensure autonomy by adhering to these simple design rules, for each visual aid:

- Figure or table number is necessary for ensuring proper viewing order and quick reference.
- Figure or table title is necessary to explain point/message (content and purpose).
- Labels and explanations are necessary to enable a viewer to find key information and understand visual aid.

You can compare the two versions of the same basic visual aid in Figure 5.2.20. On the top (a), most of the required items are omitted. On the bottom (b), the visual aid is autonomous because the title is improved, the horizontal line is labeled, as are the dashed lines, and the three curves are identified.

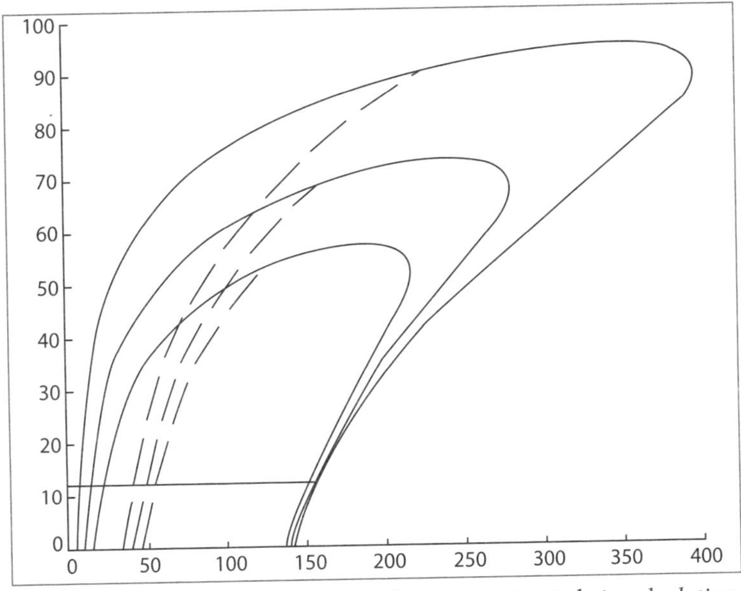

Figure 5.2.20a. Three flight envelopes that were estimated via calculations

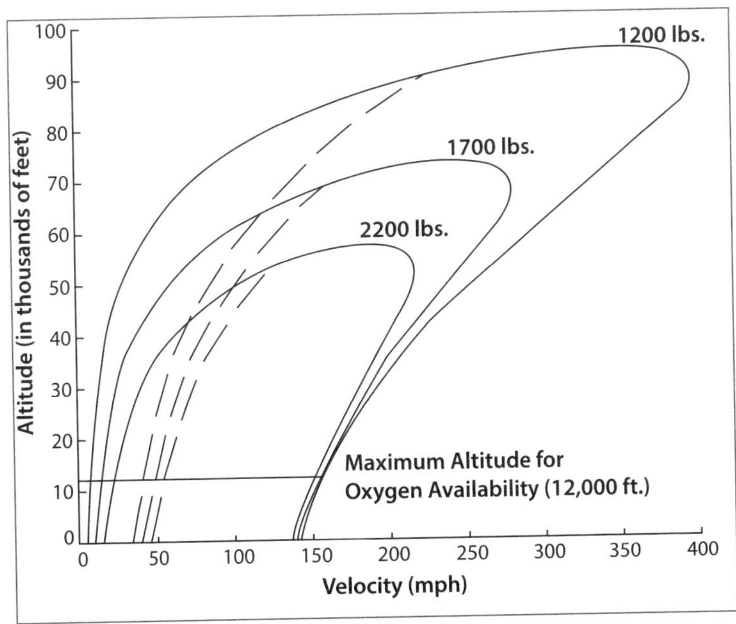

Figure 5.2.20b. Calculated flight envelopes for three weight configurations, showing the largest ranges of altitude and speeds for the lightest configuration

5.2.2.4 Simplicity

A visual aid should convey its message effectively, honestly, and autonomously. After that, it needs to do nothing more. Excessive design flourish, superfluous colors, three-dimensional portions, and other complex design elements add clutter. In the worst cases, the clutter leads to ambiguity in the message and confusion in the reader's mind. At a minimum, clutter wastes readers' time, as their eyes and mind study the portions of the visual aid that add zero content because they are unnecessary, redundant, and distracting. A visual aid should have just the right elements and nothing more. This is the quality of simplicity that you must strive for. Simplicity is elegance. Simplicity is perfection. Some scholars refer to this quality as efficiency. Efficiency has been described as using just the right amount of ink in the visual aid, and no more. With efficiency, the ratio of total ink to useful ink is 1. You will not likely make such a calculation as you design your visual aids. Nonetheless, use your own eyes and judgment to review the visual aid and ensure that each element has a useful reason, and you have made the visual aid as simple as possible.

Simplicity and the removal of clutter is an intrinsically good characteristic. But it yields an additional benefit that can be parlayed into further advantages. Namely, after clutter and other superfluous elements are removed, the visual aid has some free space that may be useful for adding new elements that improve efficacy, honesty, and autonomy. This transformation from cluttered to simple is demonstrated in the pair of visual aids in Figure 5.2.21 and 5.2.22. To say that the graph in 5.2.21 lacks simplicity is an understatement. With improvements, the new graph (5.2.22) is simple, effective, and autonomous. (Both versions are honest, to the credit of the author who produced the cluttered version.)

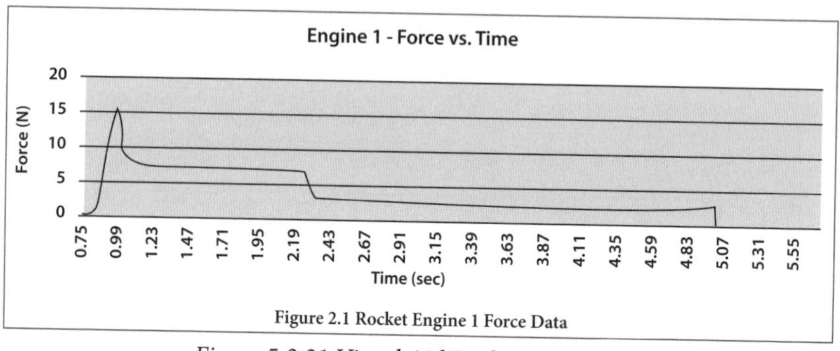

Figure 5.2.21 Visual Aid Lacking Simplicity

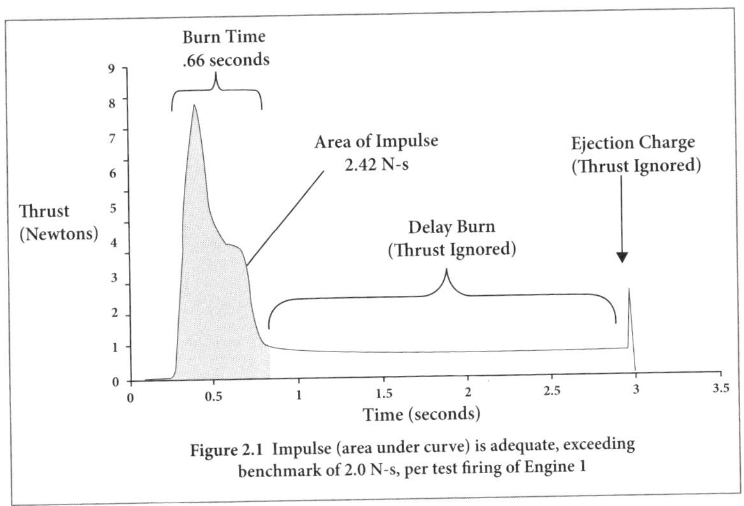

Figure 5.2.22 Visual Aid Achieving Simplicity

In the top example, you can observe numerous items of zero-content clutter. The shading within the graphing area obscures the actual data curve. The horizontal, scale-based grid lines serve no purpose. (Often a single horizontal line can be used to indicate an important standard, statistic, or benchmark, while default scale-based lines offer no information.) Along the x-axis, too many units of time are used and the series is non-standard without purpose. Two titles are provided when one suffices. The second title (center, top) merely repeats the axes labels and the engine number, so it is redundant on both counts. The visual aid boasts a top x-axis and a right y-axis, without utility. Furthermore, it is surrounded by a border box, which should not be needed given other formatting choices for the whole document. Worst of all, because the visual aid consists of numerous items of clutter, it lacks useful items that would be helpful for efficacy and autonomy, as is shown in the improved version.

I cannot anticipate all the possible items of clutter that can be added, but here is a list of possible culprits I see often and that you can be on the watch for:

Redundant titles: two or more elements serving as a title, with exactly the same or overlapping information as appears elsewhere.

Superfluous shading: backgrounds and features filled in with unnecessary shading, which means the shading provides no information linked to a key or legend.

Extra border: a second border box around a visual aid that already has an explicit or implicit box.

Additional axes: a single x axis (bottom) and a single y axis (at left) are often sufficient. An additional x axis at top or y axis at right are unnecessary. On occasion, when the y-axis at right is used to show a different unit and scale than that used on the left, a second y-axis would have utility.

Unnecessary units: Within a scale, a suitable and normal sequence of units is ideal. Any scale broken into either unnecessary or unusual units is messy.

Gratuitous scale-based grids: Horizontal and vertical lines extending across the full width or height of a graph are obstructive because they usually provide little benefit while taking away space that can be used for helpful information (labels and notes). In worst cases, they obscure clear viewing of the data or associated labels and notes. (A full grid may be useful when the visual aid is intended to be used as a look-up table.)

Avoidable legend or key: The ideal is to label parts of a visual aid directly "within" the visual aid itself, just adjacent to the part itself (using arrows when necessary). Only when this cannot be done due to insufficient space or other reasons, should a legend or key be used. Legends and keys are not absolutely prohibited. They simply are a second-best option. Use them only when necessary. The visual aid will always be simpler and easier to understand when parts are labeled directly, without recourse to a legend/key that explains the line style, shading type, color, symbol, or shape.

Extra dimensions: Three-dimensional bars, pie pieces, data curves, and such often obscure the specific values intended to be shown, so the extra depth introduces both ambiguity and clutter.

Metadata: Error bars, linear regression values or functions, plot characteristics, and similar analytical enhancements often obstruct the visual's intended simple message(s). These important analytical values must be conveyed in the report somewhere, but they should not be included where they undermine a visual's utility.

Lastly, in addition to all the aforementioned types of clutter that defeat simplicity, unused empty spaces may equally detract from a visual's quality, as it fails to appear tight and compact. Empty spaces often arise when axes scales do not correlate well with actual data, leaving a large part of the graph devoid of data. Or, the placement of elements in the visual aid may be non-optimal, producing purposeless empty spaces. White space is helpful in a visual, to keep the necessary elements separate, but excessive blank spaces are unnecessary.

5.2.3 Techniques for Common Visual Aid Types

Most specialties within ABERST have visual aid types unique to their work, equipment, and analyses. Out of the multitude, the ones below are the types I see most often:

- Bar Graphs
- Line Graphs
- Pie Charts
- Tables
- Data Matrices
- Photographs
- Line Diagrams (Sketches)
- Block Diagrams
- Flow Charts
- Decision Trees
- Chronologies
- Data Maps

For each one, I discuss some common pitfalls and demonstrate with one or two bad examples, followed by one or two good examples that meet the four design objectives introduced above. If you will be designing a visual aid that does not fall into one of the dozen categories covered, I hope you can adapt the lessons to your needs. For all types discussed below, if the visual aid does not convey a clear message and just "shows something," it must be improved to fix this shortcoming. That fundamental rule is applicable to all of them, so it need not be repeated in each and every section below.

5.2.3.1 Bar Graphs

Common pitfalls for bar graphs are the following:

- Bars are unnecessarily 3-dimensional
- Bars are overcrowded (too much data squished into one graph)
- Bars have insufficient white space between them
- Bars are in random order
- Relationship among bars is not obvious (e.g., comparison or correlation)
- Legend/key is used as default where it may be avoided
- Abbreviations or symbols are used in legend/key without definition (undermines autonomy)
- Labels on bars are difficult-to-read due to rotation, slant, or small size
- One or more bars push right up against top of graph (no head space provided)

Bad examples:

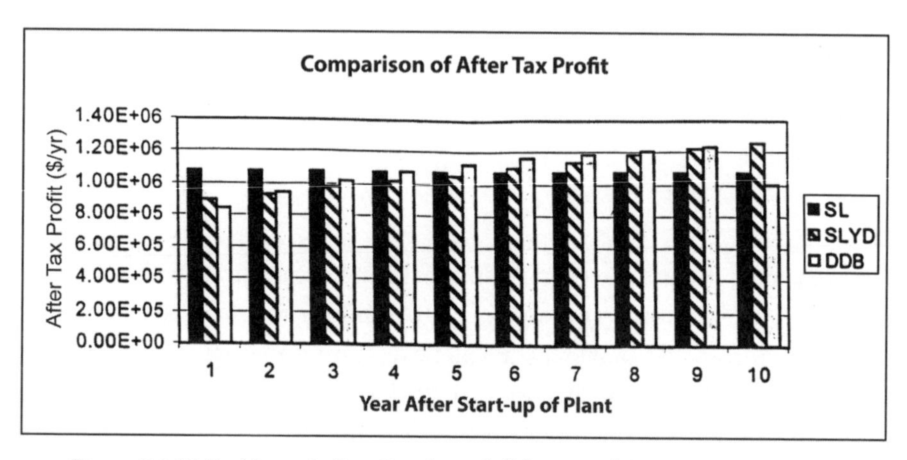

Figure 5.2.23 Problematic Bar Graph: with "phantom" bars instead of white spaces, undefined acronyms, and non-obvious dollar units and magnitudes

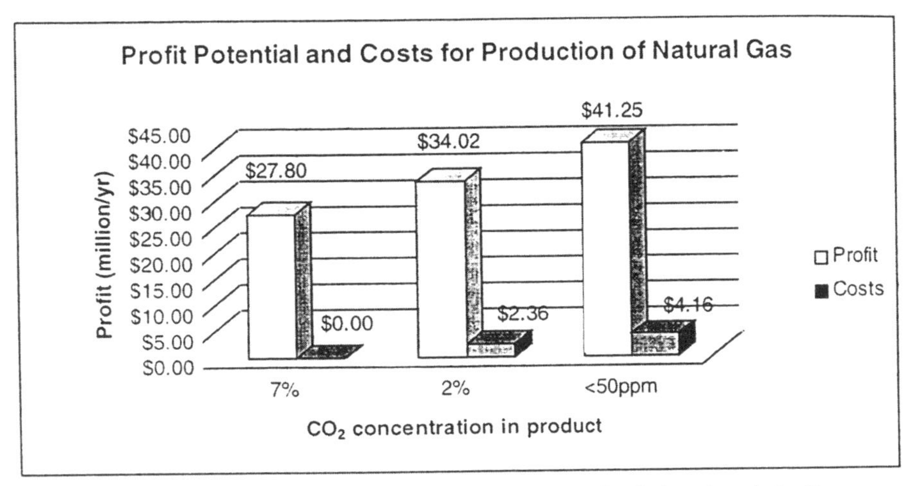

Figure 5.2.24 Misleading Bar Graph: with confusing depth (3-D) and shading

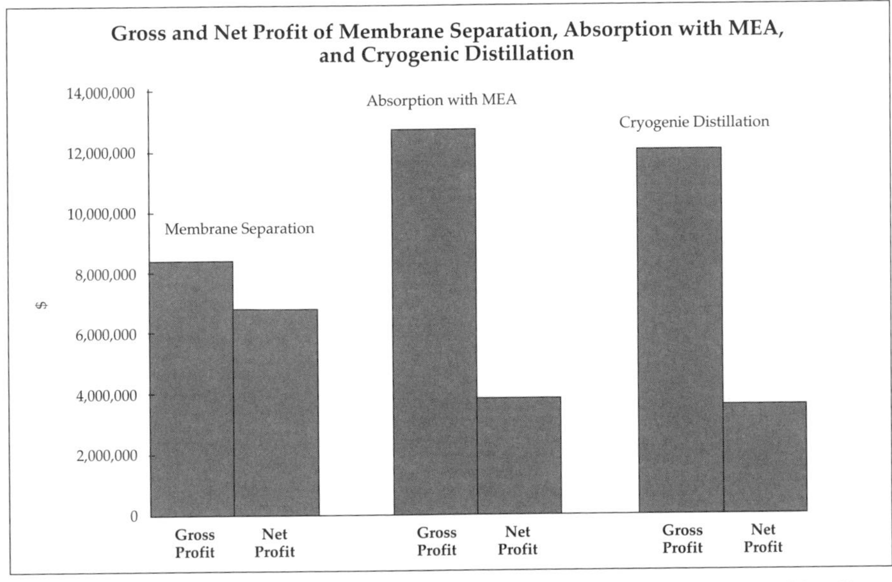

Figure 5.2.25 Crowded Bar Graph: with six, ink-heavy bars while three would suffice

Good examples:

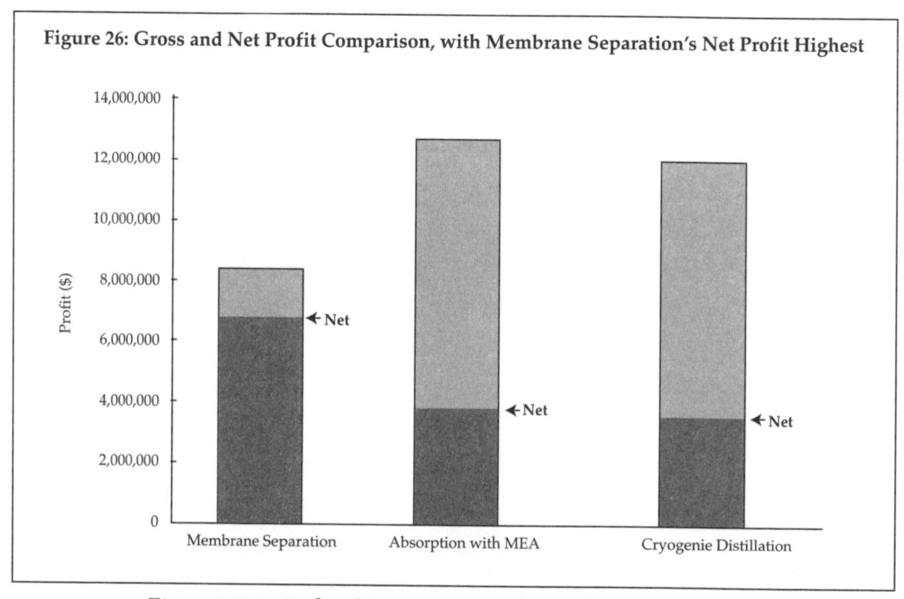

Figure 5.2.26 Refined Bar Graph: with simplicity via stacking (Improved version of Figure 5.2.25 above)

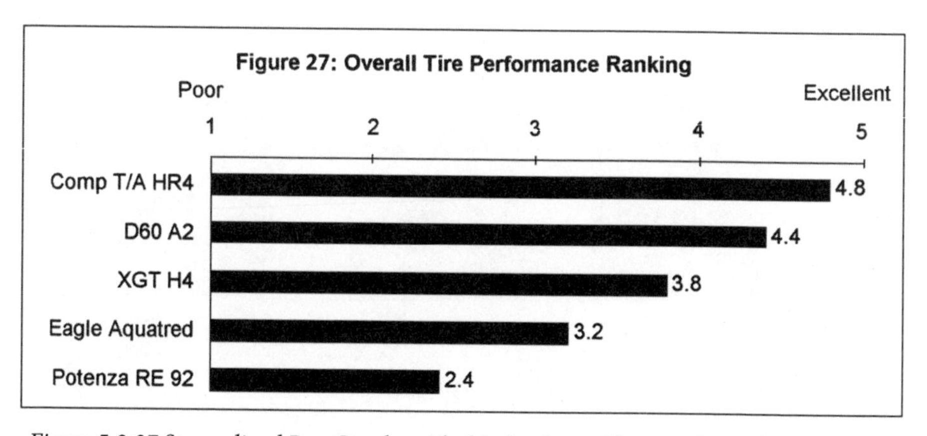

Figure 5.2.27 Streamlined Bar Graph: with thin horizontal bars and ample white space

5.2.3.2 Line Graphs

Common pitfalls for line graphs are the following:

- Either x- or y-axis is suppressed without indication (dishonesty ensues)
- X-axis continues past last data point
- Data push right up against top of graph (no head space provided on y-axis)
- Graph has unnecessary x- and y-axis on top and right side, respectively
- Default grid scale-based lines obscure data without adding information
- Titles are merely repetitions of x- and y-axis labels
- Error bars are dominant aspect, overshadowing message and other key points of data
- Comparisons are distorted when y-axis changes across side-by-side graphs
- No callout (message) is provided at significant features of data

Bad examples:

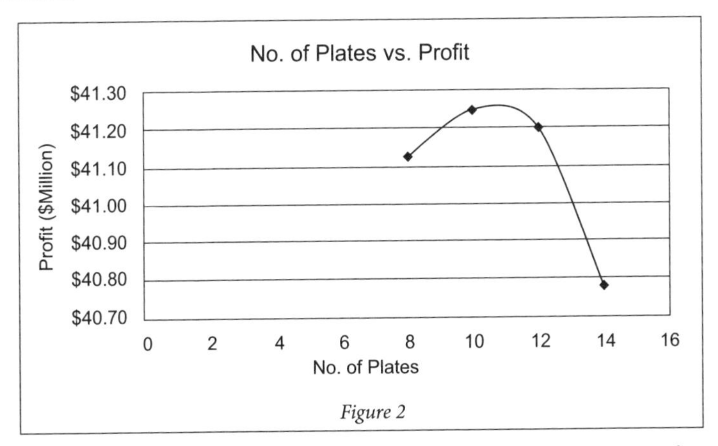

Figure 5.2.28 Problematic Line Graph Needs Improvements: area of no data from 0-7 plates could be replaced with x-axis suppression; default horizontal unit lines could be replaced with a single benchmark line; y-axis is dishonest without indicator of suppression; x-axis should stop at 14, not 16; and a title is needed instead of repeat of axes labels

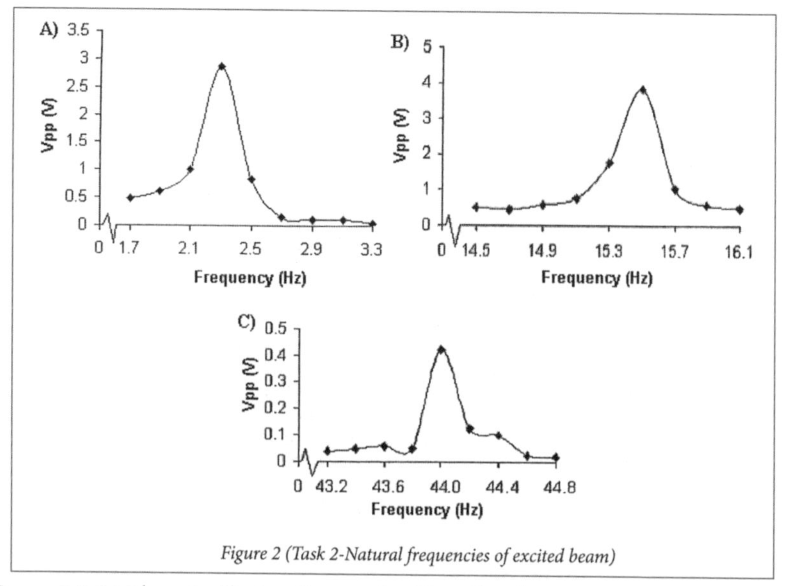

Figure 5.2.29 Three Ineffective Companion Line Graphs: comparative strengths in voltage of the three line curves are not apparent (despite the honest indication of the x-axis suppression); no message is provided; and y-axis is labeled with an abbreviation

Good examples:

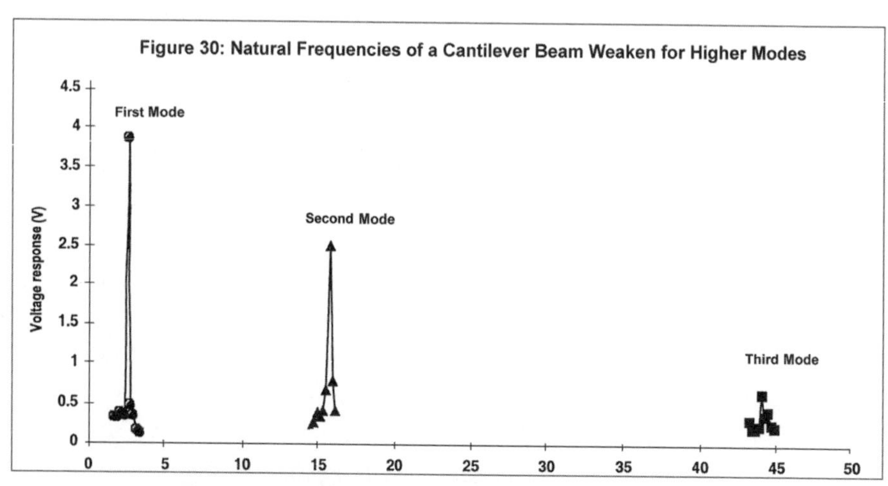

Figure 5.2.30 Three Effective Companion Line Graphs: convey message using single y-axis scale

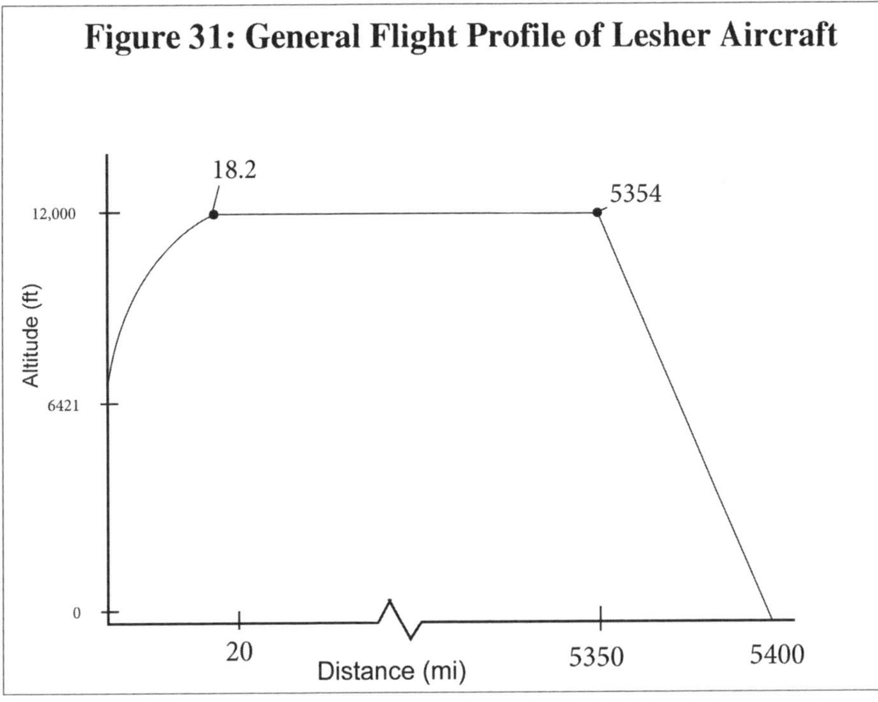

Figure 5.2.31 Clear Line Graph: with y-axis suppression to enable focus on start and end of flight data

5.2.3.3 Pie Charts

Common pitfalls for pie charts are the following:

- The items in the chart do not add up to 100%, as they should
- The size of the pie segments to not match their associated percentage value
- The chart is too simple to warrant the effort to develop and present it
- Too many small, indistinguishable pieces of pie inhibit meaningful comparison
- Fill shading or colors of pie segments are difficult to distinguish
- The items add to 100 but are not truly percentages

Bad example:

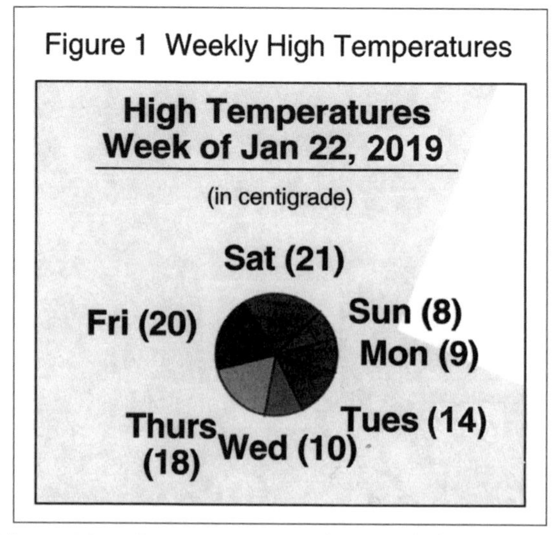

Figure 5.2.32 Inappropiate Pie Chart: with false message that values are portions of a whole (100%)

Good example:

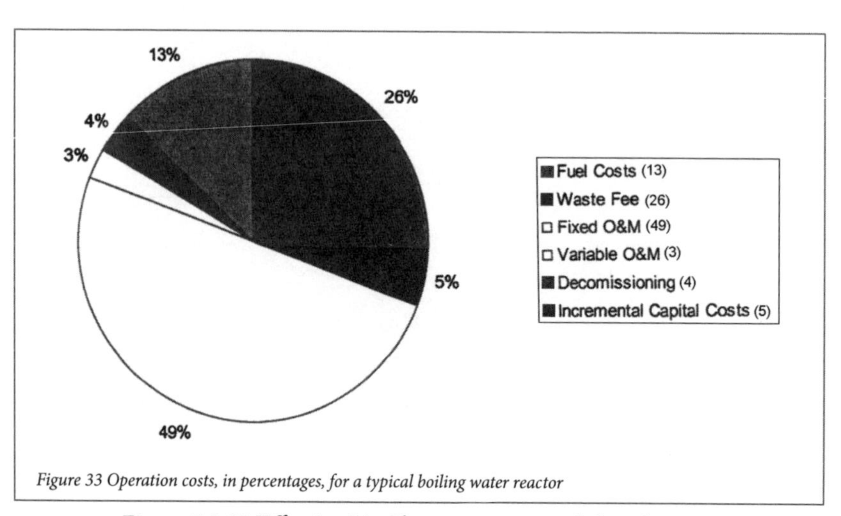

Figure 5.2.33 Effective Pie Chart: proportional distributions of all costs (100%) across six categories

5.2.3.4 Tables

Common pitfalls for tables are the following:

- Excessive use of grid lines (borders) obscures true groupings (tables are not Bingo Cards). (As stated earlier, grid lines are useful when the visual aid is intended as a look-up table.)
- If showing amounts of money, items are not aligned on decimal
- If showing amounts of money, inappropriate significant figures are used
- If showing numbers, scientific notations are used instead of decimal notation
- Empty cells exist with no explanations
- Column or row labels (headings) are absent, or undefined abbreviations are used
- Values (data) that are meant to be added or averaged are placed across a row, instead of down a column

Bad examples:

Table 34. Capital Investment Breakdown

Capital Requirement	$
Total equipment and installation	3.42E+06
Piping	3.42E+05
Total direct costs	3.76E+06
Total indirect costs	1.25E+06
Fixed capital investment	5.01E+06
Working capital	9.77E+05
Miscellaneous costs	5.26E+05
Total investment	1.53E+07

Figure 5.2.34 Problematic Table: number style obscures relative magnitude of each line item; grid lines obscure important distinctions of headings and sum from values to be added; too much space and ink

Table 35. Speedboat Comparison per Standard Features

Baja Waverunner 36	**Seabreeze 40**
8 person capacity	8 person capacity
Transom mounted ski tow point	Professional heavy duty ski post
Swim step with handles	Reversible seats
1 cooler and additional storage	Swim step with handles
Ski locker	2 coolers and additional storage
	Radio
	Sunning deck
	Ski locker

Figure 5.2.35 Poorly Arranged Table: with changing (random) data in a single row and empty cells due to missing row labels

Syllabus			
Introduction to the Universe			
Metaphysics 498			
Fall 2018			
Date		Topic	Assignments due and reading
9/9	Thursday	Course Intro; Project Intro; Questions and Answers	Vol 1, Chp 1
9/14	Tuesday	Team Formation	Handouts
9/14	Tuesday	Laws/Physics	Vol. 1, Chp 2.3
9/16	Thursday	Case Study 1; Problem solving	Handouts
9/21	Tuesday	Discussion: Laws	HW #1
9/21	Tuesday	Mechanics	Vol 1, Chp 5
9/23	Thursday	Gravity and Time	Vol 1, Chp 6

Figure 5.2.36 Poorly Formatted Table: syllabus with each day lined out (weeks not visible) due to excessive grid lines (cell borders)

Good examples:

Table 37. Capital Investment Breakdown

Capital Requirement	Cost ($)
Total equipment and installation	3,420,000
Piping	342,000
Total direct costs	3,760,000
Total indirect costs	1,250,000
Fixed capital investment	5,010,000
Working capital	977,000
Miscellaneous costs	526,000
Total investment	15,285,000

Figure 5.2.37 Sharply Styled Table: with proper spacing, lines, and number style (Figure 5.2.34 redone)

Feature	***Speedboat Model***	
	Baja Waverunner 36	**Seabreeze 40**
Capacity	8 persons	8 persons
Skiing	Transom-mounted tow point	Professional heavy-duty post
Swimming	Swim step with handles	Swim step with handles
Coolers	1	2
Storage	Additional storage	Additional storage
Locker	Ski locker	Ski locker
Deck	0	Sunning deck
Seat Options	0	Reversible seats
Media	0	Radio

Figure 5.2.38 Nicely Arranged Table: with row labels (features) and proper columns (Figure 5.2.35 redone)

Syllabus
Introduction to the Universe
Metaphysics 498
Fall 2018

Week	Date	Day	Class Type	Topic	Assignment Due	Reading
1	9-Sep	Thurs	Lect	Course Intro; Project Intro; Questions and Answers		Vol 1, Chp 1
2	14-Sep	Tues	Disc	Team Formation		Handouts
	14-Sep	Tues	Lect	Laws/Physics		Vol. 1, Chp 2.3
	16-Sep	Thurs	Lect	Case Study 1; Problem solving		Handouts
3	21-Sep	Tues	Disc	Discussion: Laws	HW #1	
	21-Sep	Tues	Lect	Mechanics		Vol 1, Chp 5
	23-Sep	Thurs	Lect	Gravity and Time		Vol 1, Chp 6

Figure 5.2.39 Effectively Formatted Table: syllabus with discrete columns and lines, showing weeks (Figure 5.2.36 redone)

5.2.3.5 Data Matrices

Common pitfalls for data matrices are the following:

- Comparative values are poorly placed in rows instead of across columns
- Abbreviations are used in row or column headings
- Inadequate cell space is provided for data to be collected
- Empty cells are provided for independent variables that should be listed in headings

Bad example:

Testing Matrix 1	Date:		
Circle all that apply:			
Honeycomb Specimen	Upstream Configuration	Motor Setting	U (m/s)
None	Clean	Hz	
Ten "	Diamond		
Twelve "	Square		
Traverse Distance	Manometer (" H2O)	HWA (v)	Data File

Figure 5.2.40 Weak Data Matrix

Good example:

Testing Matrix 1 Testing Date:			
	Testing Conditions		Options/Units
	Honeycomb Specimen:		0, 10, or 12
	Upstream Configuration:		diamond or square
	Motor Setting:		Hertz
Traverse Distance (millimeter)	Dynamic Pressure (manometer - inch H2O)	Hot Wire Anemometer (voltage)	Data File Name (.xln ending)

Figure 5.2.41 Strong Data Matrix

5.2.3.6 Photographs

Common pitfalls for photographs are the following:

- Items shown are unlabeled
- Wrong items are emphasized (too much captured in photograph beyond essentials)

Bad example:

In the example (Figure 5.2.42), the photograph includes distractions such as a shelf with plastic storage boxes and other cardboard boxes to the right. Key items are not labeled: pressure tap piping and primary valve connected to pump.

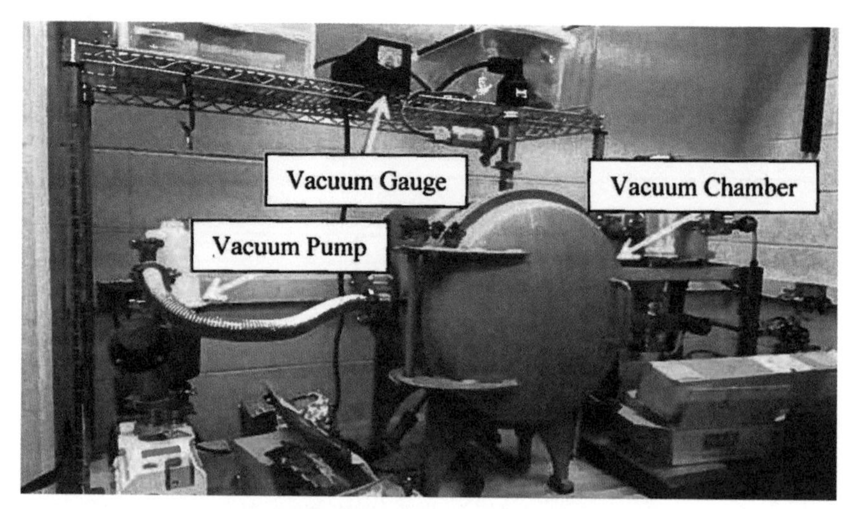

Figure 5.2.42 Crowded Photograph of Vacuum Chamber Setup: unnecessary items shown

Good example:

In the good example, the photograph (5.2.43) shows the primary devices configured into an infrared testing setup, without extraneous items. Each key item is labeled clearly.

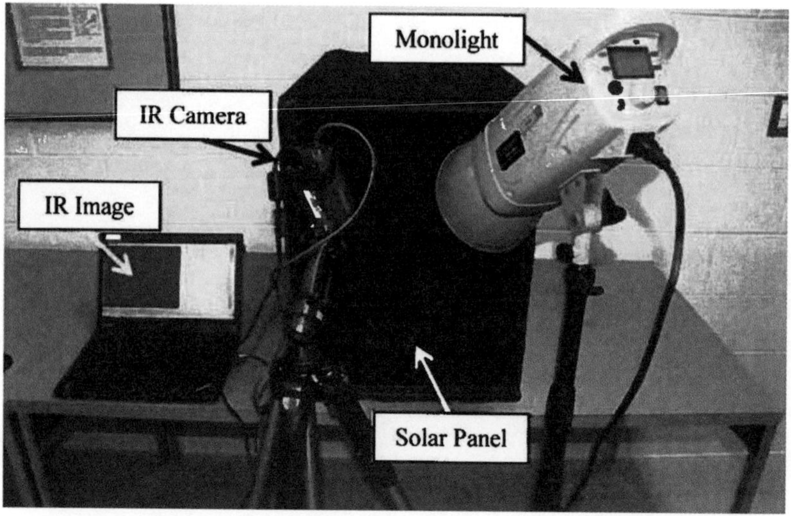

Figure 5.2.43 Clear Photograph of Testing Setup: no superfluous items

5.2.3.7 Line Diagrams (Sketches)

Common pitfalls for line diagrams are the following:

- Image creates a Moiré effect (where it appears to be wiggling on page)
- Abbreviations or symbols are used in labels without definition (undermines autonomy)
- Image overall may be incomplete, with items omitted (photographs are opposite in that they may show too much, including extraneous items)
- Image contains inconsistent shifts from iconic elements to block or symbolic items

Bad examples:

Figure 5.2.44 Visually Peculiar Line Diagram: with Moiré Effect

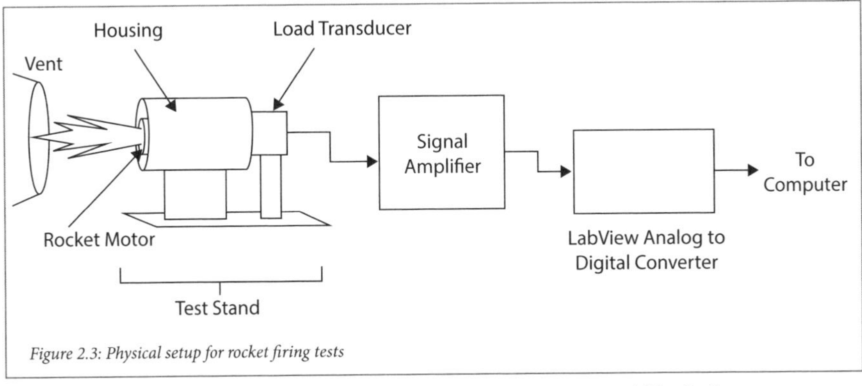

Figure 5.2.45 Test Setup Diagram: with both iconic and block elements

Good examples:

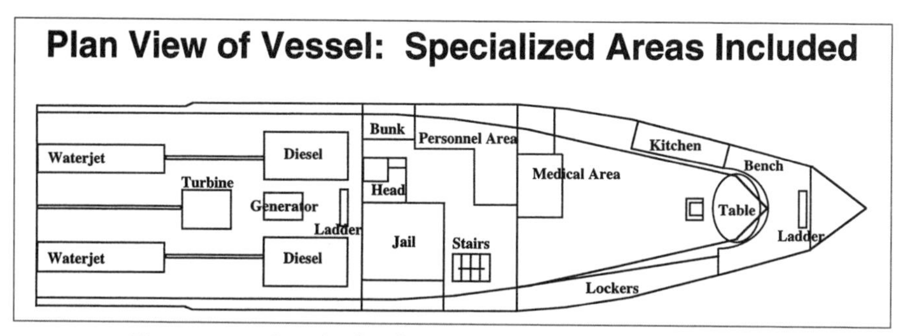

Figure 5.2.46 Simple Sketch: with iconic images and clear labels

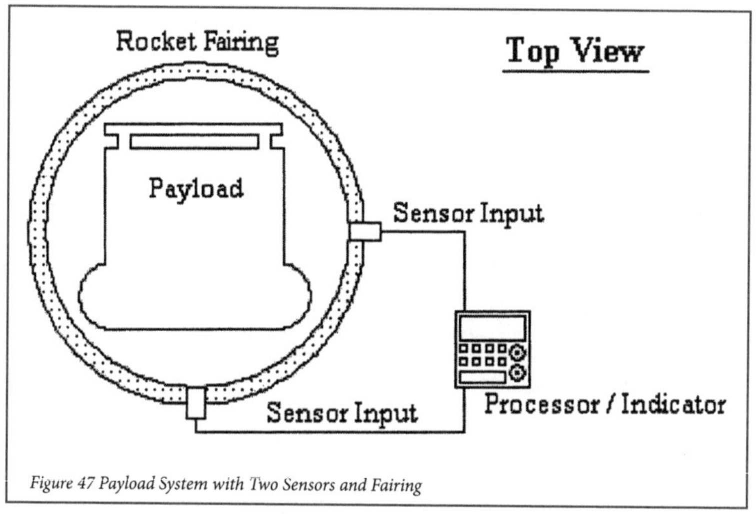

Figure 47 Payload System with Two Sensors and Fairing

Figure 5.2.47 Simple Sketch: with iconic images and clear labels

5.2.3.8 Block Diagrams

Common pitfalls for block diagrams are the following:

- The hierarchy of organization among blocks is poorly shown
- The blocks change size for no apparent reason

Bad example:

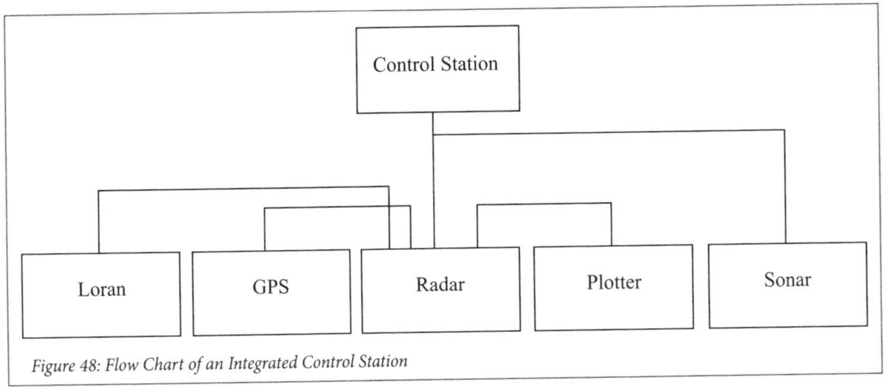

Figure 5.2.48 Rough Block Diagram: without visible hierarchy of components

Good example:

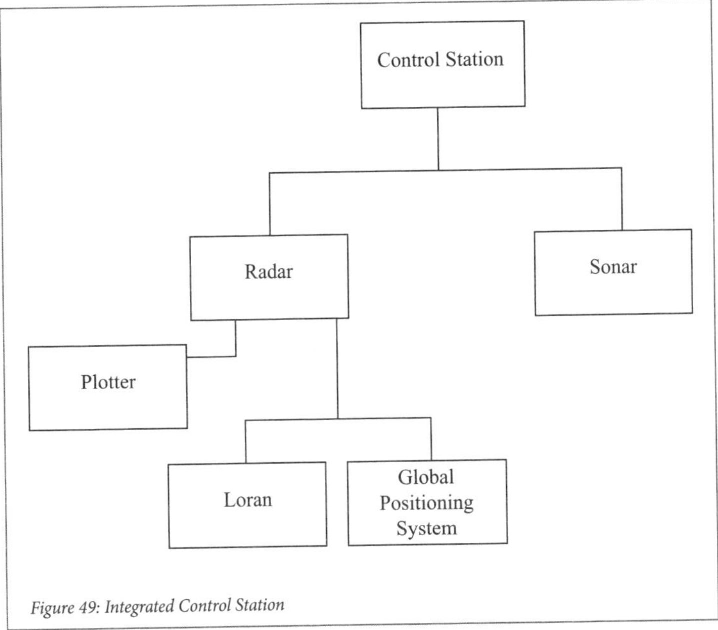

Figure 5.2.49 Strong Block Diagram: with hierarchy evident (Figure 5.2.48 redone)

5.2.3.9 Flow Charts

Common pitfalls for flow charts are the following:

- Labels of flow (streams, progress, movement) are either missing or unclear
- Labels of stages (processes, actions, tasks) are either missing or unclear
- Direction of movement is incorrect (flow should be from top to bottom, and left to right) or changes abruptly
- Components are poorly balanced (items are bunched up with either not enough or uneven white space between them.
- Description of actions required is incomplete
- Important elements are not emphasized well

Bad examples:

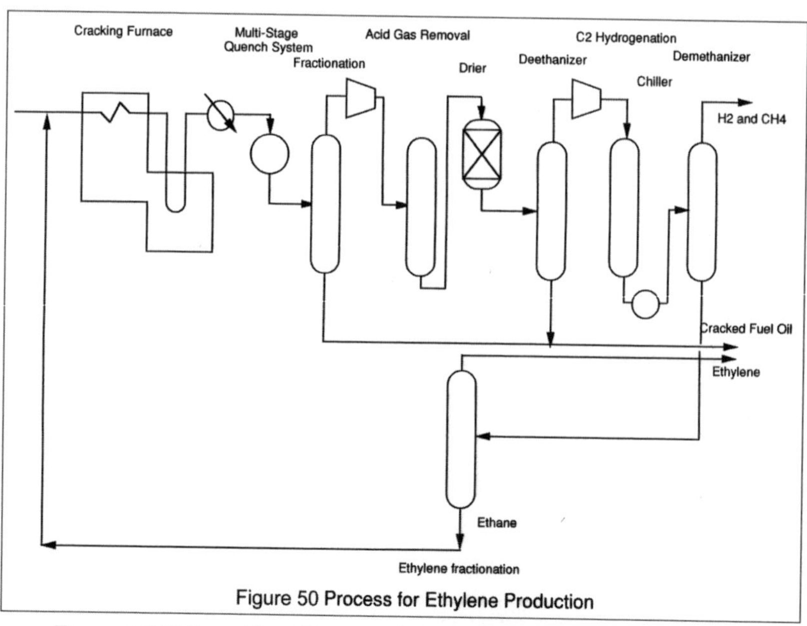

Figure 5.2.50 Poor Flow Chart: with final column only item at bottom; direction goes right to left near end before abruptly reversing to left to right; and final product, ethylene, not isolated skillfully for emphasis

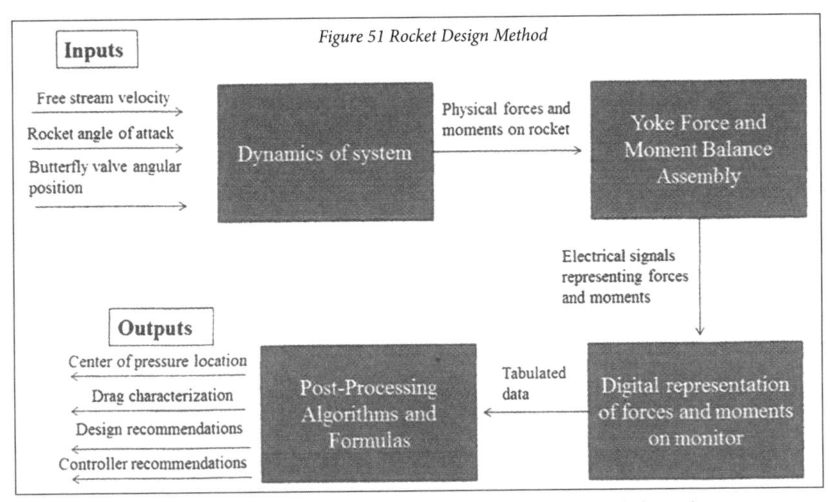

Figure 5.2.51 Poor Flow Chart: ends at bottom left and stages and streams not well organized

Good examples:

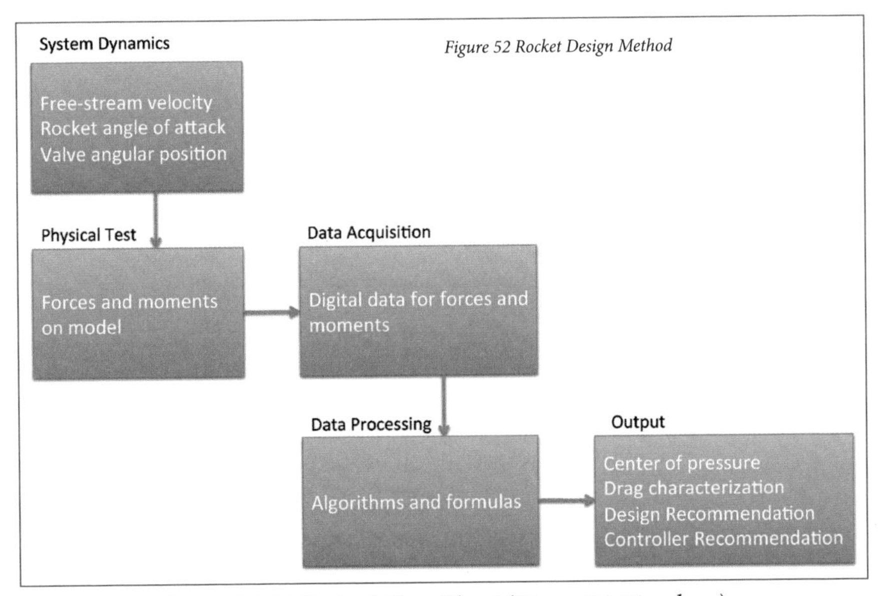

Figure 5.2.52 Revised Flow Chart (Figure 5.2.51 redone)

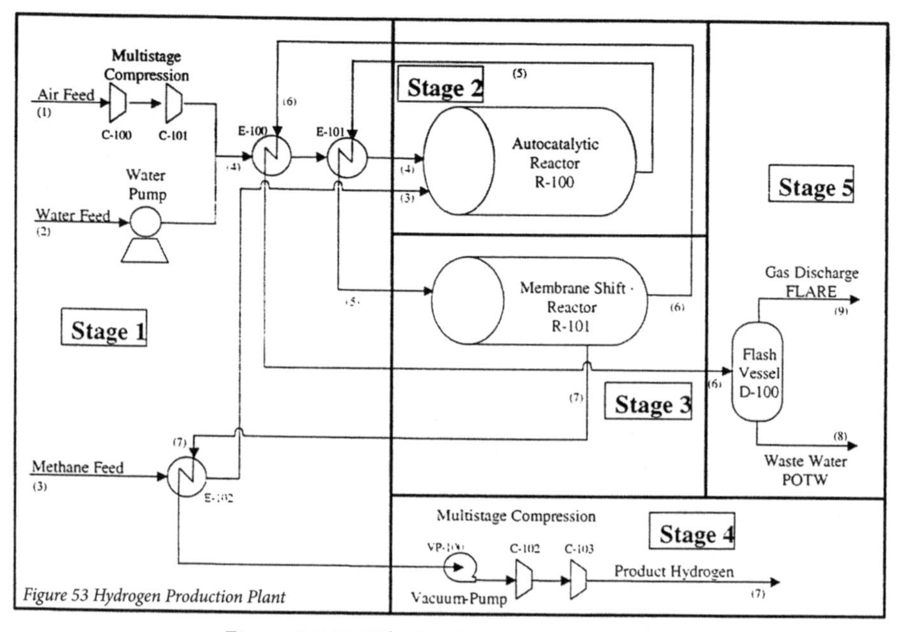

Figure 53 Hydrogen Production Plant

Figure 5.2.53 Effective Five-stage Flow Chart

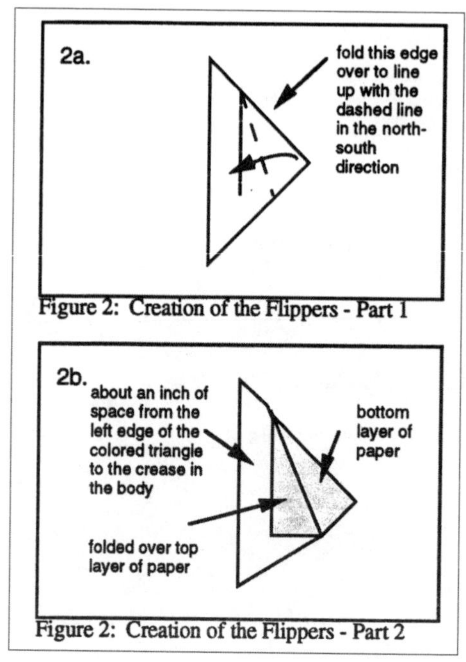

Figure 2: Creation of the Flippers - Part 1

Figure 2: Creation of the Flippers - Part 2

Figure 5.2.54 Clear Explanatory Process Flow Diagram

5.2.3.10 Decision Trees

Common pitfalls for decision trees are the following:

- They do not move from top to bottom, and left to right
- The blocks change shape for no apparent reason
- Description of actions required is incomplete
- They are simply flow charts, and not decision trees

Bad example:

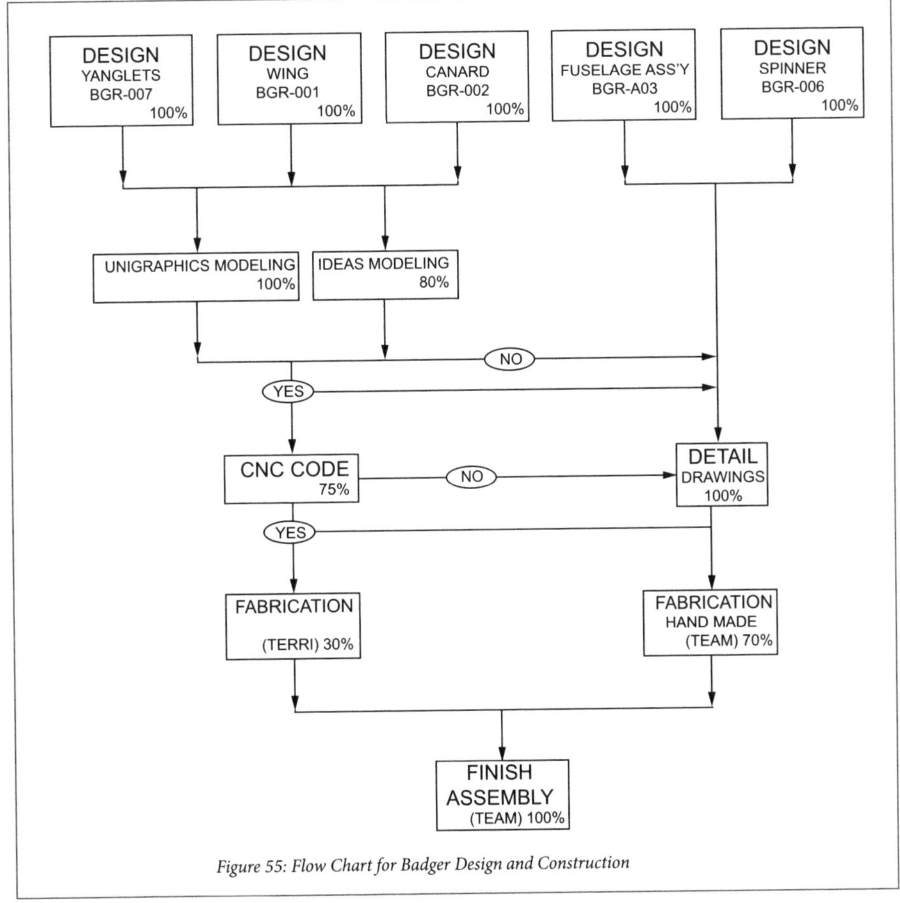

Figure 55: Flow Chart for Badger Design and Construction

Figure 5.2.55 Confusing Decision Tree

Good example:

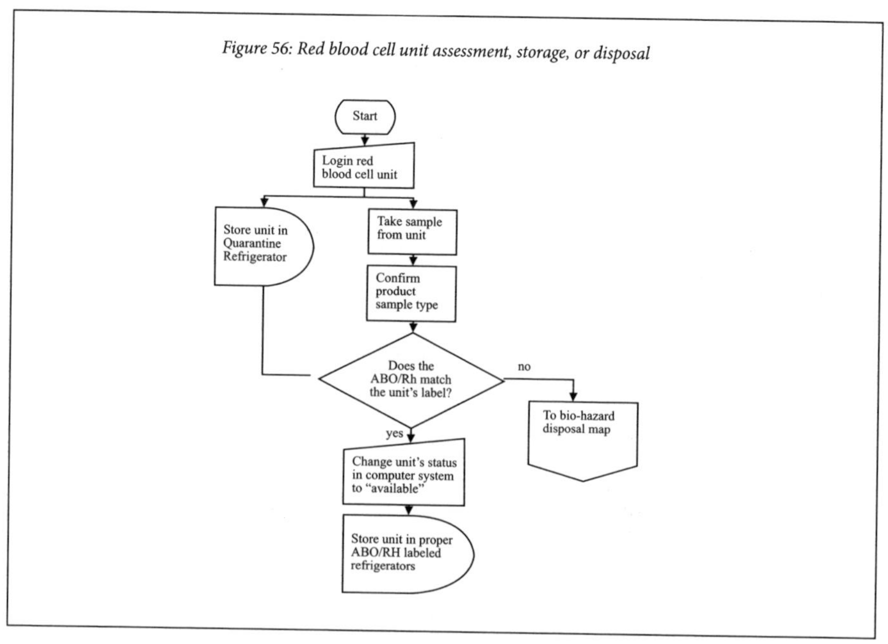

Figure 5.2.56 Effective Decision Tree

5.2.3.11 Chronologies

Common pitfalls for chronologies are the following:

- Events depicted are not clearly explained
- Graphical clutter is emphasized over useful information
- Date formats are inconsistent
- Abbreviations or symbols are used without a convenient legend/key

Bad example:

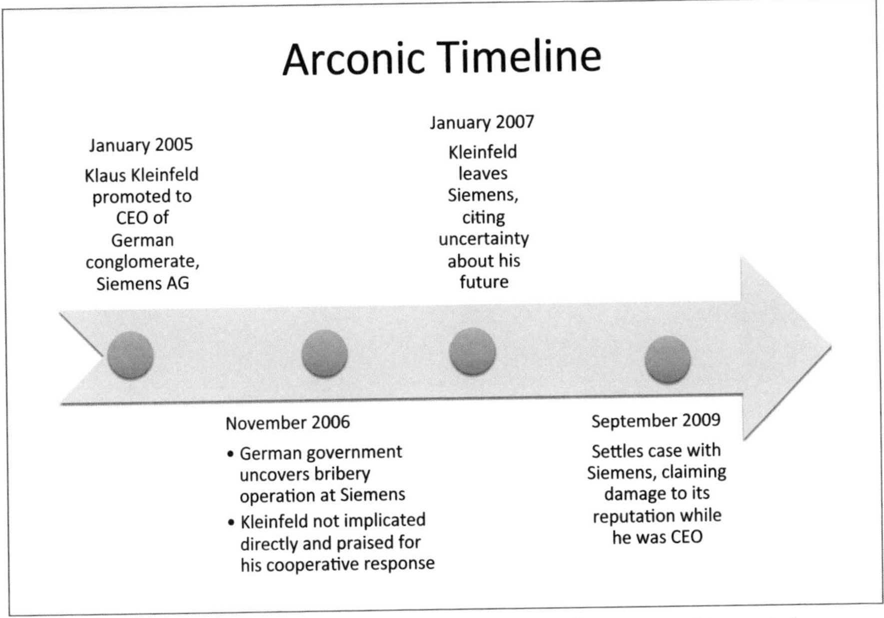

Figure 5.2.57 Poor Timeline: with emphasis on arbitrary graphic symbols

Good example:

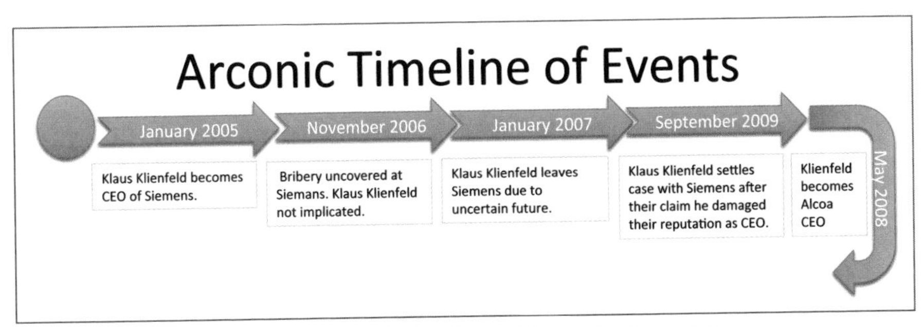

Figure 5.2.58 Good Timeline: with emphasis on dates

5.2.3.12 Data Maps

Common pitfalls for data maps are the following:

- The map is not drawn to an appropriate scale
- North is not at top
- Data are not clearly evident (distinguished) from other features of map
- Data are not defined (symbols are undefined)

Bad example:

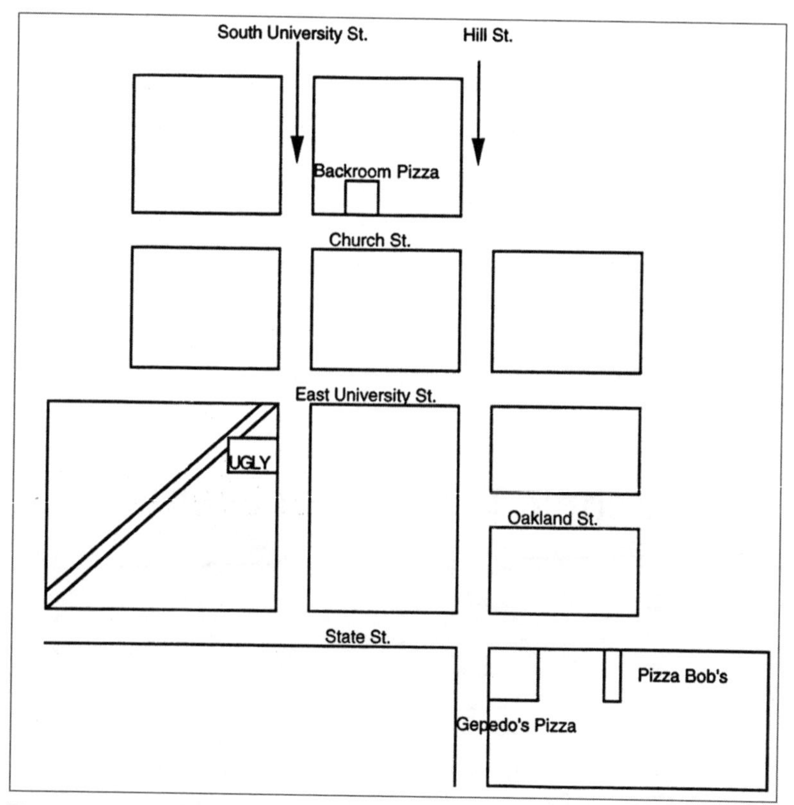

Figure 5.2.59 Bad Map of Pizza Restaurants: north neither at top nor compass indicated, data not highlighted, spelling error, and undefined abbreviation

Good example:

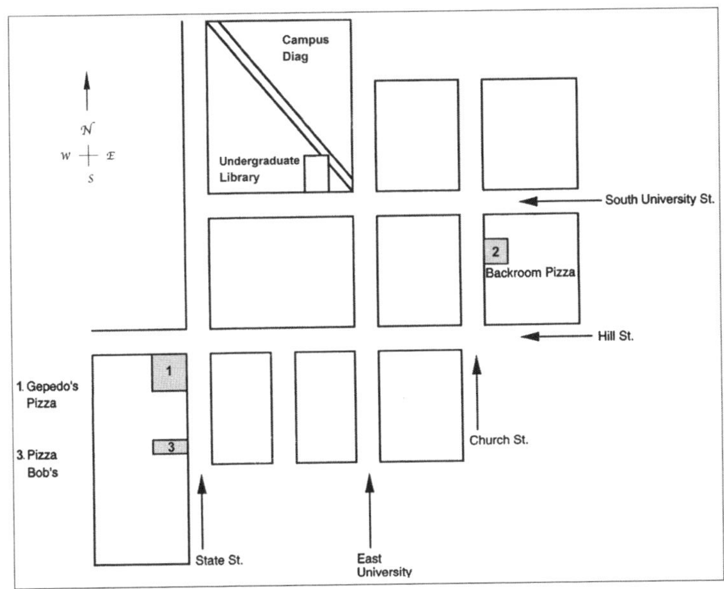

Figure 5.2.60 Good Map of Pizza Restaurants: north at top, data highlighted, and no abbreviations

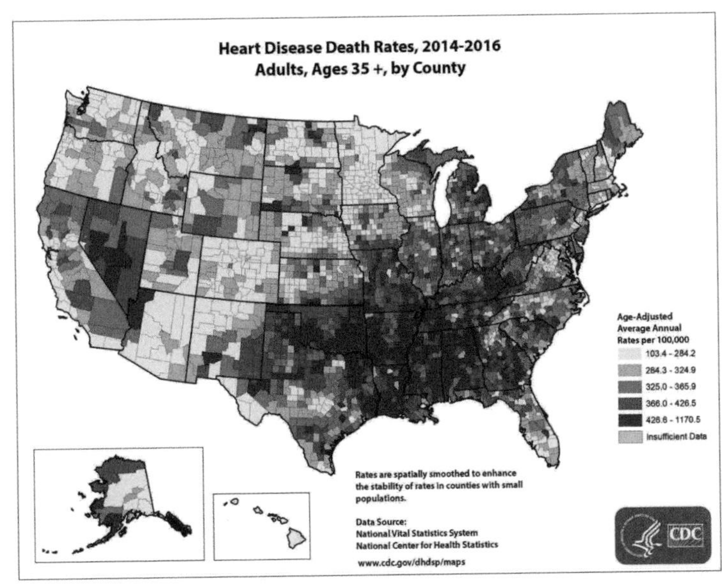

Figure 5.2.61 Classic Data Map: created by US Government to depict vital health data by county

Appendix A

Common Composition Mistakes

This appendix is dedicated to my major warnings of trouble and problems that could sneak into your documents. These warnings, or "red alerts," encompass those all-too-common mistakes and sloppy constructions that create "anguished English," as one author, Richard Lederer, has called it. The red alerts will put you on guard to watch for the common hazards. If you stay clear of these hazards, you will have the good fortune of being remembered for your document's lucidity, not ludicrousness.

A handful of red alerts are presented, and this collection should provide ample protection, as long as you are attending to the other guidelines of this book. As you read through the hazards in detail below, remember that you already have one huge advantage on your side: you have used communication purpose as a boundary line to eliminate unnecessary text. This significant cropping enables you to spend your time keeping watch for, and avoiding, the following common mistakes found in technical documents. As mentioned in Chapter 2.3, proofreading is crucial. The red alerts warn you against the following hazards:

Common Composition Mistakes

- Spelling slipups
- Vocabulary blunders
- Terminology troubles
- Personal pronoun problems
- Run-on sentences
- Sentence fragments
- Misplaced modifiers and clauses
- Which/that clause confusion
- Ambiguous demonstrative pronouns
- Faulty comparisons
- Series errors
- Wordiness
- Empty and self-evident phrases
- Verb distortions
- Imprecise phrases
- False questions via interrogative pronouns
- Redundancies
- Contradictions
- Punctuation problems

A.1 Spelling Slipups

You must avoid embarrassing spelling errors that spell checkers will NOT catch, such as the following:

- The corporation distributed 25,000 pubic notices to the members of the surrounding community.
- All moving parts will suffer where and tear.
- The dominatrix frequency sits at peak amplitude.
- Developed by the arm services, GPS (global positioning system), for all intensive purposes, would be beneficial for civilian navigation.
- The existance of ocean thermal energy conversion pants could have a armful impact on the environment and it's surrounding ecosystems.

You have heard it said before, and it is true: a spell checker will not find all your spelling errors. Many times, a word is spelled correctly, but not for the context you intend. That is, you have typed either the wrong word (*expedite* instead of *expedition*) or the wrong variation of the word (*help* instead of *helps*), though it is spelled correctly. The computer does not know which word you wish to use. Smart systems might detect some spelling errors that introduce grammar errors, but you cannot be certain. You must read your text over for yourself, every word.

Not one of the errors in the following sentence was caught by my software's error checkers:

- Threw are analysis of the mortars emissions, we can a test that the fore filters, too placed in front and to place behind, make this mortar the most clean-burning won ever.

The previous sentence should have been as follows:

- Through our analysis of the motor's emissions, we can attest that the four filters, two placed in front and two placed behind, make this motor the most clean-burning one ever.

Here is a list of word pairs (or triads) that are easily typed interchangeably. As a result, often the wrong term is used inadvertently but accepted as correct on first glance because the terms are homophones of one another or otherwise very similar.

- Whether/weather
- Their/there/they're
- Forward/foreword
- To/too/two
- Won/one
- Later/latter
- Threw/through/thorough
- Right/write
- Former/formal
- And/an
- Where/were
- It's/its

The last pair confuses nearly everyone. To distinguish them, I offer the following definitions and examples used in a sentence:

It's—
Definition: a contraction of two words: "it" and "is"
Sentence example: They wondered if *it is* possible to run a marathon under 4 hours.

Its—
Definition: the possessive version for the noun "it"
Sentence example: The chair felt soft because *its* seat had been reupholstered.

A.2 Vocabulary Blunders

The preceding examples are words most writers' brains can distinguish but not their typing fingers. A different set of words pose trouble for brains, not fingers. Indeed, these are easily confused words, leaving authors to ask, "Which word is the correct one?" Most English handbooks devote a few pages to presenting examples of easily confused words. Such lists are helpful, and one should peruse them and attempt to commit the different words to memory. This will save you the embarrassment of using the wrong word, usually because it sounds similar to the one you intended to use. I cannot claim to have discovered this *faux pas*; it has been a pitfall for writers since time immemorial. Nonetheless, I can provide a list with the ones I see as most worrisome for ABERST technical writing:

> Accept: agree to or receive something
> Except: leave out or to exclude
>
> Affect: influence (verb)
> Effect: outcome or result (noun); produce or make happen (verb)
>
> Adverse: harmful, unfavorable, or opposed to one's interests—said of conditions or events, not persons (Sentence example: We came upon adverse conditions on the trail.)
> Averse: reluctant, unwilling, or set against—said of persons (Sentence example: The manager was averse to the idea.)

Because: indicates the cause of, or the reason of, something (Sentence example: Because milkweed supports monarch butterflies, we planted it.)
Since: continuously from a prior time, during the time following an event, for a period of time (Sentence example: Since we planted milkweed, we have seen monarch butterflies return to the area.)

Comma: punctuation mark used to separate parts of a single sentence (noun)
Coma: state of unconsciousness possibly induced by reading a writing textbook for too long (noun)

Complement: suit as a companion; make something whole (verb)
Compliment: praise (verb) or a word of praise (noun)

Confident: possessing strong sense of personal ability, value, or virtue
Competent: able to accomplish a task with skill, intelligence, or knowledge

Criterion: a single expectation or requirement of something to be judged
Criteria: plural of "criterion"

Datum: a single item, fact, or figure; often of measurement
Data: the plural of datum, that is, two or more items, facts, or figures of measurement

Decent: satisfactory, acceptable, or OK (adjective)
Descent: decline in altitude, rank, and so on (noun)

Efficient: favorable use of resources to accomplish a goal
Sufficient (or, Adequate): Acceptable use of resources to accomplish goal

Exhaustive: covering an issue fully
Exhausting: tiring to the point of total enervation (energy loss)

Expended: Used up or fully activated/drained to emptiness (important in analysis of crash of Valujet Flight 592)
Expired: Past its ideal use date or beyond an allowable timeframe or termination

Formidable: difficult to overcome or prevail against; causing dread or fear
Formative: helping to shape or mold; related to development

Indigent: poor, needy, or destitute
Indigenous: native to an area or region

Influential: having potent impact on something or someone
Impressionable: easily affected, molded or influenced, sensitive to influence

Insure: to protect against loss
Ensure: to make certain

Farther: a larger measurable distance (farther down the trail)
Further: a larger degree or non-measurable quality (further into psychoanalysis)

Fewer: not as many when item can be counted (Example: fewer parts)
Less: not as many when item cannot be counted (Example: less stamina)

Ordinance: law enacted by city government or a direction from authority
Ordnance: ammunition, weapons, and related material for warfare

Principal: main or primary (Sentence example: The principal component of this system is the one-way valve.) Principal also refers to the head of a school. (Sentence example: The principal is a friend to students and parents alike.)

Principle: value, maxim, judgment, or rule to guide one's actions (Sentence example: The principle of altruism motivates her to work with orphans.)

Your: possessive pronoun for 2nd person *you*, indicating possession of object that follows (Examples: your basketball or your lemonade)
You're: contraction of *you* and *are*; a short way of saying *you are*. (Sentence example: You're the best writing student in this college.)

This final pair is not all that germane to ABERST, but I included it because it is the mistake I make the most in my writing. I really disappoint myself with this blunder, and perhaps you have seen it sneak into emails, texts, and reports too. You might have your own peccadillo in this category? If you want to send it to me, I would love to send you a note with a "thanks for your contribution," and I might include your *peccadillo pair* in the next edition.

A.3 Terminology Troubles

A similar problem to conflating two different words is to refer to something by two or more different terms without clarifying the relationship between the terms. This can be confusing for readers, unless the words are explicitly defined as synonyms. Or, if the terms are related in a way that is different than synonyms, this explanation must be provided to readers. Otherwise, the text is confusing, and readers may wrongly suppose that the items are interchangeable. In ABERST documents, the confusion usually affects an important subject, such as a concept, device, or component. In sum, the hazards are several and distinct and can be avoided as follows: (1) Limit the use of interchangeable terms for the same concept. (2) Explicitly define synonyms or clarify relationships between terms that are often used interchangeably but should not be used in that way. (3) Ensure that you do not call something by the wrong name due to a careless and liberal use of alternative terms.

Here is an example from computer science:

- The electrocardiogram records are effectively analyzed through an updated deep-learning architecture. The deep-algorithm learning is slower, but more effective, than its predecessor.

If the two key, similar-sounding nouns ("deep-learning architecture" and "deep-algorithm learning") refer to the same thing, not all readers will find that obvious. Readers could rightly expect the two items to be different concepts: Perhaps the algorithm is a component of the architecture, and it is primarily the part, not the whole architecture, that has been updated. Or, readers might interpret this as conveying that "deep-algorithm learning" is another name for the whole architecture, but they would be assuming this. It is not explained.

Here is an example from aerospace engineering:

- The rocket engine was of primary concern because previous testing suggested it was imprecise. Our testing of the rocket, however, indicates strong consistency across samples. In fact, the motor deviated only .04 percent when looking at the variation of total impulse at a 95% confidence level.

In this example, the author begins with "rocket engine," but switches in the next sentence to "rocket" by itself. These different words imply two different components in aerospace design. Usually, the propulsive component of a rocket is its engine. The rocket itself is the body that typically holds fuel, payload, controls, and aerodynamic surfaces. In writing a report about propulsion testing, an author must be careful to say the rocket's engine was designed and tested. Changing this to "rocket" is erroneous. Furthermore, perhaps "motor" might slip into a sentence, and then the reader contends with a seemingly new idea: the motor is now the issue. Are the motor and the engine one and the same, or a different item, say, some subpart of the engine? A hybrid automotive vehicle might have both a motor and an engine, but does a rocket? Some do: The Delta II and III rockets both have a single main engine and additional motors. Some do not: The Atlas V 531 has an engine and additional boosters, but no motors. In these rockets, an engine and a motor would refer to different components. So, the terms should not be used interchangeably without explicit discussion. Without any explanation of their relationship, or all of the rocket's components, a serious confusion arises.

As the examples above demonstrate, use of terminology over generously or carelessly causes confusion and creates a document that generates a sense of bewilderment in the reader. Writers may think they are showing off their large vocabulary or their understanding of the lingo and jargon in their field, but the effect on the reader is to confuse and puzzle. Look at the following example from civil engineering and natural resources:

- The treatment plant units are the following: bar rack, grit chamber, primary sedimentation basin, activated sludge basin, and secondary clarifier. We optimized the primary clarifier only.

Sadly, "primary clarifier" had not been mentioned as one of the units. We must ask: Is this a synonym for "primary sedimentation basin," or is the initial list missing this? Readers cannot know based on the text.

One last example from mechanical engineering:

- The engine's total weight must be reduced. The power plant originally weighed 125 pounds. The flywheel alone weighed 25 pounds. The cylinder block weighed 45 pounds. The other parts weighed 55 pounds. The design review team mandates that the engine come in below 120 pounds.

Without knowing that "power plant" is a synonym for the engine itself, a reader may add up the weights of all the listed components and reach a sum of 250 pounds. This will be hard to get below 120, so the problem is severe. Or, the other items, not counting the power plant, add up to 125, so the power plant is the engine, and it is merely 5 pounds too heavy, which is a small problem. The trouble with the terms obfuscates the intensity of the problem, so it results in poor communication. Readers may not follow the English, while, perhaps worse, the math is not clear either, which is abhorrent in ABERST, truly detestable.

A.4 Personal Pronoun Problems

The first- and second-person pronouns, primarily I, we, and you, are surrounded in controversy. Some writers have been told to never use these personal pronouns in ABERST. They have been told that science and engineering are objective and impersonal, and these personal pronouns have no place in such fields. Some writers have been told nothing about them, and still others have been told to use them to personalize their documents. I take some responsibility for those in the third group; Indeed, I encourage you to use first- and second-person pronouns, just as I did in this very sentence. I explain the reasons behind this advice in subsections 2.2.3.2 and 2.2.4.6, as pronouns relate to the topics of tone and word choice, respectively.

Nonetheless, the first- and second-person pronouns can lead to mistakes, especially in the subjective case (I, we, and you), while not as often in the possessive (mine, yours, ours) or objective case (me and us).

Because I suggest that the personal pronouns belong in technical reports, you might be surprised to learn that two problems I see frequently are (1) overuse and (2) inappropriate use. I also see (3) absence when the prose would be improved with personal pronouns.

Problem 1: Overuse—The pronouns are substitutes for actual nouns to be discussed, which makes the text less direct and forceful than it could be with actual nouns:

- I decided to repeat the sampling three times to reduce instrument- and human-based errors. I operated the equipment for an additional hour beyond the scheduled time to complete my testing.

Here, the writer, his decision, and his overtime hours take center stage as the sentences' subject, verb, and objects. That's not incorrect, but it detracts from the essential message of the passage, which is that multiple samples were taken to improve the integrity of the result. So, the sentences can be improved as follows:

- Sampling was repeated three times to reduce instrument- and human-based errors. This required one more hour of testing than planned.

A similar problem occurs with "you":

- You will see that we used a 1/5-scale model and adjusted fluid speed to match Reynolds number.

The direct address to the reader, via "you," is unnecessary because nothing is expected of him/her but to "see," which means nothing more than "read" or "note," which is understood by including the information in the report, in its pure form. The same is true for "we used," in this sentence, which is about testing and not opinions asserted (where "we" would be appropriate). Below is an improved version:

- Testing involved a 1/5-scale model and fluid speed adjustments to match Reynolds number.

Problem 2: Inappropriate Use—The pronouns are used sloppily and convey a nonsensical message. Look at these samples:

- We noted that the output voltage peak to peak dropped off no matter which way you went.

Here, the person addressed as "you" is not going anywhere; rather the authors are trying to say they adjusted a signal's frequency both lower and higher, so the "you" is a poor substitute for a dial on a function generator, which "went" lower and higher.

A few more problematic examples:

- Since the diode is wired in parallel with another resistor, you actually receive a decrease in resistance. (We hope the reader isn't wired into the circuit.)

- The higher up in resonance frequency modes you go, the more nodes you get in the structure.

Here, the higher up "you go," refers to acoustic waves traveling through a pipe, tube, or duct, not a person "traveling" higher or farther. A similar occurrence of peculiar human motion is described in this example:

- As you move down the centerline of the wind tunnel, the velocity increases slightly.

Again, the reader is not traveling through the wind tunnel. As air moves down the wind tunnel, its velocity increases slightly. Air should not be referred to as "you."

Problem 3: Absence—Significant content is not expressed in the sentence without suitably using personal pronouns.

When the personal pronouns convey crucial information regarding persons making decisions or changing protocols, they should be included so readers can understand that specific persons have taken actions or reached conclusions. Look at this example from a post-testing report with a new device recommended:

- Good: Previous experiments involved hot wire anemometers, but I thought of using a hot film anemometer.

As happens often in ABERST, the author is a researcher with a problem to solve and an innovative solution to report. The author needs explicitly to convey that, in contrast to his predecessors' efforts, her new approach is different. If "I" were removed, the meaning would be ambiguous, as in this poor example:

- Bad: Previous experiments involved hot wire anemometers, but this one involved a hot film anemometer.

Here, the sentence expresses no contrast between the others and the author; it suggests that either (1) the author had no part in either of the experiments or (2) the author performed both types. The absence of personal pronouns leads to an incomplete message and ambiguity.

Here is another example with a meaningful and important use of personal pronouns:

- The propulsion engineers are grumbling that maybe our medium-size wind tunnel's computerized data acquisition system (DAS) is obsolete. We talked with them about the delays this causes in their work, so we are writing to request approval to purchase a new DAS.

Without the first- and third-person pronouns, the sentence does not reveal the important people who are motivating and handling this concern. The equipment does not simply "get old" and "becomes updated" on its own. People have a role to play in these opinions and decisions. The pronouns reveal the "players" and their "parts." Here is a revision without personal pronouns:

- Perhaps the computerized DAS is obsolete, as some say, and a new one has been requested.

We can only ask, "Who thinks the DAS is obsolete; who has requested a new one; and what is the author's relationship to these other, mysterious persons?"

Similarly to using the first-person pronouns effectively, authors can use "you" directly to address the reader: This is sometimes vital for expressing your writing purpose: provide information to someone in particular to ask that person to do something specific, as in this example:

- The decision remains for you. You need to select one of the three vendors presented in this report.

A.5 Run-on Sentences and Sentence Fragments

Run-on sentences are like your embarrassingly tacky cousins from the backwoods who camp out in their rusty trailer in your front yard for 2 weeks after the family reunion, and sentence fragments are your pompous, elite, east-coast relatives who show up to family events only for a split second, in between their jaunts to other, important and fashionable events in cities more cosmopolitan than yours. Both are the opposite of the ideal: natural, easy-going, courteous complete sentences that neither overstay their welcome nor stop in too briefly to provide any lasting meaning in your life.

A.5.1 Run-On Sentences

The rule in English is that a sentence is, at minimum, one independent clause. An independent clause requires a noun (the shoe) and a verb (fits). A sentence can expand to include either additional subordinate clauses or another independent clause if joined with a proper coordinating conjunction. That is the basic rule. So, you have three options:

1. Single independent clause: Kumar drives his car fast to impress girls.
2. Independent clause plus subordinate clause: Although she had a fear of heights, Elin agreed to try skydiving.
3. Two independent clauses joined: Ang likes film, but Lee enjoys television.

In ABERST, the sheer volume of information that needs to be put into reports overwhelms the ordinary author. The information is not only complex and difficult to explain, but it just goes on and on forever. Writing becomes a marathon. As a result, in an effort to shorten the route, and get finished sooner, authors try to reduce the number of sentences required. It makes sense. The sentence seems to be the primary building block of a report. Thus, the fewer sentences, the sooner the writing work will be done. This is misguided but understandable. Mathematically, it might seem clever. If I have 50 ideas to convey, my work might be easier if I pack 10 ideas into each of 5 sentences. Thus, I need to write only 5 sentences rather than 50. An author reduces the number of sentences by packing lots of ideas into one long sentence. The problem with this approach is that those five sentences will be difficult to read and insufficiently clear. They will be run-on sentences, certainly. Here is an example:

- The feedback loop may create an overdamped system, or an underdamped system can occur, however the goal is to reach an ideal critically damped system where the steady-state error is zero, or close to it, which can be on the order of hundredths of a volt and this is observable and to be measured.

How do you prevent run-on sentences? The answer is twofold: 1) Know the rules of independent clauses and complete sentences. 2) Proofread all your sentences after you have written your first draft, and fix any run-ons. Fixing the preceding run-on is done as follows:

- The feedback loop may create either an overdamped or an underdamped system. Neither is ideal, which is to reach a critically damped system, where the steady-state error is zero or close to it. This small error, which can be on the order of hundredths of a volt, is observable nonetheless. Both reducing and measuring this error must be done.

A.5.2 Sentence Fragments

In this subsection, I present several typical sentence fragments and explain the reason they are fragments. Fragments slow readers down, making them pause to deduce how the fragment connects to either the preceding or subsequent sentence. Sometimes the connection is obvious, so the reader's pause is brief. Sometimes the connection is less obvious, and the reader's pause is long, or it is interminable when no certain meaning can be determined. Here are passages from reports that contain sentence fragments, and you can see for yourself how you must stop up and try to figure out the sentence problems and, eventually, the author's meaning. In the final example, you will not be able to determine the author's message because of the fragment.

Short pause:

- We evaluated several important performance parameters for each filter. Specifically, the frequency response limit, the linear response of the filter beyond its limit, and the time-dependent voltage response.

Within a few moments, a reader can see that the fragment, beginning with "Specifically," is simply the list of performance parameters evaluated. Nevertheless, a reader expects a verb to appear somewhere after the list (of nouns) because it is presented as a full sentence. (The fragment begins with a capital letter S following the period at the end of the previous sentence.) After reaching the final period, the reader realizes that no verb is included in the fragment, so the fragment is merely a list, which should be presented like this:

- We evaluated several important performance parameters for each filter: the frequency response limit, the linear response beyond such limit, and the time-dependent voltage response.

Long pause:

- The train runs at high speeds. Because it is aerodynamically advanced. Production models achieve nearly 92% of the theoretical and simulated maximum speeds.

The seemingly independent sentence starting with *Because* is a fragment: on its own, it makes a typical reader wonder if an additional message will be provided, such as "it is quiet" or "it burns fuel efficiently." Readers proceed to the next sentence to see a new piece of information, then they re-process the two prior clauses as necessarily linked. The first two ideas fit together nicely and most readers will see the connection, but only be certain after reading the third sentence. Thus, readers must process and re-process three sentences to absorb the message the author should have delivered with one sentence. In the one, corrected sentence, the "because" fragment is an adverbial clause that is dependent upon, and enhances, the first independent clause preceding it:

- The train runs at high speeds because it is aerodynamically advanced. Production models achieve nearly 92% of the theoretical and simulated maximum speeds.

Interminable pause:

- The flange immediately underneath the reverse-clad housing put into place to improve durability.

This fragment is incomprehensible because readers are unsure whether the flange or the housing improves durability. If the flange improves durability, the author simply left out "was" in front of "put into place." It would be this:

- The flange immediately underneath the reverse-clad housing *was* put into place to improve durability.

But if the housing improves durability, then the verb that is missing must come after "durability," and the fragment is missing some statement about the flange, which is the noun subject of the fragment. In some way, the corrected sentence must be primarily about the flange:

- The flange immediately underneath the reverse-clad housing put into place to improve durability . . . is made of dual-phase steel? is an ovoid? is 2- millimeters thick?

As you have seen from the examples above, fragments seem to be created inadvertently by authors, who leave out one word or use incorrect punctuation. The ease of error underscores the importance of proofreading, especially considering that a fragment can be a fairly detrimental mistake yet quickly fixed. Thus, although fragments are often results of unintentional error, the following review can be helpful in preventing them. The following are typical phrases and clauses used to enhance meaning and improve your message. They are fine to incorporate into sentences, but they are not, on their own, complete sentences.

Prepositional phrase:
- After the storm
- Under the radar
- In the response
- At the turn of the century

Gerund phrase:
- Averaging the five values
- Visiting the laboratory

Participial phrase:
- Carrying a parasite
- Getting dimmer

Noun-modifier phrase:
- Inter-cooled, jacketed heat exchanger
- Light-weight composite material

Subordinate Clause:
- Whoever runs the test
- Because the unemployment rate is steady

Linked together with an independent clause, they create meaningful complex sentences. On their own, they are dependent on a partner, that is, an independent clause. The short phrases and clauses are likely obvious to everyone as incomplete sentences, but when a few more words are added, phrases and clauses can seem to be full sentences when they are not, as in these examples:

- After the storm, which had not been forecasted.
- Light-weight composite material in tests and experiments with small variations upon error analysis.
- As we increase the frequency, a steady decrease in the peak-to-peak amplitude response.

Readers must ask about "the storm," the "composite material," and the "steady decrease." With the latter, readers may wonder: Does a "steady decrease" appear or occur? Was it expected? Would it give one reason to worry? In other words, the fragment is incomplete and readers must guess as to the author's intended meaning.

Again, sentence fragments may result from either the author's (1) belief that the fragment is a full sentence or (2) a missing or incorrect word, phrase, or punctuation mark. Only authors can fix these hazards before the report is submitted. After that, readers can only pause, make the mental corrections themselves, or remain scratching their heads in confusion. Here is one final example that involves a tiny mistake and a large confusion.

- The principal investigator became dizzy. With the project schedule in his head; he was disappointed he had to delay the pilot study two more days while he went for medical testing for possible exposure to neurotoxins.

A fragment is produced with a tiny mistake: incorrect use of a semicolon instead of a comma. This little error creates the vastly wrong impression that the PI became dizzy because the project schedule was swirling around in his head. However, the schedule combined with the delay made him disappointed, not dizzy. His dizziness derived from possible exposure to neurotoxins. An improved version without a semicolon fragment, and with a correctly used comma, follows:

- The principal investigator became dizzy and had to delay the pilot study two more days while he went for medical testing for possible exposure to neurotoxins. With the project schedule in his head, he was disappointed with this delay.

A.6 Misplaced Modifiers and Clauses

Modifiers must be placed immediately before or after the noun or clause they are intended to modify. If not placed correctly, an unintended meaning is produced. In particular, when placing a subordinate clause in front of a noun, be sure to place the correct noun after such clauses to avoid a common grammatical mistake. Look at the examples below. I enjoy a good joke, but this is not the way to infuse your documents with humor, although that is often the inadvertent consequence of misplaced modifiers and clauses. You can try fixing each one in the space provided.

- Having tested seven software programs, the faulty keyboard was identified as the root cause. (Keyboards cannot conduct testing.)

____

____

- Thinking about saving money, the Epson projector offers our company the best combination of features and price. (Projectors do not think.)

____

____

- Though unsightly and non-aerodynamic, engineers could quickly apply duct tape to the flapping photovoltaic cells to resume flying the solar airplane. (Engineers might be introverted but it is unkind to say they are unsightly and non-aerodynamic.)

____

____

- Once vibrating, we learned that by adjusting the frequency and voltage we could find the natural frequencies in the beam where resonance occurs. (I picture the authors sitting in one of those vibrating massage chairs at the mall to conduct this mechanical testing of a beam.)

____

____

- Weighing 250 pounds, the hamburger hardly filled me up. (If the hamburger weighed 250 pounds, it should fill you up.)

____

____

- Being inclined to bark, whine, and noisily lick themselves, business unit managers should not bring their dogs to the office. (Managers need to behave themselves.)

____

____

The solution to a misplaced modifier problem is to rearrange the phrases, as in the following:

- When working at the office, business unit managers should not bring their dogs, which are inclined to bark, whine, and noisily lick themselves.

A.7 Which/That Clause Confusion

Some people feel that obsessing between "which" and "that" is an unnecessary pedantic quibble. They say use either one you feel like, interchangeably, and stop worrying. Here is the problem: without careful selection of the word AND punctuation, your message may be misleading and, worse, erroneous. You do not want to state an incorrect fact because of this English pitfall.

The twofold which/that rule is simple:

1) Use commas with "which" and non-restrictive relative clauses, indicating it is an optional clause and not necessary for full understanding of the antecedent. The information between the commas is optional and not essential in identifying the antecedent.
2) Do NOT use commas with "that" and restrictive relative clauses, indicating the clause must be included in the sentence for full understanding of the antecedent. A restrictive clause is essential in identifying the antecedent.

Practicing with some examples helps to comprehend this important rule and avoid the hazards of which/that mistakes.

- We designed a jacketed reactor, which allows the bacteria to ferment and create beta-carotene.

This construction implies that all jacketed reactors allow bacteria to ferment and create beta-carotene; in other words, the purpose of all such reactors is to do that, and only that. But, that is not the case. Reactors come in all shapes and sizes, and they serve a myriad of purposes, depending on the industry. Try this improvement:

- We designed a jacketed reactor that allows the bacteria to ferment and create beta-carotene.

With this construction, which is the correct one, you are conveying the point that your designed reactor (the antecedent) serves the very specific purpose of allowing the bacteria to ferment and create beta-carotene. The restrictive clause is crucial for identifying the jacketed reactor specifically.

Here is an appropriate use of "which" after "reactor":

- We designed a reactor, which is a tank or vessel wherein a chemical, thermal, or biological reaction takes place, to facilitate beta-carotene production.

In this construction, the clause containing "which" serves to be an unrestricted relative clause, as opposed to a restrictive relative clause with "that." It is optional, and it is not required to identify the antecedent but rather explain it.

Here is another problematic which/that example:

- First used by the U.S. military, microwave antennas which operate in the 4.2 to 6.3 gigahertz band are health risks with prolonged exposure.

This is about human health. It is important. The sentence suggests, because the "which" clause is not encapsulated with commas at start and finish, that ONLY those specific microwave antennas that operate in that 4.2 to 6.3 range are health risks, while other microwave antennas are NOT health risks. A proper "which" clause with commas, however, tells readers that ALL microwave antennas operate in that range, and, thus, ALL pose health risks. So, which is it? Some subset or all? To be clear in identifying the antecedent, you must follow the rule, especially when you are the expert and the person explaining the specific situation. Here are the possibilities:

- **Option 1:** First used by the U.S. military, microwave antennas, which operate in the 4.2 to 6.3 gigahertz band, are health risks with prolonged exposure. (ALL)

- **Option 2:** First used by the U.S. military, microwave antennas that operate in the 4.2 to 6.3 gigahertz band are health risks with prolonged exposure. (ONLY the subset)

One more example:

- This manual explains in five stages how to detect and fix a bicycle's flat tire using equipment which can be found in most households.

Recall the rule: The "which" clause requires a comma to precede it, indicating that it is an extra piece of information that can be removed from the sentence without significantly altering the message or confusing readers as to the precise antecedent (equipment). Using a "that" without a comma is appropriate when the extra piece of information following "that" cannot be removed from the sentence without significantly altering the message. It is essential to the meaning of the preceding words. So, here, again, the mistake is to use a modification that hybridizes both versions and leads to ambiguity. The resolution is as follows: "Equipment" is really meaningless (unidentified) in this context until the additional clause elucidates that the manual demands nothing but "equipment found in most households." It is necessary and restrictive. The correct version is this: This manual explains in five stages how to detect and fix a bicycle's flat tire using equipment that can be found in most households. Same with this example from aerospace:

- The 8V8 aircraft will have an engine which boasts 70,000 pounds of thrust.

You can see the problem now without any explanation.

Look at the following examples from an imagined public health report. Anyone working in ABERST would certainly want to make sure these statements are unambiguous so the public could be properly educated. All three need correction for accuracy and certainty. Feel free to correct the sentences directly on the page:

- We recommend eating primarily vegetables which are low in acid.
- We recommend eating primarily vegetables, that are low in acid.
- We recommend you supplement with whole grains, that are low in acid.

Practice fixing these problematic which/that clauses:

- I like vegetables which are high in soluble fiber.

____

____

- I enjoy Haiku which is a form of poetry with Japanese origins.

____

____

- Our only OSHA inspection team which is operating under a new protocol plans to be on-site for 3 days.

____

____

- The OSHA inspection which is responsible for auditing the coatings laboratory arrives tomorrow.

____

____

- It is a pressure vessel which contains an agitator, cooling jacket, and dip tube.

____

____

- Invertase is an enzyme which is capable of catalyzing the inversion of sucrose.

____

____

Here is my analysis of the last one. This sentence suggests all enzymes (by definition) are capable of catalyzing the inversion of sucrose, but no comma is present to confirm this. In fact, it's an incorrect interpretation. The capability of invertase is a restrictive clause required to understand the subject (invertase) fully (not enzymes generally).

If you are getting hungry and thirsty after extensive study of this subsection, I will end this topic with an incorrect example to help you think of refreshment:

- Wrong: He enjoys ales and lagers which are handmade in small batches in microbreweries.

Placing a comma before the "which" would render the sentence non-factual because not all ales and lagers are handmade; some are made in huge breweries using automated equipment. The only way for this sentence to be accurate is to be rewritten with a restrictive clause using "that," as follows:

- Correct: He enjoys ales and lagers that are handmade in small batches in microbreweries.

A.8 Ambiguous Demonstrative Pronouns

The red alert here is to make sure that your demonstrative pronouns ("that" "this" or "it") are explicitly linked to a definite antecedent. Without careful attention, these demonstrative pronouns could refer to one or more items previously introduced, as in this example:

- We adjusted the potentiometer as well as the angle of the antenna to get the measurements we needed to finish our field work and submit our report, and it was easier than we had imagined.

Which item was easier, the field work or the report? Here are three possible correct revisions:

- **Correct 1:** We adjusted the potentiometer as well as the angle of the antenna to get the measurements we needed to finish our field work, which was easier than we had imagined, and submit our report.
- **Correct 2:** We adjusted the potentiometer as well as the angle of the antenna, which was easier than we had imagined, and we thereby got the measurements we needed to finish our field work and submit our report.
- **Correct 3:** We adjusted the potentiometer as well as the angle of the antenna to get the measurements we needed to finish our field work and submit our report; and it was easier to write than we had imagined because we followed the guidelines from a great book on technical communication for college and career.

A.9 Faulty Comparisons

Many years ago, I learned an important grammar lesson from a standup comedian. Sadly I can't remember his name, but he was funny. And here was his joke: "What's with tobacco advertisements on billboards? Carlton has less tar. Less tar than what? They don't tell you, do they? It might not be such a great cigarette if you knew... less tar than your average freeway on-ramp."

At its most basic form, a comparison always requires two, or more, items to be compared, say A and B. In comparisons, therefore, the reader must be clear about both the A and the B terms. The mistakes usually concern these terms, and I address the two most common ones I see in technical writing in the sub-sections below.

One common flaw is to leave out the B term (as in the Carlton example above). Another one is to put in an incorrect B term so that A is being compared to something nonsensical.

A.9.1 No B Term

Let's look at a serious example that has a missing B term:

- In terms of braking, the Camry stops more slowly on wet pavement.

A reader is unsure of the Camry's performance. Does it stop more slowly than the other vehicles tested and compared to the Camry? Or is this "more slowly" than the Camry's stopping time on a surface other than wet pavement, such as dry pavement? So, to complete and clarify the comparison, the other term must be added. Here is a possible solution, although only the author can fix this comparison definitively. The rest of us must guess.

- In terms of braking, the Camry stops more slowly on wet pavement than on dry, loose gravel.

You can fix this one:

- The hospital administrators were worried about the health of x-ray technicians, so they introduced a procedure that would be safer.

A.9.2 Nonsensical B Term

Let's look at an example that comprises mismatched items for comparison:

- The standard chiropractic coursework is not as comprehensive as a medical doctor.

Maybe a doctor would like to be compared to an encyclopedia, but even doctors would have a hard time matching up. Here, a doctor is compared to coursework. Instead, to fix this nonsense, you must compare two sets of coursework, in one of the following acceptable constructions:

- The standard chiropractic coursework is not as comprehensive as that of a medical doctor.
- A chiropractor's coursework is not as comprehensive as a medical doctor's coursework.

You can even leave out the second "coursework":

- A chiropractor's coursework is not as comprehensive as a medical doctor's.

The rule applies to anything (noun) that is compared: people, places, and things. Here is an example with power, followed by one that is personal:

- That engine's efficiency is higher than this engine.
- John has a bank account larger than Pete.

You cannot compare efficiency to engine. Either efficiency to efficiency, or engine to engine:

- That engine's efficiency is higher than this engine's (efficiency).
- That engine is more efficient than this engine.

John might be larger than Pete, but his bank account cannot be compared to Pete's physical stature. So, it is fixed as follows:

- John's bank account is larger than Pete's.
- John has a larger bank account than Pete has.

A.10 Series Errors

Presenting a series of related items is an excellent and useful approach to providing information to your readers. All too often, however, authors trip up on series constructions. A series might start out fine with two or three items fitting nicely together, but authors tend to pack additional information into the series that no longer matches the pattern and style of the original items. In such instances, the sentence becomes difficult to read, and the information is often unclear at best and incomplete, confusing, or redundant at worst. Here is one example:

- I will show you how to sauté zucchini, steam broccoli, and carrots that have been peeled and washed carefully.

Is the reader going to steam carrots along with broccoli, or do something else with carrots, which is missing? Boil? Shred? And has the broccoli been peeled and washed carefully, or just the carrots? Neither the intended meaning nor the exact problem is obvious to readers. The series could be fixed in various ways, only one of which is consistent with the author's intent. Thus, only the author can fix this, perhaps as shown in the following corrections:

- I will show you how to sauté zucchini, steam broccoli, and shred carrots, which must be peeled and washed carefully.
- I will show you how to sauté zucchini and steam broccoli and carrots. The carrots must be peeled and washed carefully.

"Carrots" was added on to the series somewhat too casually. In the next example, all the items do belong in the series but they are presented in varied style, which slows down readers:

- The company has modified the technology to be applicable to finding stolen cars, missing pets, lost adult recovery, and identifying distress calls.

It starts fine, with "finding stolen cars and missing pets," but it makes no sense to "find" the next item: lost adult recovery. Here the author has added a new term "recovery," which replaces "find." This addition applies to the last item also: "identifying distress calls." The solution is to have just one gerund: "finding" and eliminate the other two. When looking for the best term for all items in the list, perhaps "finding" can be switched to "locating":

- The company has modified the technology to be applicable to locating stolen cars, missing pets, lost adults, and distress calls.

Or, if a unique verb is important for each noun (direct object), you can use four verbs:

- The company has modified the technology to be applicable to finding stolen cars, tracking missing pets, recovering lost adults, and identifying distress calls.

Mixed-up series constructions are a common problem, and my speculation is that it derives from the need to provide readers with a single comprehensive, multi-faceted package of information. This is a laudable and worthwhile effort; you have a lot to explain to your readers, so why not serve up some information-packed sentences? But, this approach can backfire. Authors become overly ambitious and try to pack more information into a series than it can hold. Here is another example:

- The U.S. Armed Forces are using the global positioning system (GPS) for foot soldiers, surface and air vehicles, rocketry, missile navigation, and precision landings.

Looking at the series, we can see that foot soldiers and vehicles are plural nouns, but rocketry is a field of aerospace engineering (not a plural noun). Furthermore, "missile navigation" is not a plural noun but, rather, the combined expression of the purpose of "using GPS," namely navigation, along with "missile" (a particular usage), and precision landings is a task accomplished by one of the previous nouns with the help of GPS, likely air vehicles or missiles, or both. In sum, this series is a potpourri of ideas all related to GPS but not a good, clear list. The following is a correction:

- The U.S. Armed Forces are using GPS for foot soldiers, surface and air vehicles, rockets, and missiles.

This is all too typical. Authors start off down one path, then they meander in a new direction, then reverse direction, only to end up repeating a thought from the beginning, yet in a different form that slightly masks a redundancy. A pragmatic solution might be to shorten the series and use another sentence to explain the additional point(s):

- The U.S. Armed Forces are using the global positioning system (GPS) for foot soldiers, surface vehicles, airplanes, and missiles. GPS improves both navigation and precision landings.

In fact, quite often the last item tacked onto the end of the series, which appears at first to be the last item in the series, is actually the author's next thought. The solution is to present the series carefully, and create a new sentence for any item that represents a related but separate idea from the series.

- Our report discusses the filter's parameters, limitations, advantages, and conclusions regarding a purchase recommendation.

This series implies that "conclusions" belongs to the filter, along with parameters, limitations, and advantages, but conclusions come from the authors. The last item in the series must be removed and placed into a new, independent clause, following the coordinating conjunction "and":

- Our report discusses the filter's parameters, limitations, and advantages, and it includes our conclusions regarding a purchase recommendation.

One final example reinforces the main lessons of this section:

- This literacy program addresses students with diverse backgrounds including living in shelters, single-parent homes, having trouble with the law, enrollment in English as Second Language programs, and cultural heritages from Latin America or the Caribbean.

Readers expect the series to follow after "living," with a list of places students are living: shelters, single-parent homes, and . . . another location. Instead a new issue entirely is introduced: *having trouble with the law*, which might work if "living," "having," and "enrolling" were used as the key items (adjectival participles) of the series, but the last two items are "enrollment" and "cultural heritages," which are not both participles to match *living* and *having*. So, at this point, a fairly strong correction might come to mind:

- This literacy program addresses students with diverse backgrounds including living in shelters and single-parent homes, having trouble with the law, enrolling in English as Second Language programs, and possessing cultural heritages from Latin America or the Caribbean.

The solution uses the adjectival phrases as the items, to allow more information to fit into the series than is possible with only one adjectival participle: "living." Now there is "living," "having," "enrolling," and "possessing." Also this fixes the first two items, which are objects of "living," where a comma between shelters and single-parent homes is misplaced and confusing. This is similar to erroneously placing a comma between apples and oranges in this sentence:

Wrong: The doctor suggested eating more apples, and oranges.

No comma is necessary when only two items are included. The comma is used when the series has three or more items.

Correct: The doctor suggested eating more apples, pears, and oranges.

A.11 Wordiness

Public enemy #1 wreaking havoc on technical writing is excess verbosity, or, simply, wordiness. The ideal is to be concise and clear, using the minimum amount of words required to convey your message and fulfill your communication purpose. Other types of writing do not have this requirement, and the language itself may be the message from author to reader. Words, words, and more words might be appropriate for novels, memoirs, and poems. But working in ABERST, people are busy and time is of the essence. Reports must be concise. Thus, you must seek out and destroy all instances of wordiness in your writing. If you miss a few loquacious passages here and there, readers will forgive you, as long as the bulk of your prose is tight and trim.

Most English style and grammar books cover this topic and offer some excellent guidelines. I learned myself from reading this principle in textbooks and observing it first-hand in my work as a technical editor. Below I address the key pitfalls you are likely to struggle over so you can steer clear of them.

A.11.1 Self-Evident Subordinate Clauses or Phrases (empty, vacuous, or platitude)

Wordy: In the field of medicine, as well as in the field of law, the classroom training is rigorous.

Concise: In medicine and law, the classroom training is rigorous.

Option: In medicine, as well as in law, the classroom training is rigorous.

The option enables you to retain the additional thought, if it's important: "On second thought, my observation applies to law, as well." You still eliminate the first wordy phrase, "in the field of."

Wordy: Hewlett-Packard is a company that sells scientific equipment and computer peripherals.

Concise: Hewlett-Packard sells scientific equipment and computer peripherals.

Wordy: My manager is a person who oversees all machining operations of our domestic factories.

Concise: My manager oversees all machining operations of our domestic factories.

Wordy: Currently, the employees on the developer team in our company are developing a mobile app for person-to-person money transfers.

Concise: Our company is developing a mobile app for person-to-person money transfers.

Wordy: At this time today in this country, American automobile sales have surpassed pre-recession levels.

Concise: American automobile sales have surpassed pre-recession levels.

Enhanced: According to last month's data as reported by the *Wall Street Journal*, American automobile sales have surpassed pre-recession levels.

The enhanced version is helpful if you need or want to add a single word or phrase of introduction to link this sentence to the prior sentence or to emphasize an angle indicative of your message, such as "Remarkably, or "Unexpectedly," or "Indeed," or "According to data compiled by J.D. Powers & Associates."

Wordy: Appendix A is an attachment that shows the primary circuit schematic.

Concise: Appendix A shows the primary circuit schematic.

Using "is an attachment" is both obvious and wrong, for Appendix A is an appendix, not an attachment.

Wordy: As per your request in a letter of August 16, 2016, the meeting in Geneva will be extended by 1 day.

Concise: Per your request in a letter dated August 16, 2016, the meeting in Geneva will be extended by 1 day.

Concise: As you requested on August 16, 2016, the meeting in Geneva will be extended by 1 day. (If the medium of the request is not essential.)

The expression "per X" is a good one, but it is often incorrectly corrupted into "as per." That corruption is wordy and unnecessary. It seems to stem from a misguided motivation to use both "as" and "per," as if an author starts out with an "as" phrase in mind, then changes plans and truncates the intended phrase with a quick "per." One can use either an "as" or "per," but not both combined.

A.11.2 Verb Conversions

Another wordy blight on ABERST documents stems from the conversion of original verbs into multi-word, noun-heavy pseudo verbs. The ideal is to emphasize the crucial concept in a single verb, but often confusing verb conversions are common. For this topic, let me show you examples that were shown to me as a graduate student:

- Wordy verb phrase: Put emphasis on
- True verb: Emphasize

In the wordy version, the verb is "put" with a direct object (noun) of "emphasis." "Put" is a pseudo verb because the true verb is "emphasize," which you can derive from the noun that follows.

- Wordy verb phrase: Increase by a factor of two
- True verb: Double

- Wordy verb phrase: Give an explanation of
- True verb: Explain

The improvement makes sense because it is a concise, single word (verb) that has the same meaning as the phrase, which has a misleading style. Instead of "giving," we are "explaining." OK, so here is my list of these collected over the years, with the first and last ones done for you. You can do the others:

- Give the ability (enable)
- Develop a design ( )
- Be a success ( )
- Participated in a competition ( )
- Measured the weight ( )
- Conduct research ( )
- Complete surveys ( )
- Established communication with ( )
- Worked to solve problems ( )
- Give the ability ( )

- Take into consideration ( )
- Provide certification of (certify)

A.11.3 Imprecise Phrases

Closely related to multi-word, noun-heavy pseudo verbs are imprecise phrases. With these handfuls, you need to clarify your meaning, which will usually be done by finding an exact word as a replacement for the ambiguous phrase, as in the following:

- Figuring out the test setup (decide, determine, select, or identify)
- Ran into a problem (encountered)
- Negatively impact (harm or disrupt)
- The reason is that.... (because)

Many authors use sentences beginning with "The reason is that...." This tends to be both wordy and imprecise. A better approach is to properly place a clause starting with "because" either before or after the true subject of the sentence, which is not "reason." Look at this example:

- The reason the research team believes the catamaran design would work is that the design is much more stable and maneuverable than the conventional monohull tanker.

In a rewrite, either the design or the personal conclusion can be emphasized, rather than "reason." Emphasize the design as follows:

- The catamaran design should work best because it is much more stable and maneuverable than the conventional monohull tanker.

Emphasize the conclusion by the team as follows:

- The research team believes the catamaran design would work best because it is much more stable and maneuverable than the conventional monohull tanker.

The most complicated and incorrect version of this "reason vs. because" dilemma is the instance when both terms are used together. Here is an example:

- The reason I added an I-beam to the structure is because I anticipated its maximum load to be 5,000 pounds.

In this sentence, the reason for adding an I-beam is stated as a fragment, which is incomplete. A reason must be a complete idea (an independent clause), not a fragment. Look at this simple contrast:

- The reason I closed the door is dog.
- The reason I closed the door is the dog might get out otherwise.

Here, "the dog might get out otherwise" is a complete idea, which explains the closing of the door. "Dog" is a single-term fragment that explains nothing.

Returning to the I-beam example, the reason stated beginning with "because" is a grammatical fragment instead of an independent clause (complete idea). The true reason is the full idea, "I anticipated its maximum load to be 5,000 pounds." Also, after making the reason a complete idea, it is best linked to the consequence itself (the action taken), rather than the reason behind that action. Linking the reason to the reason is circular. If one reverses the order of the clauses, the confusion is evident. Look at this simple example:

- Because I eat my vegetables, the reason I run fast.
- The reason I run fast is because I eat my vegetables.

Looking at the first example above, when beginning a sentence with a "because" subordinate clause, the consequence follows naturally, not a second clause discussing the reason.

- Fixed: Because I eat my vegetables, I run fast.

Similarly, looking at the second example, if some condition or explanation is a reason, it does not need to be preceded by "because" when "the reason" is used to preface it:

- Fixed: The reason I run fast is I eat my vegetables.

Returning to the I-beam sentence, two options are available for eliminating the grammatical fragment and circularity:

- Option 1: I added an I-beam to the structure because I anticipated its maximum load to be 5,000 pounds.

- Option 2: The reason I added an I-beam to the structure is I anticipated its maximum load to be 5,000 pounds.

A.11.4 Negative Phrases

I have reiterated in this book the need to be concise. I have used, however, an awful lot of words myself, to communicate the many lessons of this textbook. In this section, I have an opportunity to emphasize conciseness once again, and be concise in doing so. The problem and solution are simple ones: Often, a wordy negative phrase has a concise counterpart that is better to use. Thus, here is a straightforward rule: Transform the negative into positive (a recurring theme of this book), as in the examples below.

Replace negative with precise term:

Not many	Few
Not unaware	Aware
Not continue	Discontinue
Not efficient	Inefficient
Not feasible	Infeasible
Not disapprove	Approve
Not reliable	Unreliable
Not tardy	Punctual
Not durable	Flimsy

A.11.5 False Questions

Pronouns are wonderful little words. They help us ask questions, as in these:

- Who wrote that fantastic report?
- Which test comes next?
- Where is the property's northern border?

They help us add subordinate clauses to independent clauses, producing a complex and informative sentence that breaks up the monotony of continual simple sentences:

- This part is manufactured at our Midwest transmission plant, where the cast metal processing was streamlined to improve production volume.

- The optimization of design falls under the responsibility of the chief engineer, whose team of 25 creative operations engineers has produced 22 patents in the past 3 years.

So, while all these pronouns—who, whom, whose, where, what, which, why, and when—have lots of benefits for both interrogative sentences (introduce a question) and declarative sentences (add more information), they yield a diction problem when used imprecisely in pseudo questions that are actually declarative sentences with poor wording. Here are some examples:

Weak: Why I earned an MBA was to make myself more employable.
Improved: I earned an MBA to improve my job prospects.

Weak: The settling pond is where gravity is harnessed to promote separation of oil and other contaminants from the water.
Improved: The settling pond uses gravity to separate oil and other contaminants from the water.

Weak: What you do next is disinfect the laboratory equipment.
Improved: The next step is disinfecting the laboratory equipment.

Weak: What centaurs do is allow us to learn more about the stability of orbits.
Improved: Centaurs allow us to learn more about the stability of orbits.

A.11.6 Wordy Idioms

Many standard phrases are inherently redundant and, therefore, verbose. You can shorten such common wordy idioms when you encounter them. Typically, you will find redundant pairs, as in the following examples:

- advice and counsel
- due and owing
- assist and help
- first and foremost
- prompt and immediate
- help and cooperation

Similar to the pairs above, redundant modifiers are unnecessarily wordy.

- final outcome
- close proximity
- red in color
- past experience
- share in common
- advance planning

The following sentence can be made concise:

> Wordy: We asked the research team to conduct an analysis of soil that was reddish-brown in color, sandy-clay in composition, and 20% in moisture content percentage.
>
> Concise: We asked the research team to analyze reddish-brown, sandy-clay soil having 20% moisture.

A.11.7 Verbosity Summation

Putting together some of the common indicators of verbosity discussed so far, you can appreciate this example, from a famous document, intentionally edited, by a highly motivated group of intelligent writers, to be concise and cogent, namely, the second paragraph of the American Declaration of Independence:

- "We hold these truths to be self-evident, that all men are created equal, that they are endowed by their Creator with certain unalienable Rights, that among these are Life, Liberty, and the pursuit of Happiness. —That to secure these rights, Governments are instituted among Men, deriving their just powers from the consent of the governed, . . . "

Had this been written by a busy professional working in ABERST today, without time to produce a second draft by revision, it would have turned out as follows:

- "At this point in time, there are truths that are held to be self-evident and obvious by us, and the first of these maxims being that all men are created equal; that their Creator has endowed all men with certain unalienable Rights is the second principle, that among these are Life, Liberty, and the pursuit of Happiness, in our opinion.—That to secure these rights, among Men have been instituted Governments, deriving their just powers from the consent, approval, and acquiescence of the governed, . . . "

A.12 Redundancies

Verbosity has a companion topic, to which I turn now: redundancy. Redundancy is not to be confused with repetition. Although strategic repetition of important ideas at the right places is a strong characteristic in a report, redundancy is a weakness. Redundancy is providing the same information more than one time when a single time is sufficient. Capably distinguishing between repetition and redundancy comes with mastering all the guidelines of this book. Sadly, this is not easy, so often ABERST reports possess instances of redundancy yet lack strategic repetition. This needs to be reversed. Elimination of redundancy must be done at the proofreading stage, if not earlier.

Here is how to eliminate redundancy, in a nutshell: You read through your report, looking for places where you realize you have already told the reader this information. I give you an example of this below. Also, I point out one common construction that is inherently redundant, as well as secondary items that often are duplicated unnecessarily in reports. These can serve as some particularly useful red alerts.

A.12.1 Redundant Example

In a report, let us say, the author has a section (section 2) that is used to describe the original machine part that has been improved. Other sections (3 and 4) introduce the improved version of the part. Well, if the original part is described as having five primary components in section 2, and those components are listed in bullet form, the list does not need to reappear in sections 3 and 4. Those sections are for new information. The bullet list of components in section 2 is sufficient; additional listings are redundant.

A.12.2 Double Introduction

I see double introductions frequently. These are a redundancy you can seek and destroy, as demonstrated below.

- Redundant: We were asked to meet three criteria. These three criteria are few parts, durable materials, and permanent lubrication.
- Fixed: We were asked to meet three criteria: few parts, durable materials, and permanent lubrication.

- Redundant: The final equipment-related cost to be considered is the cost of utilities, which has three components. These components are as follows: cooling water, electricity, and steam.
- Fixed: The final equipment-related cost to be considered is the cost of utilities, which has three components, as follows: cooling water, electricity, and steam.

A.12.3 Redundant Secondary Items

Certain items in ABERST reports are of secondary nature but often are given a prominent position, front and center, throughout reports. This is mistaken emphasis, which amounts to little more than redundancy. Recalling the difference between primary and secondary information (Chapter 3.4), authors must remember that your main message almost always concerns a recommendation, conclusion, finding, request, proposal, design change, solution, and so on. This is the information that matters, and the other items are less important (though still useful). I can offer several reasons that authors may emphasize such secondary information to the point of redundancy.

Secondary items are familiar and integral to one's daily tasks. Indeed, most of one's time is spent with these items, whereas innovations and solutions are discovered only at the end of a project and are often not that "present," but more hypothetical, experimental, and innovative in nature.

Secondary items are certain—real, physical, practical, tangible, visible, reliable, and so on—whereas primary items are controversial, conceptual, and theoretical.

Secondary items are usually personal—about oneself and one's tools, work group, equipment, laboratory, and offices—whereas primary information is intended for management and people higher than the authors and elsewhere in the organization as well as in other divisions, sections, and other organizations entirely, so this information seems to "belong" to another place in the organization or another organization entirely.

Lastly, routines are emphasized when doing work, which is good, so the routine nature infiltrates the report, with lots of redundant presentation to match the repetition of work tasks, but one does not want reports to seem full of monotony, repetition, and routine of doing tasks repeatedly, as is normal and suitable for work in ABERST. In a report, routine and repetition becomes redundancy.

Typical secondary items that are duplicated fall into these categories: esoterica, minutae, jargon, procedure, software, tools, and key people, as you will see in the examples below. These should give you the flavor of this type of redundancy. Perhaps you recognize it from reports you have read or written, so you will immediately connect with this red alert. If these are new to you, they should alert you to a potential pitfall you are advised to avoid in your writing. Notice the repetitive routine of work infiltrating the samples below and be wise to avoid this work-infused, inductive, redundant style.

- Using Excel, we converted the force values into lift and drag coefficients. The Excel tables of our raw and converted data can be found in Appendix B.
- In Matlab we were able to simulate 500 scenarios of control. Matlab found the optimum combination of gains through these simulations.
- Looking at the plot of velocities, you can see the error bars on most of the data points. Error bars were calculated to depict maximum error in our measurements due to human and equipment imprecision.
- The data were collected in a single 14-hour test session. With 14 hours of testing completed, we felt the testing was thorough, so we canceled the secondary testing day scheduled for the following day.
- The theoretical calculations were simplified following the work of Bergson and Durey (2010). Bergson and Durey approached conflagration from the perspective of rocket fuel, but our work is similar enough that Bergson and Durey's derivations can be used.

A.13 Contradictions

Repetition of primary ideas several places in the document is required, but this can lead to contradictions if a main message (or secondary message) changes over time and across different drafts of the document. Also, as a document is improved, a formulation or analysis might be refined, and thereby modified, from its construction in early drafts. This process of reworking and rewriting, albeit necessary and helpful, can lead to contradictions from one place to another. The following two sentences were several pages apart in a report, but I juxtaposed them here so the contradiction can be seen easily:

- Our recommended distance between equipment is 18.5 meters. From our testing, we conclude that 16.5 meters is the optimal and recommended distance between equipment.

I provide a few different examples in this subsection so I can alert you to potential situations that may come up in your writing. In the next example, the authors developed a new flight control system, and they changed the description of the system as it evolved. In the report, the system was described in different ways, which was confusing, because the report had remnants of the superseded design, as follows:

- Page 3: Our proposed system has three components. The second component has two main parts.
- Page 8: We propose a four-component system.

A similar situation arose in this example:

- Our data suggest insignificant drag reduction from the three techniques tested, leading us to conclude that none of the techniques is worth the additional investment. We recommend two of the three techniques as drag-reduction methods.

Sometimes the contradiction is so blatant it is painful. Apparently, during initial drafting a certain conclusion was formulated, which, upon additional work and analysis, was reversed. So, one or more places in the document present the new conclusion but the previous, superseded conclusion is not removed from all locations. Therefore, a reader comes upon the exact opposite point of view in the same document, as in this example:

- We recommend against adopting the new method that uses vortex shedding frequency measurement. Another reason we recommend using vortex shedding frequency measurement to calculate velocity is that the frequency sensor has an extremely fast time constant.

A.14 Punctuation Problems

A review of the most helpful punctuation rules is provided in Chapter 2, Section 2.2.8, but here you will find four red alerts for common mistakes that are easily made and easily overlooked during proofreading.

A.14.1 Comma Incorrectly Inserted between Subject and Verb

A serious and common mistake with a comma is to insert one incorrectly between the major parts of a sentence, namely the subject and verb, which means the sentence's subject and predicate are separated for no good reason. Here is an instance of the mistake, and I hope you can see that it is in fact wrong:

- Wrong: The Chicago Bears, are the best team in football. The Bears, rule!

This mistake is corrected as shown below, with the correct Chicago accent added for emphasis:

- Correct: Da Bears are da best team in football: Da Bears rule!

I would wager that no one reading this book would ever make the exact mistake shown above. But in ABERST the subjects are often so long and stretched out (see subsection 2.2.6.3) that commas find their way between them and their verbs on a surprisingly regular basis. It is nearly a norm, so let's put it to rest.

The lesson here is simple: Do NOT ever use a comma to separate subject and verb; this is the dreaded "comma splice," which I call "Da Bears Rule." Below are some examples:

- Wrong: An analog-to-digital converter chip testing laboratory power supply and regulator to enhance selective micro diamond sputtering deposition electrode pattern creation, has been implemented in our laboratory.
- Wrong: Each of the three types of chemical sensors that were recently assembled onto 3-millimeter high, multi-layered silicon chips, can be used to accomplish air pollution monitoring inside automobile passenger cabins.

- Wrong: The reliability ratings of the Garmin and Magellan GPSs, are distinguishable by their important power sources.

Removing the incorrect comma placed after the subjects fixes all three of the examples and every sentence ever written with a similar mistake. Remember: Da Bears Rule.

A.14.2 Comma Misused with *However*

- The distillation columns are clearly described, however the number of stages required in the columns seems conservative.

However is not a coordinating conjunction; instead, it is a sentence adverb. So, it cannot be used with a comma to link two independent clauses. It must be inserted into an independent clause:

- The distillation columns are clearly described. The number of stages required in the columns, however, seems conservative.

A.14.3 Colon Wrongly Separating Parts of a Phrase

- The research team took four biological samples including: breath, saliva, blood, and urine.
- The research team took four biological samples including breath, saliva, blood, and urine.

The participial phrase starting with "including" has four objects that must follow the participle directly with no separating punctuation. Colons are used only when the preceding phrase or clause is complete.

A.14.4 Semicolon Wrongly Used Instead of Comma

The semicolon is correctly used in place of a comma in certain instances, but it cannot be used interchangeably with a comma at an author's whim. A common mistake is to use the semicolon when a simple comma is required. This demands that you know the rules for commas and semicolons, which are covered in Section 2.2.8. Even before you have memorized every rule, you can keep an eye out for the most frequent incorrect substitution of semicolon for comma, namely, when a comma is needed to link a subordinate clause to its companion independent clause, as in the example below:

- Wrong: The Marketing Team plans to implement several promising strategies derived from broad survey data; unless follow-up focus groups fail to corroborate these initial conclusions.
- Correct: The Marketing Team plans to implement several promising strategies derived from broad survey data, unless follow-up focus groups fail to corroborate these initial conclusions.

Appendix B

Resumes, Job Cover Letters, and Thank-you Notes

This supplemental appendix is not intended to be your sole source for instructions on writing a resume, a job cover letter, or a thank-you note. Many good reference books are, in fact, dedicated to job-seeking skills, encompassing preparing the necessary written materials, as well as interviewing favorably and dressing appropriately, among other issues. I read and consulted lots of materials and resources when I started my professional career. Most of the reputable reference books and materials on this topic offer good advice and provide the basic "how to" instructions. I suggest you consult one or two such reference books, seek out advice from career centers, and gather suggestions from friends and family. In this appendix, I assume you have some initial understanding as to the required information in a resume and job cover letter, and you have reviewed some guidelines as to the pros and cons with such materials. You have also seen some examples of best, good, and bad resumes and job cover letters.

If you are using this book as your first and only source of expertise on this topic, you can skip ahead to Section B.11 and complete a quick study of the required information on a resume, then return to the beginning here and work your way through my ten specific guidelines for creating successful job cover letters and resumes. At the end, you will find Section B.12, with assistance for writing thank-you notes after having an interview.

In brief, my contribution is to comment on controversies and provide expert tips. The goal is to help you take an unfinished or mediocre resume or letter and convert it to an excellent one. I resolve issues of controversy with explanations that I hope you find convincing. I offer my best tips based on reviewing these materials (as both a hiring manager and a college instructor) for over 25 years. One of these best tips, and I begin with it, is to leverage basic technical communication principles to create a letter-resume combination that greatly increases one's chances of getting a job. I call this the "twofold strategy," and it is covered shortly below, as part of the first of my ten guidelines on this subject.

At the outset, let me say that there is no single correct way to design a resume or letter. Although this is a document genre that has its conventional approach, some creativity is necessary as well, otherwise all resumes and letters would appear as clones of one another. This means that while you must follow conventions, you must also add your own personal twist on these materials to catch a reviewer's attention in a subtle and positive way. This is no small task. The samples in this appendix indicate various design options, not all of them executed perfectly but all equally feasible.

People will update and revamp their resumes and letters many times over a career. Your resume after college will not likely look the same 10 years into a career. Nevertheless, whether you are reading this as a college student, recent college graduate, or seasoned professional, the advice below applies just the same. It is intended to be timeless and, therefore, especially helpful to college students who might find they are making an unexpected good impression on employers by presenting themselves in much the same way that a seasoned professional would go about designing and delivering her resume and job cover letter.

B.1 Guideline #1: Use Structure and Purpose to Your Advantage

To begin my presentation of the "twofold strategy," I must take a moment to explain the reason this appendix exists in this book at all. This is a technical writing instructional book, as you know. Without any doubt, moreover, letters, resumes, and notes are part of the larger universe of technical writing. They are required in ABERST on both individual and organizational levels. For the individual, they facilitate career placement and advancement. For the organization, they are instrumental in hiring and transfers, team formation, and contract procurement. Indeed, when an organization seeks a contract or funding through a proposal, it includes resumes of key personnel in each proposal to present its staff to the decision makers; the resumes help prove the organization can assemble a qualified team of individuals to do the sought-after work.

Resumes and letters are, therefore, crucial to the process of moving individuals into and among organizations so that both sides prosper. To achieve this vital communication task, each letter and resume presents a single, clear point: the individual described is fully qualified. In addition to having a very clear point, as should any well-written technical document, the structure for a letter-resume combination is none other than the standard pyramid you have seen throughout this book (Figure B1).

B.1.1 Structure of a Letter-Resume Combination

Applying this structural understanding to job-seeking materials enables you to commandeer all the advantages of a well-written technical document to your own cause of job hunting. You leverage the lessons learned rather than shut this book, forget things, and tackle the task of writing a letter and resume devoid of all that you have mastered for technical documents. It is not a separate category of writing. On the contrary, if you see a resume from the perspective of a technical report, you can take full advantage of the strategies and structures used for all successful ABERST documents, and you will have a leg up on your competitors applying for the same jobs and career advancements, not to mention you will have an easy-to-use formula for organizing and writing your letters and resumes. All of which leads to a writing task that is manageable and readily mastered.

So, to begin, let me map the standard structure onto a job letter and resume (Figure B1). The first paragraph of the letter serves as the Overview. The subsequent paragraphs of the letter compose the Discussion. And, finally, the resume is the Documentation.

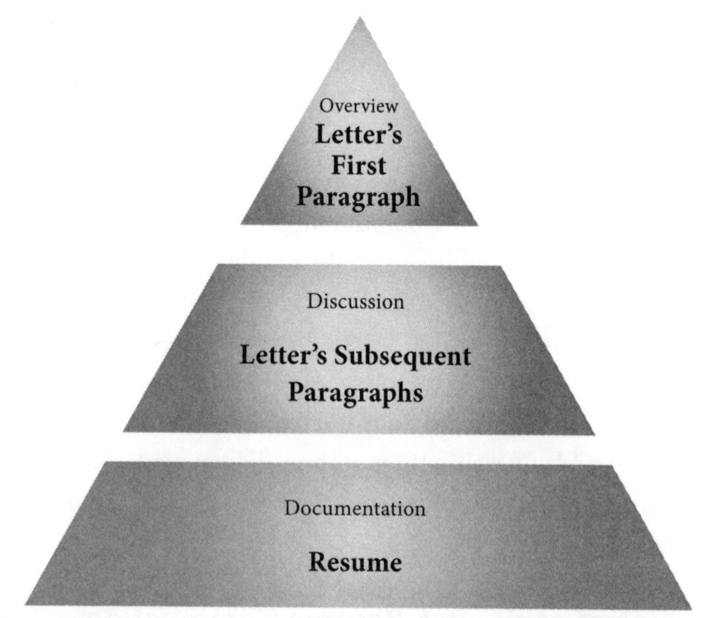

Figure B1 Three-part structure of letter-resume combination for job seeking

You should discern that the letter is the primary item, while the resume is of secondary importance. Like any report documentation, the resume has support details and supplementary information that the reader may consult to acquire further details beyond the information in the Discussion, but the reader need not consult the Documentation if the information in the Discussion suffices. This may shock you: as much as the venerable resume is emphasized for job hunting, it is merely support to the letter. Am I actually demoting the resume and labeling it as mere support? Yes, I am. It matters, but letters can be even more valuable to your efforts. With a letter, you can direct your efforts at a specific person who has the power to hire you. You can define a job as you wish and propose it to an employer. With a letter, you can be the only applicant to send in a resume to an organization not actively hiring, and you can persuade them to interview you. Resumes cannot speak directly to an employer you wish to work for. Letters can. Combining a letter and a resume is a twofold strategy that greatly increases your odds of procuring an interview and, ultimately, being hired. Thus, not only can a letter be more versatile than a resume, it can be sufficient on its own to prove your qualifications without anyone needing to see the resume. See Figures B2 and B3 for a sample letter-resume combination.

This does not mean you can disregard the resume. No. You must make an excellent resume. It may well be reviewed very carefully. It must provide elaboration of the points expressed in the letter. It can also serve as a backup item in case a particular employer does not put much stock in letters and prefers to skip to resumes (some do that), or if the letter seems weak to the reader but she remains interested enough to look over the resume.

Here are the three possible outcomes this twofold strategy may produce:

1. The letter makes a great impression and secures an interview invitation.
2. The letter is so-so but entices the reader enough so that she looks over the resume, which secures the interview.
3. The letter is skipped (you have no control over that), but the strength of the resume alone secures the interview.

The only time the twofold strategy is inappropriate is when an employer has specifically asked for resumes without letters (recall Chapter 1.1). This demand must be honored. In all other instances, the twofold strategy is optimum: provide a letter in cases when either nothing is said about a cover letter or when one is requested.

When you classify the letter-resume combination as simply one more genre of technical writing, you parlay that perspective into a clear advantage as a job hunter. This perspective suitably emphasizes structure (as stated above) and purpose, which is fundamental to all technical documents. Indeed, suffusing your letter-resume combination with purpose enables both flexibility and utility. The combination can serve a few different purposes, all of which have the utilitarian objective of landing you a desired job. Purpose deserves, therefore, some additional explanation.

B.1.2 Purpose of a Letter-Resume Combination

To see the power that purpose can lend to your letter-resume combination, first consider a resume alone. A resume on its own has no stated purpose, as it is predominately a summary of a person's experiences. (Including an Objective statement on a resume provides a hint of purpose, but it does not convey sufficient content-based purpose.) Combined with a letter, however, a resume becomes a document with a purpose that can be pinpointed to a specific recipient to produce an intended result: employment.

1234 Ash Street
Ann Arbor, MI 48105
(617) 321-9876
September 2, 2017

Russell Brady
Extreme Blue Lab Manager
IBM
10330 Wimbledon Dr.
Charlotte, NC 28262

Dear Mr. Brady:

I am writing this in response to IBM's online post about the Extreme Blue internship program. I am a junior at the University of Michigan studying Computer Science Engineering with a minor in Business, planning to graduate with my BS degree in December 2019. After talking with an IBM representative at my university's career fair, I believe that the Extreme Blue internship would provide me with an unparalleled opportunity to create and design an innovative software solution. I am confident that my background in programming, big data management, and team leadership makes me particularly well suited for the Extreme Blue position.

Through coursework, I have developed the necessary web development and programming skills that would enable me to effectively contribute to an Extreme Blue team. Within my Entrepreneurial Design Class, I collaborated with a group of fellow students to develop an application by using a standardized design methodology, such as interviewing anticipated customers and conducting user research. I not only gained valuable user-interface design techniques, but also specific programming skills in HTML, Angular JS, Flask, and several third-party APIs. This project gave me experience in solving a problem with programming strategies.

Furthermore, through my position as a Laboratory Research Assistant, I have obtained experience with big data analytics. Specifically, I analyzed over 50 papers and articles on various topics in machine learning, particularly natural language processing and computer vision. Building on this literature research, I created a script using Microsoft Computer Vision API to generate feature data from a set of photos. I utilized Python to create a visualization of the data in order to communicate the results to non-technical stakeholders. During the Extreme Blue position, I will be able to incorporate these innovative technologies into any project.

In addition, I have experience as a team leader, with strong interpersonal, planning, and communication skills, as evidenced by my leadership roles in Theta Kappa fraternity (TK) and Palomar Emerging Leadership Experience (PELE). As TK's recruitment chair, I organized events for over 100 members. Furthermore, through PELE, I learned how to work effectively with new colleagues by identifying and leveraging each team member's strengths.

I believe the Extreme Blue team position will benefit from my ability to create software products that leverage big data management. I have enclosed my resume for your consideration, and I will check with you in three weeks' time to confirm receipt of my resume and answer any questions you may have. I hope to be invited to an interview at your earliest convenience.

Sincerely,

Benjamin Farley

Benjamin Farley

enc.

Figure B2 Sample letter-resume combination (letter)

Benjamin Farley

(617) 321-9876 • bfarley@umich.edu • 1234 Ash Street, Ann Arbor, MI 48105

OBJECTIVE

Summer internship as software developer, starting May 2018

EDUCATION

University of Michigan, Ann Arbor, MI

Bachelor of Science in Computer Science Engineering, December 2019
Minor in Business, Ross School of Business
GPA: 3.32/4.00

COMPUTER SKILLS

Proficient in C++ and Python
Knowledgeable in HTML, CSS, JavaScript, Angular, and Java

EXPERIENCE

Kronos Incorporated, Chelmsford, IA, April 2016 – July 2017
Software Performance Intern

- Developed protractor testing framework using page object model for maintainable E2E testing
- Debugged code written in Agile software

University of Michigan, Ross School of Business, Jan. 2015 – June 2015
Research Assistant

- Conducted and presented survey on computer vision and machine learning algorithms for natural language and image processing
- Implemented various computer vision algorithms using Python

PROJECTS

The Related Artist Network, Jan 2017 – April 2017
Programmer

- Created graph visualizations using Spotify API, Networkx and D3

University of Michigan, Nov 2016 – April 2017
Project Team Member

- Utilized design methodologies to interview, design and develop web app that alerts users when friends are within certain distance
- Implemented prototype using HTML, Flask and Google Maps API

Accenture U.S Innovation Challenge, Jan 2016 – Feb 2016
Team Leader

- Led interdisciplinary team to research, construct, and present business strategy for prevention of food waste by grocery stores
- Selected as runner up in national quarterfinals

City of Newburgh, June 2015 – Aug 2015
Student Researcher

- Provided data analysis using Python, Pandas, and Google Maps API to estimate CO_2 produced by students' commute
- Developed and presented visualization to support proposal of bike lanes and sidewalk repair

ACTIVITIES

Accelerate CS (Google), September 2015 – Present
Lead weekly lesson at middle school to teach students computer science skills

LEADERSHIP

Palomar Emerging Leader Experience, September 2015 – December 2015
Selected as 1 of 34 students to participate in sophomore-year college leadership program

Theta Kappa Fraternity, September 2014 – Present
Recruitment Chairperson: planned and organized many events

Figure B3 Sample letter-resume combination (resume)

Furthermore, a letter-resume combination can be used flexibly for two different purposes: to apply for a posted position or to apply to an organization that is not overtly hiring or posting job openings. The former situation is typical: If a job opening has been posted, advertised, circulated, and so forth, the letter is the appropriate vehicle to introduce yourself and your resume, using the twofold strategy described above. The latter situation may be less familiar to you but may be more beneficial than the former: if you seek to apply for a job that is not posted or advertised, that is, you simply want to introduce yourself to an organization where you would like to work, you can and must use a letter to announce your purpose in writing. You cannot simply send a resume to an organization that is not actively hiring. It will not speak for itself.

The letter, however, can explain all the reasons you believe the recipient should hire you. You can state your general qualifications, overarching enthusiasm, and particular areas of expertise that are useful to the organization. Submitting a letter out of the blue, similar to making a "cold call" for sales, is a strong move and may earn you respect. (It may not, but then you cannot control this; all you can do is try, if it means enough to you.) If the letter-resume combination has merit, you may indeed secure an interview from an organization that is making an accommodation just for you. You have created a job opening for yourself, and you are the only applicant. No better situation can be found for a job applicant. In sum, letter-resume combinations work for both posted job openings and "cold call" personal introductions.

Many college students ask me if they are allowed to send a resume into an organization that is not overtly hiring, such as by advertising a job opening. Perhaps some recent graduates in their first or second jobs have this same question, having found employment to date only by applying for a posted job opening or engaging with company recruiters. My answer is emphatically simple: Yes, of course you can communicate your interest to any organization you wish. This sets you apart from others. This puts you in the position of being the only applicant. Your odds are better for getting hired when you are the only applicant rather than one of hundreds of applicants. Figures B4 and B5 are a letter and resume sent upon the author's own initiative.

You may not get an answer to this letter (an event-initiated document), but you are allowed to try. An organization may discard your materials without any reply to you, and that is its prerogative. But, quite possibly, you will get a reply.

The organization may appreciate your initiative. Moreover, if you offer something of value to them, which presumably you do or you would not be wasting your time with the effort, they will be impressed with your qualifications. In that way, you will combine initiative with capability, and that should get the attention of the letter's recipient. You have done your best, and the rest is left up to factors outside your control. Importantly, when you do choose to pursue employment on your own initiative, my instructions on submitting the document correctly (Chapter 1.5) are germane and crucial. An uninvited letter-resume is a classic event-initiated document, so attending to the three steps of submission should prove helpful: identification, attention, and distribution. You may wish to revisit those concepts with job seeking in mind.

Now that you have an understanding of how the fundamental principles of structure and purpose apply equally to a letter-resume combination as they do to any technical document, you can grasp the ideal content of both, and we will start with the letter.

B.2 Guideline #2: Use a Standard Letter Layout and Persuasive Content

Writing a successful one-page job letter constitutes my second guideline on this topic. Importantly, the letter is customized to the recipient. That is, it highlights the parts of the resume most applicable to the recipient's needs. In addition, it is cogent, concise, clear, and courteous. Let's look at each of these in turn:

- **Customized:** you address the letter to a specific person at an organization, and you express interest in being hired to do something specific at that organization; you selectively highlight your qualifications as germane to the specific role or task you wish to assume there.
- **Cogent:** you must have a clear message that you are qualified via a few select strengths, and you must present proof of those selected strengths; this makes your letter convincing.

987 April Street
Southfield, MI 48149
(734) 585-8285
October 3, 2017

E.M. Chung, Professor
Electrical Engineering and Computer Science Dept.
University of Michigan
2195 EECS Bldg., 1400 Beal Avenue
Ann Arbor, MI 48109-2411

Dear Dr. Chung:

I am writing to express my interest in contributing to your microdevices lab as a part-time research assistant, available immediately. I am an undergraduate majoring in electrical engineering planning to graduate in May 2020. I am interested in the lab's work on advanced neural probes, which I read about in the October issue of *The Record*. I have relevant laboratory research experience as well as an aptitude for effectively communicating my ideas to fellow team members in an academic context.

Over the summers of 2014 and 2015, I assisted professors at Western Michigan University with their research projects, and this work prepared me to contribute to your lab's efforts. For example, in the summer of 2014, I assisted professors Vivek Krishnan and Charles Pinter in their development of soy-based coatings. The purpose of the soy-based coatings was to create a "green" alternative to the petroleum-based coatings that dominate the market today. While working in their lab I helped them prepare many different mixtures of coatings according to their specifications, and then I tested them accurately and precisely through many different trials. I learned to execute test after test to make sure that the results are statistically solid, and I would certainly come into your lab with this mindset. Additionally, in the coatings lab I was introduced to many new concepts and apparatuses on an almost-daily basis, and I think that this experience will help me more quickly acclimate to your lab.

Over the summer of 2015, furthermore, I worked in the lab of Dr. Bob Timshal to research different methods of developing nanoparticles and the effect those methods had on the nanoparticles' ability to transfer from water to various organic solvents. My contribution to this lab was that I repeated experiments and took careful measurements in order to assure the validity of our results. In doing so I gained experience working with small-scale systems that required patience and caution. I understand that a lot of the nano machinery that is used in your lab is also created via chemical processes like my nanoparticles were, and as such I already have a certain familiarity that would allow me to begin contributing to your lab even faster than I would otherwise be able to. During my time working with nanoparticles, moreover, I became familiar with different types of data analysis software, photo spectrometers, and a wide array of basic chemistry apparatuses. I feel that my familiarity with lab procedure and the research environment will allow me to contribute to your lab in a significant way.

Finally, being able to effectively communicate one's findings in the laboratory setting is of utmost importance, and at the end of each of my research assignments, I had to prepare a report of my findings and present them to a group of people who were unfamiliar with what I had done. My presentations were meant to explain my professors' research in a succinct manner and capture the interest of other members of their departments, while also convincing them that our research was yielding useful results. Enclosed you will find my resume, and I will call you next week to check to see if you have questions or want to schedule an interview with me. In the meantime, I will be reading further about your new 12-site probe.

Sincerely,

Victor Marcus

Victor Marcus

enc.

Figure B4 Letter sent upon the author's own initiative

VICTOR MARCUS
987 April Street
Southfield, MI 48149
(734) 585-8285
vrmarcus@umich.edu

OBJECTIVE
Part-time research in engineering or science, over multiple years while studying full-time

EDUCATION
University of Michigan- Ann Arbor
Bachelor of Science in Engineering, Electrical Engineering, May 2020
Minor: Physics

RESEARCH EXPERIENCE
Project SEED - Western Michigan University, Research Assistant — Summers 2014 & 2015
- Followed guidelines issued by sponsor, American Chemical Society
- Developed "green" soybean-based coatings, as subsitutes for petroleum-based ones
- Identified options for transferring nanoparticles from inorganic solvents into organic ones

Technical University of Berlin — Summer 2016
- Implemented proportional-integral-derivative (PID) controller for wheeled path-finding robot

WORK EXPERIENCE
Self-Employed, Southfield, MI
Personal Tutor — June 2014 - Present
- Tutor people young and old in math and science

University of Michigan, Ann Arbor, MI
Teaching Assistant — Summer 2017
- Coordinated three co-instructors in teaching high-school students from many countries basic physics, geology, and math
- Developed methodss to facilitate effective and fun learning
- Organized laboratory activities and field trips

Michigan Union, Ann Arbor, MI
Dishwasher — September 2015 - May 2016
- Cleaned dishes, washed floors, emptied trash, and cleaned cooking equipment

University of Michigan, Ann Arbor, MI
Transit Coach Operator — July 2015 - November 2016
- Obtained commercial driver's license and completed training
- Serviced stops and drove safely for 20-30 hours weekly

Western Michigan University, Kalamazoo, MI
Physics Help Room Tutor — November 2014 – May 2015
- Helped students understand specific problems and physics concepts

SPECIAL SKILLS
- Moderate in Spanish
- Conversational in German
- Fluent in Romanian
- Competent in C++, C, and Matlab
- Familiar with Adobe Photoshop

Figure B5 Resume sent upon the author's own initiative

- **Concise:** the letter can be no longer than one page; you must be succinct and limit repetition to one area only: your interest in the job sought, which is stated at both the beginning and end, for emphasis.
- **Clear:** you must review and edit the letter to ensure broad comprehension by people who have never met you, attended your college, or done the same work; you achieve this by avoiding jargon, abbreviations, proprietary and arbitrary nouns, and overly obscure technical terminology. Here are some examples of unclear text unless explicitly defined: DaDT, ARM-M4, beat receiver, MOM, and synchronization primitives.
- **Courteous:** you use only a professional tone and you must show deference to the reader; you eliminate any elements that may seem demanding, disparaging, audacious, or confrontational.

Meeting the five requirements stated above necessitates a standard layout of paragraphs and persuasive content. I suggest a 3- to 5-paragraph letter, with the paragraphs having the content, as follows:

First Paragraph: (a) your purpose to apply for a particular position (either advertised or self-asserted)—casually "inquiring about openings" is not purposeful to my ear; it is acceptable to overtly state, "I am writing to apply for position x"; (b) where you learned of the advertised position (if applicable); (c) your status professionally or academically; (d) preview of your qualifications in a few expertise areas—generalizations—to be elaborated on subsequently; and (e) your available start date or timeframe, if a constraint of some kind. The letter to IBM above (Figure B2) has an effective first paragraph. The letter to Fitbit (Figure B6) has an unusually personal and effusive first paragraph. It, nonetheless, succinctly sneaks in the first paragraph's required elements. The subsequent paragraphs, fortunately, add the details and balance necessary after the effective yet risky company flattery (discussed in Guideline #3) at the start. In this way, it is an original letter, providing a strong sample to emulate but one that cannot easily be cloned.

777 E Madison St., Apt #4
Ann Arbor, MI 48107
(313) 242-5566
billgrat@umich.edu
April 9, 2017

Karen Richardson
Programming Lead
Fitbit Inc.
150 Spear St
San Francisco, CA 94105

Dear Ms. Richardson,

I am a senior in Electrical Engineering at the University of Michigan writing to apply for the Firmware Engineering position posted on your company's website. More importantly, however, I am writing to tell you how much I love your company and want to contribute my digital signals processing skills to your mission.

As I write this letter, with my black Charge HR on my wrist, I think about the tremendous impact this little device has had on my personal health and fitness. I love to run and was a varsity athlete in high school, but it is often difficult to balance time for working out with a rigorous engineering course load. My Fitbit changed that immediately, reminding me that I should seek progress whenever possible. The chance to contribute to Fitbit's firmware design and thus improve the health and well being of individuals around the world would be a tremendous opportunity.

This past semester I completed my Senior Design course in Digital Signals Processing, which culminated in a project that my team based on a Fitbit wearable device. My biggest contribution to the project was to model some of our device's specifications after Fitbit's successful technology. By comparing the form-factor, power needs, and weight to the Charge HR, we developed a very high level of performance criteria for our own device. Our project, the Danger Detector for the Deaf, used a microphone to "listen" for fire alarms or smoke detectors and notify a user of the danger via haptic feedback. This project was an opportunity to apply knowledge of digital signals to embedded systems programming. I hope to continue working in the firmware area of engineering.

I believe my enthusiasm and experience make me an excellent candidate for the Firmware Engineering position. Please feel free to contact me with any questions either at (313) 242-5566 or billgrat@umich.edu. I will call you in two weeks to inquire about the next stage in the application process. I hope to meet you in person soon.

Sincerely,

Bill Gratz
Bill Gratz

enc.

Figure B6 Sample letter with an unusually personal and effusive first paragraph yet all required elements

Middle Paragraphs: (a) examples to prove possession of first expertise area listed in your preview with indication of place(s) where experience was gained; (b) examples to prove possession of second expertise area listed in your preview with indication of place(s) where experience was gained; (c) examples of all remaining expertise areas listed in your preview with indication of place(s) where experience was gained. This is your demonstration of your qualifications. These are the particulars to elaborate upon the generalizations stated in the preview. The letter to Steelcase (Figure B7) has strong middle paragraphs.

Final Paragraph: (a) reiterated interest in job sought as natural progression from all aforementioned experiences in middle paragraphs; (b) next step you will take as follow-up to this communication; (c) some statement of courtesy; (d) extra offer to help organization per a trending situation within your awareness, to indicate knowledge of either organization or its wider business environment.

Two more examples of job cover letters are included here (Figures B8 and B9). As you consult these and the other examples in this appendix, please remember my warning about carefully walking the line between following standard conventions and avoiding direct duplication (cloning) of sample materials. You must find your own voice and create a unique letter. My overall comment on the letter to Veracode (Figure B8) is that it contains format problems, and, more importantly, its content is neither persuasive nor professional, demanding better adherence to both Guidelines #2 and #3, covered immediately below. The flaws should be evident in comparison to the other samples provided. My overall comment on the letter to Apple (Figure B9) is that the author makes good use of previous work experience as a qualification, so it is consistent with my guidelines. I also provide the resume (Figure B10) to go along with the letter to Apple (Figure B9), so you can see another example of a good letter-resume combination.

1423 E Washington St.
Greenville, NY 95001
(213) 754-2486
August 28, 2017

James Kantney
IT Department Director
Steelcase, Inc.
901 44th St SE
Grand Rapids, MI 94508

Dear Mr. Kantney:

I am writing in response to the employment opportunities listed on your company's website, in particular the internships in information technologies for 2018. I am a sophomore at the University of Michigan College of Engineering working towards a Bachelor of Science in Computer Science, anticipated in May 2019. I wish to contribute my application development experience and teamwork skills to the goals you have in the IT Department in the areas of applications development and systems analysis.

Over the past two years, I have gained considerable experience working with application development software by developing mobile apps. While I was a part of the University of Michigan's Solar Car team, I helped create a communications app for the race crew to signal upcoming issues on the route to the driver. Also, I developed a mobile timecard app for North Michigan Builders, where I was employed for the past few summers, to replace the paper system they were using to record hours worked at various locations.

I am also very comfortable working with a team to accomplish projects. During my employment at North Michigan Builders I had to coordinate several teams and communicate with widely displaced people to ensure construction was staying on schedule. In addition, I was partially responsible for training new employees and monitoring their performance to confirm work was done efficiently and correctly.

I have enclosed my resume for your further review of my experience. I plan to contact you in the coming weeks and possibly arrange a visit to your offices. Perhaps it will be convenient for you that I will be in Grand Rapids the week of October 14th. Beyond that, I am committed to following through on the hiring process to completion and affirming my enthusiasm and qualifications for assuming an internship under your direction.

Sincerely,

Connor Nitro
Connor Nitro

enc.

Figure B7 Sample letter with strong middle paragraphs

105 Singer Road
Duluth, MN 62505
(943)-321-4326

6, December, 2016

Steven Christiansen
Chief Scientist
Veracode
625 Network Drive
Burlington, MA 01803

Dear Steven Christiansen:

I am writing to inquire about a software engineering internship at Veracode in the future. I am a sophomore at the University of Michigan double majoring in computer science engineering and economics. I have a strong interest in software engineering with a focus on data analytics and security. I am hoping to gain experience in these areas by finding an internship that will allow me to explore these interests through work on a real-world project.

I am especially interested in Veracode because of your company's approach to software engineering. I have always prided myself on thinking outside of the box and thrive in environments where others do the same. Veracode's revolutionary cloud-based approach to data storage seems to embrace this idea and, to me, is indicative of a company who believes in looking beyond existing ideas and creating products that are new and revolutionary.

During my time here, I have taken courses in data structures and algorithms and am currently taking a course in databases. I am also enrolled in a directed study in which I and three others are working together on a semester-long project to re-design and re-implement a help queue used during office hours by multiple classes. This has given me experience in UI design and mobile application development. I am proficient in several programming languages, including C/C++ and Java, and have always been able to pick up new languages and software quickly. I believe these skills and abilities give me the technical background necessary to be an innovator.

However, just as important as my technical background, I also possess the skills to work efficiently as a leader. My background as a community assistant in a residence hall has given me experience in collaborating with people from all different backgrounds, and my experience as a tour guide has taught me how to communicate clearly and engagingly. My combined intrapersonal and technical skills will allow me to contribute new ideas while also helping continue your history of cutting-edge innovation in data handling and security.

If you have time, I would appreciate a chance to talk with you regarding any opportunities at Veracode during this upcoming summer. Feel free to call or email me any weekday morning or weekend evening.

Thanks for your time,

Linus

Linus Engels

enc.

Figure B8 Sample letter that contains format problems and lacks other strengths

B.3 Guideline #3: Strike a Professional Tone

One characteristic that all the letters share, except Figure B8, is a professional tone, and this is my third guideline. A professional tone is marked by what it is *not* as much as by what it *is*. Indeed, the pleasant absence of sophomoric, simple, and sensational statements contributes to professionalism. Professionalism is neither obsequious (sucking up) nor vacuous (empty talk). It is neither demanding nor diffident. It is confident, sensible, straightforward, and courteous.

Here are some unprofessional phrases culled from actual letters that fail to advance the author's objective in securing employment:

- My careful attension to detail has been recognized by previous employers.
- I want to join your famous company.
- My proven ability to solve the most complex problems and excel in higher mathematics will be an asset to your company, as my GPA of 3.12539 attests.
- Internship sought at your firm so I can help bring the world together through music.
- Seeking a dynamic position working on challenging problems in a fast-paced setting with top-flight scientists, available summer.
- In researching your company, I found that your products are admired around the world, so I have become interested in working for your team.
- I seek a position where creative people are valued and the company fosters personal development, with flexible work hours.
- I have always been a team player, and as you peruse my resume you can see the five personal software development projects I have worked on during my free time over the past four years.
- Seeking a progressive position where I can learn about hardware design from the best engineers in the field.
- I know I will fit into your fast-growing company producing cutting-edge products.
- I am a natural problem solver and a very fast learner; so far at college, I have completed four courses in economics.

1945 Geddes Ave.
Ypsilanti, MI 49107
(848) 402-4000
February 28, 2017

John Heisinger
Engineering Project Manager
Apple, Inc.
1 Infinite Loop
Cupertino, CA 95013

Dear Mr. Heisinger:

I am writing to apply for the position of Operations Test Engineer, as posted on Apple's website. I am currently a computer engineering student at the University of Michigan, and I will complete my Bachelor's degree in April 2018. I spoke to one of your company's recruiters, Ashley Baker, at a recent career fair on my campus, and she encouraged me to contact you directly and highlight my recent job experience. For the past two years, I have been a smartphone technician, and I am very interested in applying my experience servicing and repairing these phones to the design and production of these and related products, such as tablets and laptops. I also have strong communication and team-work skills.

In my work at QuickFix, Inc., I helped customers who had damaged products. Often, they were not sure what exactly was wrong with their phones or tablets, so I would troubleshoot these devices. After determining the cause or causes, I put great effort into explaining the situation to the customer, patiently discussing technical issues with people possibly unfamiliar with the details of a functioning smartphone. This demanded strong communication skills. In all, I diagnosed problems, explained them, coordinated a solution with a customer, and implemented the necessary repair in a timely manner. This skill in quickly solving problems while working closely with a customer should be applicable to your needs in device development and product testing. Additionally, I am able to collaborate with other team members on all types of tasks. As a loan servicing representative, I had to be in constant communication with members of other departments in order to complete the daily workload. I am committed to working cooperatively with many colleagues.

Along with this letter, I am enclosing my resume. I will contact you next week to follow up on my application. I feel I am a good fit for the open position, and I am very excited to apply my hands-on smartphone troubleshooting experience to development. You can contact me by phone (above) or email: andywass@gmail.com.

Sincerely,

Andy Wasserstein

Andy Wasserstein

enc.

Figure B9 Sample letter that makes good use of previous work experience as a qualification

Andy Wasserstein

(848) 402-4000 — 1945 Geddes Ave.
andywass@gmail.com — Ypsilanti, MI 49107

OBJECTIVE
Full-time engineering position in hardware design

EDUCATION
University of Michigan — Ann Arbor
BSE in Computer Engineering — April 2018
GPA: 3.5/4.0

Course Highlights:
Microprocessors, Programming and Algorithmic Thinking, Digital Logic, Physics—Electricity and Magnetism

EMPLOYMENT
QuickFix, Inc., — San Carlos, CA
Smartphone Technician — Nov. 2014–Dec. 2016

- Performed fast, effective repairs on damaged phones, tablets, and computers
- Assisted clients in comprehending technical aspects of repairs
- Completed sales and refunds using point-of-sales system
- Oversaw daily operations of retail and service location, including opening, closing, and taking inventory

Wells Fargo Credit Corp — San Carlos, CA
Loan Servicing Representative — Jun. 2013–Aug. 2013

- Processed customer payments and managed accounts
- Contacted insurance companies and oversaw verification of customer's policies
- Updated loan information in computerized record system

RESEARCH
University of Arizona Water Quality Research Lab — Tucson, AZ
Student Researcher — Sep. 2013–Dec. 2013
Assembled moisture sensor to measure water content in soil
Conducted field research using sensor, while working with team
Analyzed effects of extreme drought on arid ecosystems

COMPUTER SKILLS
Languages: C++, Matlab, Verilog
Application: Altera Quartus, Mathematica

AWARDS
College Honors: Winter 2016
Dean's List: Winter 2016

Figure B10 Sample resume to nicely complement its letter (B9), illustrating another example of a good letter-resume combination

- During this internship, I aim to have a concrete impact in the fields of financial services, currency trading, and global industry.
- Seeking an internship where I can gain experience.
- I learned from your representative that your company strives to be the best at everything it does, and I know I would fit in with that culture.
- I want to work at your company because it is one of the world's most well-known and popular software companies.
- I researched five companies with job openings and decided to apply for this one because it fit my background better than the other four.
- Your corporation is a decent place to gain some real academic experience.
- I believe in your product: television streaming is an excellent idea.

These actual examples compel me to clarify some common misconceptions about the purpose of a job letter and resume. Recall that you are trying to prove, primarily, that you are qualified for employment. Secondary objectives of a letter-resume combination include demonstrating you can write well and format a professional document. In contrast, you do not need to prove you idolize the organization, though you should convey sincere interest in working there. Similarly, you do not need to prove other applicants are incompetent. You do not need to prove that you have huge dreams for yourself in terms of salary and advancement. You do not need to prove you are an enthusiastic consumer. Therefore, I can summarize the following as rules:

- ▶ Wanting personal rewards is not a qualification.
- ▶ Flattering employer is not a qualification.
- ▶ Stating the obvious to an employer is not a qualification.
- ▶ Self-proclaimed virtues are not qualifications.

As the examples above reveal, rather than boast about oneself, demand ambiguous benefits or the plainly obvious, or rave about the organization where one seeks employment, a job seeker need only speak plainly and sensibly about one's qualifications and the applicability of those qualifications to anticipated work effort that needs to be accomplished at the receiving organization. The letter to Fitbit (Figure B6) pushes the boundary of this guideline, but it avoids a full-on violation. The letter to Veracode (B8), however, breaks the boundary in too many ways.

I will concede that the deep truth may be that you do very sincerely feel excited, as an adult on a career path, about the opportunity to work with experienced people at companies doing great things one hears about in the news. Maybe you are eager to share your innermost thoughts: "I can learn so much from the cool scientists and engineers at your organization, and I will be getting paid, which thrills me, and the possibilities for me are endless, and I get to test cutting-edge products and ideas before they hit the marketplace, so I will be inside the trend-setting world, on the ground floor of tomorrow's technology." Sure, all that is true, but 99 other applicants are thinking this also. So, you need to be the 1 applicant out of 100 who skips all this drivel and speaks at the level of the professionals you are hoping to join as a team member, not at the level of the neophytes you are leaving behind.

A professional tone avoids hyperbole. It is calm and balanced. It is modest but accurate about one's qualifications, as in the following positive examples:

- I have been a runner for 10 years, and I have experience with wearable technology. I would like to apply my insights from personal experience and my electrical engineering education to your company in designing and manufacturing wearable technology.
- I am interested in working in your artificial intelligence division, and I have experience in that area from three university courses and one internship, as described below.

Because a letter-resume combination is by definition a self-focused document intended to display one's strengths, the difference between right and wrong is really important. It is a fine line between being conceited and being professional, so this issue deserves a little more attention. One way that may be helpful is to distinguish between self-asserted virtues and empirically driven experiences. This hinges on understanding that it is easy to speak but harder to do. You cannot just "talk the talk," but you have to "walk the walk." Remember these adages:

- *Talk is cheap.*
- *Actions speak louder than words.*

To elaborate this difference, I put the concept into formulaic terms, which I call the *Qualification Formula*: Instead of stating you have a particular virtue or personality trait that may be entirely subjective and hard to verify, such as easy going, punctual, humorous, flexible, and so on, you need to demonstrate that you have verifiable experiences and proven abilities.

The Qualification Formula: Do not merely *state* a virtue or trait; rather *demonstrate* a qualification or ability.

Examples

Self-proclaimed virtues:

- With my coursework experiences, innovative mindset and affinity for problem solving, I know I can add to your company.
- With my relentless drive, boundless enthusiasm, and required level of resilience, I am certain I can fulfill the requirements of the posted position and meet all deadlines.

Demonstrated qualifications:

- I have attempted to improve my team leadership skills by serving as the vice president for new membership for my university's human-powered helicopter team.
- As a smartphone technician at PhoneFix Corp., I was a problem solver. Customers arrived with non-functioning phones, without any clear certainty as to the problem, and my responsibility was to diagnose the problem and fix it. I did this job for 4 months last summer before entering my final year of college.

B.4 Guideline #4: Express Some Knowledge of the Hiring Organization

As emphasized above, any toss-away assertion about the organization where employment is sought will come off as obvious, cliché, flattery... in short, not a qualification. You have no reason to waste precious space in your letter telling the company something that everyone knows or assumes: great place to work, wonderful products, and leader in the field. Even if you take away the flattery and indicate you know something about the company from your job-seeking research, it usually is only skin-deep information:

- The organization is hiring for positions x and y and locations a and b.
- The organization has business segments q, r, and s.
- Their headquarters are in Podunk, and testing is done in Plainville.

Instead, try to dig deeper, like a dog, and find the bone. This is usually information you know beyond the basic job-seeking data. You might know this because you pay attention to events in the related industry where you want to work. You can learn things from personal contacts, newspapers, broadcast or on-line news media, magazines, career centers with research materials, professional organizations including student chapters, company websites or media materials, and any other in-person, written, or on-line sources. You might spend years of your life working for this organization; why not find something out before you commit?

Indeed, you should do some research and show you know about the field. Is the company truly a leader or is it looking to buy a competitor to increase its market share? Is the organization dealing with a large recall of its product or a failure to obtain FDA approval for a new drug when it was widely anticipated? If you can show in your letter that you know something about the organization that proves your true interest in the field, you rise above the other applicants. Here are some examples:

- Edwin Wilcox, a senior programmer with your organization, told me that your inventory management system uses MySQL as the backend and AWT/Swing for its graphical user interface (GUI), and I have experience with both of those programs from my volunteer work at Habitat for Humanity.
- I know your company is considering a purchase of a blog-site company, and such a purchase is highly likely in the next 6 months. I can contribute to the integration of that new business unit because I helped to integrate Sarcos after its purchase by Raytheon.

B.5 Guideline #5: Combine Generalizations with Particulars

This is used throughout the letter-resume combination, just as it permeates any technical document, at various levels, as you have read in the main parts of this book and in B.2 as it pertains to the letter's paragraph structure.

The relationship between generalization and particulars is a one-two punch. First, you open with a generalization to let the reader know exactly how you want her to interpret the particulars. Second, you follow with particulars to support the generalization in just the way you have telegraphed to the reader through your preceding generalization.

Example from a letter:

- I have significant experience in rocketry avionics, gained while serving as lead engineer for my college's student-based rocket team. I was personally responsible for hardware development for the team's first-generation flight computer, as well as a proprietary global positioning system (GPS) receiver for high-speed operation, which incorporates radio frequency (RF) communication components.

In this example, the author begins with a general qualification of rocketry avionics. She has some experience with this technical area. These particulars follow: hardware development, flight computer, GPS receiver, and RF components.

Two examples from a resume:

> EDUCATION
> University of California
> Masters of Science, Computer Engineering, May 2018
> Dean's List: fall 2016, winter 2017
> Course highlights: linear algebra, data structures and algorithms, and logic

> EMPLOYMENT
> Phoenix Laboratory for Renewable Energy, Chicago, IL
> *Laboratory Assistant*
> Separated DNA . . .

In the Education example, the university attended is the most general qualification: It says broadly, "I went to this university." The items underneath are particulars that elaborate upon that general qualification: I earned a Master's, in Computer Engineering, and I made the Dean's List, twice. Moreover, I took three courses I want to highlight. Starting with particulars would not be as effective:

> EDUCATION
> Course highlights: linear algebra, data structures and algorithms, and logic

Readers will necessarily ask, "What about those courses? When and where did you take those? What did they lead to?" But, listed after the generalizations that precede them, they add value and embellishment.

In the Employment example, the place of employment is listed first; it is a general qualification. It makes a good impression on its own, without any particulars: the applicant worked at the Phoenix Laboratory for Renewable Energy, Chicago, IL. Wow. That trumps restaurants, lawn care, summer camp, and all other decent, but temporary, work sites for a rising ABERST professional. After the listed organization, particulars are given: position of laboratory assistant and duties in that position, such as separating DNA. Without these particulars, the general assertion of employment with Phoenix Laboratory is also weak in itself.

B.6 Guideline #6: Remove All Superseded Items

Job seekers might be tempted to provide an entire work history, listing all experiences and educational efforts of one's life. This is unnecessary. A letter must be only one page, and the same is true for a resume. You simply must leave some items out; instead, you select the most valuable and meaningful qualifications. This allows you to remove superseded items that you, by your continued efforts, have made less valuable.

Let me give specific suggestions relative to education, employment, and activities (hobbies, clubs, and similar).

B.6.1 Education

One of your greatest strengths is your education, assuming you have one or more advanced degrees from a college, university, or graduate/professional school. You serve yourself well by listing, for each degree, the granting institution, the degree earned, the study area, and the month and year graduated. These few crucial items just listed, as well as any study abroad and specialized training, are the only educational data that are necessary. Simplicity emphasizes the key achievements.

Any and all superseded information is unhelpful. If you have earned a Bachelor of Science (BS) degree, you do not provide information about your high school. The BS degree implies you finished high school. The employment you seek requires education beyond high school, so it is inconsequential, self-evident, and valueless. This is true of community colleges or schools that contributed to your transfer to a 4-year college or university where you took your terminal degree. Those courses and credits helped you transfer, and no reviewer/employer needs to know that you started elsewhere on your college journey. Again: all data from schools that merely contributed to a terminal degree at another institution are superseded. All that is required are your terminal degrees.

If you have a PhD, however, you still want to indicate your other college degrees: Master's and Bachelor's. These degrees are strengths on their own that did not just "fold into" the higher degree. (If you have a Master's, you also list your Bachelor's.) They each have separate curriculum and challenges that indicate your readiness and preparation for employment.

People often resist this advice, feeling that they must mention they were valedictorian of their high school class. My answer is twofold: One, nearly everyone who moves on to college has some high school honor in their past; Second, if you stopped your education after high school, you would not be a qualified applicant. Only because you moved on to college are you even being considered for professional employment. Thus, only upon your college accomplishments will you be compared to other applicants, not your high school glories. If you did great in high school and not so well in college, you will have an uphill battle (not insurmountable), and you must fight that battle with your college record and all other post-high-school activities.

A final exception to the superseded rule for education: employment during high school can be listed on your resume, especially if it lasted for a good duration or shows responsibility, diligence, expertise, and maturity. You will eventually discard such employment as you replace it with new jobs and experiences that fill up the space of the resume, but until then, it can be a good strength. Such information is not merely "folded in" to your college education because it is part of a separate category, "Employment," which I cover next.

B.6.2 Employment

For each organization listed under headings such as "Employment," "Work Experience," or "Experience," you need only indicate the last position held, not a series of job titles. If you started as dishwasher and ended as kitchen manager, the latter goes on the resume.

Here is a logical derivation to support omitting all superseded job titles at the same place of employment: Imagine each employer (interviewer, recruiter, and screener) reading a section of your resume and asking, "That's good? What else do you have to show? Can you top that?" That is how reviewers look at resumes. Well, if your next item is a lower-level job (dishwasher), it doesn't top the first item (kitchen manager). So, you have slid backwards, instead of moving forward, in the reviewer's estimation. You always want the reviewer to learn something new about you, in a cumulative sense. After reading about one job you have had, the reviewer should move on to either another job or a fully different section/category of information. If you make the reviewer move on to a previous job chronologically, even if your position there was low, that is acceptable because it is another place of employment altogether, which is a strength. Showing that you worked at a different employer previously is a new detail, as it shows you have varied experience and more than one employer has thought you worthy of employment. If you have finished providing your employment highlights, the reviewer will move on to another area, for example, "Awards" or "Hobbies." That is also good, as those sections provide new information. They "top" the employment by showing another dimension to you. But, if the employer moves from kitchen manager to dishwasher, at the same employer, the effect is a let down.

Should the employer wonder whether the current or last position you held at a job is the one you started at, they can ask you about that. If you have worked for 5 years at the same place, and your job title is "Manager," and they are not sure if you have been a manager for all 5 years, and that's important to them, they can ask. Most people get promoted and change job titles throughout their careers, so this is not your core strength, and it is common to everyone. Thus, it is not required to document all those details on a 1-page resume. I would suggest removing fairly typical specifics such as the following, which are taken from resumes I have reviewed:

- Started working 5 hours per week, then was requested to work 10 hours
- Promoted to counter manager after 3 months as sandwich maker
- One of only three assistants asked to continue working in the summer

In fact, rarely will your resume list all your jobs, experiences, and successes, and it does not have to. That would be impossible. Resume writers drop off items as they are superseded by more relevant, more recent, and more impressive alternative items. Each author has a choice of which items to include and exclude. If you place an obviously superseded item on your resume, you are not selecting items for optimal impact.

I know someone who started at a software company in California as a receptionist and, over time, rose up the ranks to become a vice president for customer relations (VPCR). She lists this company in her "Employment" section with just this final job title and a 5-year duration, although she was not the VP for all 5 years. Both are nonetheless true: She worked there from 2007 to 2012, and she held the position of VPCR.

One exception to this rule may be if you have worked for a company for many, many years, and the two (or three) positions you have held during this time involve distinctly different skills and responsibilities, not just progressively more seniority. For example, if you started as a chemical engineer and ended as a vice president for research and development, you might want to list both under a single line for employer name, as follows:

- Acme Industries, Detroit, MI
- September 2001 to Present, Vice President
- June 1995 to August 2001, Chemical Engineer

Think of this as two separate jobs with an extra formatting bonus: you only have to list the employer once, saving you at least a line of type.

B.6.3 Hobbies, Clubs, and Other Activities

These items, which are not the crucial ones on a resume, must be listed with utmost brevity in mind. Redundancy in a hobby section is anathema to your purpose with a resume. Thus, you will not want to list, for example, every marching band you have performed in. To say you were in your college's marching band, trumpet player, is adequate. Similarly, if you are on your school's robotics team, you do not need to list all the different positions held (unless you are including this somewhat prominently in the resume as crucial project experience).

A strong word of caution, therefore, is warranted for this resume data: As if added as an afterthought, activities often are the section on a resume with the most formatting and typographical errors. Although activities can add strength and serve as ice breakers at interviews, their presentation can undermine a good impression in terms of conciseness and professionalism, which brings us to the next section.

B.7 Guideline #7: Fix All Format Flaws

A resume's general look should be neither too sparse nor too crowded. It must be neither too blocky nor too loose. As in film, the *mise en scene* is all that lies within the borders of the screen or the frame of film; similarly, the resume must look unique, attractive, well organized, inviting, and neat.

Resumes often have these flaws: too sparse, too crowded, poor layout, and messy use of fonts, text sizes, and highlights. Solutions are presented below.

B.7.1 Fill or Add White Space

A resume is a conundrum, for it must contain as much biographical information that can fit on one page while simultaneously seeming spacious, nicely segregated into sections, and easy to scan. In other words, a resume must contain some white space but not too much. It cannot seem half finished, nor can it be packed from edge to edge with non-stop text. This is a difficult challenge. The various examples throughout this appendix demonstrate different ways to organize the information into sections, enabling a reader to both see the separations and find related data placed in similar, repeating patterns. Most formats are acceptable, as long as the white space is helpful. The first example in this appendix, the resume sent to IBM (Figure B3), provides a good baseline for appropriate use of white space. It is not, however, the only allowable format style. Examples below show both positive and negative use of white space while presenting various formats that potentially can be fine for resumes. While reviewing the examples, pay special attention to the use of white space, with either too much or too little.

Figures B11 and B12 are examples of resumes that are light on information and seemingly unfinished. Other examples illustrate resumes that are too jam-packed with text and, thus, appear overwhelming, disorganized, and intimidating (Figures B13 and B14). So, to help you maximize white space without sacrificing substantive data, I have two tips, as follows:

1) Remove one-word continuation lines
2) Maintain adequate margins

B.7.1.1 Remove One-Word Continuation Lines

If a bullet point or a segment of information continues to a second or third vertical line, you are advised to use that whole line, or most of it. If not, you are wasting the vertical line for little data. Figure B15 has numerous instances of one-word continuous lines. Here are two bad/good examples, one with a one-word continuation line, and one with a full use of the second line:

<u>Skills</u>
Project management, mobile development, user-experience design, agile methodology

<u>Skills</u>
Project management, mobile development, user-experience design, agile methodology, computer game design, emerging technologies, and security

EMPLOYMENT
Wealth Management, LLC, Chicago, IL, Summer 2016
- Applied knowledge of Angular JS to develop application for mobile devices

EMPLOYMENT
Wealth Management, LLC, Chicago, IL, Summer 2016
- Used Angular JS to develop mobile application targeted at travelers

FRANCIS LOUDON MACINTOSH
4094 E. Cross Street
Lansing, MI 46532
+1 (254) 490-8609
frldemac@msu.edu

OBJECTIVE: To obtain an entry-level job at General Electric

EDUCATION
Michigan State University School of Engineering
Will Graduate Spring 2020, Majoring in Electrical Engineering
Present college GPA: 3.292

AWARDS/HONORS
2016: Dean's Honor list for Academic Distinction, fall semester
National Merit Scholar

EXTRA-CURRICULAR:
2017-Present CHISL Design group—tech team
Biggby Coffee project team—market research and web development

EMPLOYMENT:
2015: TestTech Solutions—Intern, then hired as Electrical Engineering Technician
—Assembled and learned about design of wire harnesses for vehicle test bunks

2016,17: Founder's Brewery Grand Rapids—Server's Assistant/Barback
2017: EMU Libraries—Library staff member

INTERESTS AND TALENTS
Piano—Studied for 12 years
Guitar—Studied for 4 years, performed in band

REFERENCES
Ralph Ellison, CEO of TestTech Solutions, LLC:
Phone: +1(324)454-9843
Email: rellison@testtechsolutions.com

Figure B11 Sample resume light on information and seemingly unfinished

cfragu905@gmail.com
315.343.8234

CHRIS FLANAGAN

OBJECTIVE	Software developer interested in building and maintaining Android and web applications
SKILLS	• Java (Android, RxJava) • Python (Django) • Web (HTML, CSS, JavaScript, SQL) • Testing (Espresso & Appium) • Salesforce (Apex, Visualforce, SoQL)
PROFESSIONAL EXPERIENCE	**CRM CONSULTANT, MADISON PARTNERS I MAY 2016 – JULY 2016** Developed web applications on the Salesforce platform configured to clients' CRM organizations. **QA INTERN, INNOVATION AI I SEPTEMBER 2015 – APRIL 2016** Wrote automated tests for a mobile application (Android and iOS) in various frameworks.
PERSONAL PROJECTS	**MySpotify** An Android application which uses the Spotify Android SDK and web API to provide the same services as the official Spotify app **Linear Algebra** A Java library for performing basic operations in linear algebra **NBA Scraper** A Python program used to scrape & structure data about the NBA from the web
EDUCATION	**UNIVERSITY OF COLORADO** – BOULDER • BSE in Computer Science • Expected graduation: June 2018

Figure B12 Sample resume light on information and seemingly unfinished

ARJUN KRISHNAKUMAR

ADDRESS: 1001 Hayward St, Ann Arbor, MI 48109
MOBILE: +1 (269) 999-1717
EMAIL: arjunkk@gmail.com

WEBSITE: arjun.xyz
GITHUB: github.com/arjunkk
LINKEDIN: linkedin.com/in/arjunkk

EXPERIENCE

DEPOT TECHNICIAN
University of Michigan
SEPTEMBER 2018-PRESENT
Working as a Computer Consultant at the Tech Depot division of Information Technology Services.

INTERN
Air Zoo
JUNE 2018-SEPTEMBER 2018
Worked as a web developer to create an AdSense revenue-heavy product showcase website with HTML5, CSS3 and JavaScript.

INTERN
Black Lab Five
JULY 2017-MAY 2018
Worked as a web developer to create Wordpress websites with themes made from scratch, and fixed the backend of some Wordpress sites.

TEACHING ASSISTANT
Western Michigan University
SEPTEMBER 2016-JUNE 2018
Helped students of the AP Computer Science A class with their projects and provided support as needed every Sunday for one to two hours.

INTERN
Technology Solutions of Michigan
JUNE 2016-AUGUST 2016
As a part of a professional team, worked as a backend web developer and created a phone billing system with a simplified user experience.

ACCOMPLISHMENTS

- Winner of IBM Master The Mainframe 2016, 2017, and 2018—Parts 1 and 2.
- Intel International Science and Engineering Fair (ISEF) 2018 Finalist.
- Co-programmer for Team Strykeforce, a qualifier at 2016 FIRST Robotics Competition (FRC) World Championship, Winner of 2017 FRC World Championship, and Winner of 2018 FRC World Championship.
- National Merit Scholarship Recipient from 2018 to 2022.
- Intel International Science and Engineering Fair 2018 Finalist for "PEROVSKITE NANOSTRUCTURES AS LEDS: TOWARDS FLEXIBLE DISPLAYS" with guidance from Dr. Ramakrishna Guda at Western Michigan University.
- Winner of the Intel Excellence in Computer Science Award in the 2016-17 Southwest Michigan Research Fair for "PERFORMANCE OF PARALLEL SOLUTIONS FOR THE N-BODY PROBLEM" with guidance from Dr. Elise deDoncker at Western Michigan University.

EDUCATION

B.S.E. - COMPUTER SCIENCE
University of Michigan
SEPTEMBER 2018 - JUNE 2020
Freshman in the College of Engineering, starting with 52 of 128 credits done. Taking EECS 280, EECS 203, MATH 215, PHYSICS 140, and TCHNCLCM 300. Part of the Solar Car Team and Project RISHI.

DUAL ENROLLMENT
Western Michigan University
SEPTEMBER 2014 - JUNE 2018
Academically Talented Youth Program (ATYP) with 3 years of English and 1 year of Computer Science, and Differential Equations (MATH 3740).

LANGUAGES

ENGLISH
MARWARI
HINDI
SPANISH
MARATHI
SANSKRIT
KANNADA

SKILLS

C/C++, Java, Python
HTML5, CSS, NodeJS
PHP, .NET & SQL
Android & iOS Apps Design

HOBBIES

</> CODING
ORIGAMI
MUSIC
MECH-ING
BIKING
LANGUAGES

Figure B13 Sample resume with small text, unnecessary format flourishes, and inconsistencies

PAMELA BREE KANTOR

1015 Church St. – Ann Arbor, MI – 48104 // 7346804836 // pambkan@gmail.com

EDUCATION

- **Bachelor of Science in Engineering (BS Eng): University of Michigan, Ann Arbor, Michigan** *Aug 2015 – May 2019*
- **Cumulative College GPA:** 3.167/4.0
- **Major in Data Science Engineering; Minor in Program in Sustainable Engineering**
- **Honors:** Dean's List for academic excellence *Jan 2017 – May 2017*
- **Relevant courses:** Machine Learning by Andrew Ng, Stanford University, California
  Introduction to algorithms, Machine Learning, Data Structures, DBMS
- **Study Abroad in China:** Program in Clean Energy, Xiamen University, China *May 2016 – July 2016*

WORK EXPERIENCE IN DATA SCIENCE

Intern – Chhattisgarh InfoTech and Promotion Society (CHiPS) under PwC consultants *May 2018 – July 2018*

- Conducted an end to end project on Social Media Analysis of CHiPS' Facebook and Twitter pages by web scraping published material using Graph API Explorer and Twitter API in Python
- Created predictive models, and performed sentiment analysis using Google Cloud NLP and text analysis to highlight the difference between the sentiments of the government versus that of the public discussed on social media posts
- Leveraged Microsoft Power BI to perform analytical data visualizations and storytelling to derive insights pertaining to the sentiments of users to improve the schemes and increase audience involvement by 35% using sentiment analysis
- Reported recommended changes to Government Public Schemes by conducting data analysis of data gathered from 120,000 public beneficiaries across the state directly to the Chief Minister of the State

Research Assistant – Undergraduate Research Opportunity Program (UROP), U of M *Sept 2016 – June 2017*

- Collected Michigan dining inventory data from 2012 to 2016 and conducted comparative analysis using databases and personal networking
- Increased the yearly purchasing and composting of sustainable food products across campus from 65% to 75% by presenting analytical data visualizations to showcase the impact

Intern – M76 Analytics, Mumbai *June 2017 – July 2017*

- Developed machine learning algorithms on training data sets to reduce client's expenditure by 15% increasing profit margins, optimizing solutions to financial problems faced by brokerage firms
- Analyzed client's past recorded data to improve the financial performance of the firm within the industry by creating an Economic Opportunity Model (EOM)™ using machine learning algorithms in Python

WORK EXPERIENCE IN SUSTAINABLE ENERGY

Research Assistant – Multidisciplinary Design Project (MDP), U of M *Jan 2017 – Dec 2017*

- Leveraged the art of Kirigami to develop dynamic structure of solar cells driving efficiency of solar panels by 200%
- Participated in a 15-person team to create convex hexagonal solar concentrators that increase the surface area of solar cells and concentrate light intensity to create the more optimal solution

Intern – Silicon Valley Renewable Energy Development Agency, Los Altos, CA *Jan 2015 – Feb 2015*

- Presented a paper to the director – "*Reduction in the efficiency of solar panels due to the effect of dust*"
- Investigated the performance of solar panels through electrical test experiments conducted daily for a month for the strategies to be successfully adopted in government offices

EXTRA-CURRICULAR / LEADERSHIP POSITIONS

Peer Advisor – Undergraduate Research Opportunity Program (UROP), U of M *Aug 2017 – May 2018*

- Conducted bi-weekly seminars instructing a class of 35 undergraduate students on scientific research
- Facilitated weekly discussions with researchers and industry professionals on scientific research, its methodologies and techniques while helping provide mentorship opportunities to the students
- Graded and judged undergraduate research work at the annual UROP Symposium with 500 participating students

Senior Vice President - Michigan International Student's Society *Sept 2016 – Present*

- Managed the only international student's society on campus with approx. 550 members, recruited a new class of 25 students each semester with an acceptance rate of 10% and organized social networking events involving alumni

Vice-President of Public Relations – Society of Global Engineers *Jan 2016 – Present*

- Organized monthly information sessions with companies focused on recruiting international engineering students
- Facilitated discussions, resume workshops and professional photo-shoots for the members of the society

Sustainability Chair – Bursley Hall Council, UM *Sept 2015 – Aug 2016*

- Administered sustainability practices for 1200 university housing residents by organizing social and professional events to bring about awareness and collaboration

TECHNICAL SKILLS

Technical: Python, C++, SQL, MATLAB, R, Microsoft Power BI, Microsoft Office

Figure B14 Sample resume with too much text and too little white space

Amanda Morningstar

333 Colorado Road
Apt. 44
Louisville, KY 40206

Email: Aman.Morning@louisville.edu
Cell: (353) 997 - 4359

EDUCATION

University of Louisville Anticipated Graduation: August 2016

- MEd in Counseling Psychology and Personnel Services with a Specialty in Counseling Psychology: GPA 3.93
- Clincal Practicum at The Cardinal Success Program at the Academy at Shawnee
- Assessment Practicum at The Cardinal Success Program at the Academy at Shawnee, Nia Center, and Roosevelt Perry Elementary School
- Graduate Teaching Assistant for the Psychology Department- Psych 365 Child Development (Spring 2016)

Denison University Graduated May 2014

- Awarded the Denison Alumni Award scholarship for Leadership
- Bachelor of Arts in Psychology: Major GPA 3.2

WORK EXPERIENCE

Graduate Teaching Assistant- University of Louisville Psychology Department December 2015- present

- Attend and assist the professor for all classes of Psych 365, Child Development.
- Grades and manages online exam administration, as well as all homework and extra credit assignments
- Holds online and in-person office hours, and stands as the first line of contact for student questions

Counselor in Training- Cardinal Success Program at the Academy at Shawnee, Louisville, KY August 2015- present

- Conduct individual sessions with middle and high school students referred by teachers, staff, and parents
- Implement TIM&SARA, a bi-weekly group therapy program developed for depression prevention in ninth-graders
- Create treatment plans, clinical notes, intake interviews, and parent/staff interviews for all referrals
- Administer assessments at local clinics for adults, adolescents, and children and create psychological reports

Mental Health Technician- Baptist Health, Louisville, KY July 2014- August 2015

- Assists nursing staff by performing select nursing procedures and routine duties in caring for Adult and Geriatric patients with a variety of psychiatric diagnoses, particularly those who are in need of immediate crisis intervention
- Support therapeutic activities by providing emotional care
- Conduct community sessions for patients involving goal setting and planning

Therapeutic Riding Intern- Equine Assisted Therapy, Centerburg, OH Summer of 2013

- Assisted instructor during therapeutic riding lessons for children and adults with disabilities including horse handling, side walking, lesson preparation, coordination of scheduling and lesson plans, as well as assist with volunteer training.
- Assisted with fundraising projects, including apparel sales, as well as research and propose grant options, as well as write and distribute various press releases.
- Assisted with the creation and distribution of the monthly electronic newsletter and other marketing/advertising endeavors including social media and web site maintenance.

RESEARCH EXPERIENCE

Family Roles in Mental Health, University of Louisville Fall Semester 2015

- *An Examination of Social Support, Family Religiousness, and Family Idealization and Their Role in Mental Health:* An examination of family idealization, social support, and family religiousness on mental health and illness in adolescents
- Data is currently being processed

Figure B15 Sample resume with numerous instances of one-word continuation lines and inappropriately small margins

B.7.1.2 Maintain Adequate Margins

You should start with 1-inch-wide margins all around. I strongly encourage you to keep these. Margins add a pleasant and conventional frame around the text, and they allow for 3-hole punches, gripping fingers, and other "edge" uses. The example above with one-word continuation lines (Figure B15) has inadequate margins all around. However, if you must fiddle with the margins just a little to enable you to include all the critical information you have your heart set on, and you like the format and overall look, you can start by adjusting the right margin a bit, then the bottom, but only rarely the left. Also, the top can be adjusted if you have your name in large text at the top, which helps to offset the slightly smaller top margin. If you reduce the top margin and leave small text up there, the effect will be to violate the space intended for a margin, as in Figure B16.

Elaine O. Crabtree

2345 Overlook Dr. Apt. 4
Toledo, OH 45734

Mobile: 415.898.5432
email: eocrab@umich.edu

EDUCATION:

University of Michigan – Ann Arbor, Michigan — April 2016
Bachelor of Science degree in Biopsychology, Cognition, and Neuroscience
Coursework specializing in neurology

VOLUNTEERISM:

Program Heal of Northwest Ohio — 12/2015 to Present
Organize community events for a non-profit to educate and raise money for men and women undergoing treatment for eating disorders. Engage with community members to encourage involvement and raise awareness across the area

Girls in the Gym – Flint, MI — 2/2014 to 5/2014
Taught the Girls in the Gym curriculum to girls in 3rd through 5th grade, focusing on emotional, physical, and mental health. Encouraged teamwork amongst participants and other mentors.

Positive Self Club – University of Michigan, Ann Arbor, MI. — 9/2010 to 12/2013
Trained to facilitate open discussions focusing on body image, mental health, and self-care. Led other students through these trainings in an effort to expand the program. Engaged with different groups and organizations throughout campus to endorse the program, lead discussions, and promote connections amongst members.

Ozone House – Ann Arbor, MI — 12/2012 to 8/2013
Staffed the crisis-helpline. Utilized skills in listening, empathy, and communication to counsel youth in crisis and provide them with community resources.

University of Michigan Health System – Ann Arbor, MI — 5/2010 to 2/2011
Volunteered on the pediatric oncology unit. Provided comfort and support to patients and their families. Supported medical staff in providing the best possible care.

WORK EXPERIENCE:

***Nanny*, Private Family** – Ann Arbor, Michigan — 1/2015 to Present
Responsibilities: Trusted to care for three children ages 4 and 6. Constructed engaging activities to aid in development, education, and enjoyment for the children. Communicated with parents on a daily basis to ensure seamless care and consistent family policies.

***Manager*, Café Zola and Zola Bistro** – Ann Arbor, Michigan — 1/2013 to Present
Responsibilities: Process payroll weekly. Collaborate with executive chef to create weekly specials and change the menu seasonally. Update the Aloha POS system as needed. Manage the training program for staff members. Host and manage the floor to ensure customer satisfaction and teamwork amongst staff. Expedite food on busy weekends. Communicate with other supervisors, managers, and restaurant owners.

COMPUTER: Apple and PC systems proficient
Microsoft Office proficient
Aloha POS system
Open Table reservation system and programming

CERTIFICATIONS:

Stewards of Children — 2015
Sexual abuse of children prevention

Crisis Text Line Counselor — 2015

Figure B16 Sample resume with small top margin and poor column spacing

B.7.2 Improve Layout

One way to improve the layout, in addition to ensuring adequate white space and clear spatial divisions between sections, is to ensure you do not cross implicit column dividers. This is a subtle issue. It is not always immediately apparent to some people. The best way to see the problem of crossing implicit column dividers is to compare two examples (look at Figures B16 and B17).

Figure B17 includes an effective use of a three-column layout. At the far left are the first-level headings, the middle column has the details, and the far-right column isolates the dates and locations nicely. Importantly, the text in the middle does not cross over into either of the other columns, and the overall layout does not contain a shifting structure of columns.

In contrast, in Figure B16, the resume's format is rough and irregular. It relies on many indentations to create a sense that items are aligned in columns, but the columns are not maintained throughout. The top section, Education, has a neat three-column layout. That layout is undermined in the section below, where text in each column overlaps the other columns in rows above or below, in particular, the descriptions of volunteer accomplishments, which overlap both the heading, Volunteerism, and the dates at the right. Overall, all the dates are meant to stand apart at the right, but the text lines above and below the dates extend across the "white space" gap and into the margin beyond the dates. Similarly, the heading "Certifications" crosses over the horizontal gap between the "Computer" heading and its details. Although the "column" alignment approach to both the Education and Computer sections is laudable, it is inconsistent with the layout of the other sections. Lastly, the middle column containing details of Responsibilities in the Work Experience section does not align with any columns elsewhere. This author's experience is impressive, but the resume's layout is not.

CLAY W. MANHEIM

734.887.7383
claywman@umich.edu

244 Division Ave.
Ypsilanti, MI 47108

OBJECTIVE: Permanent research position in cognition and learning

EDUCATION: **University of Michigan – Ann Arbor**
Bachelor of Science in Sociology, April 2018
Coursework specializing in learning development

EMPLOYMENT:

Rochester College — Sept 2014 – June 2016, Auburn Hills, MI
Peer-Assisted Learning Manager
- Collaborated with course professor and other tutors to produce effective learning session structures
- Facilitated group discussions outside of classroom
- Recorded students' progress by conducted exams
- Set academic goals for students
- Provided guidance in courses involving natural sciences, computer science, business, mathematics, and economics

Cross-Zebra Group, Inc. — May 2012 – Aug 2015, Southfield, MI
Programmer
- Designed and implemented manufacturing parts catalogue program
- Developed searching function and quantity updating
- Implemented complex data structures

Landmark Properties — Feb 2010 – April 2012, Southfield, MI
Maintenance Supervisor
- Oversaw four commercial properties to ensure continued cleaning and good repair
- Researched and enforced codes for sanitation and building upkeep
- Organized waste and recycling program

VOLUNTEER SERVICE:

Rochester College — Jan 2015 – June 2016, Auburn Hills, MI
Civil Rights Action Team
◊ Supervised advertising and marketing sub-teams
◊ Designed all forms of advertisement
◊ Planned and organized many campus events
◊ Conducted seminars regarding current events pertinent to team's mission

HONORS: **University of Michigan**
Dean's List, Winter 2017 and Fall 2017
Mills Scholar in Sociology (top 5% of class), Fall 2017

Figure B17 Sample resume with effective use of a three-column layout

B.7.3 Be Clean and Consistent with Fonts, Sizes, and Highlights

Across your cover letter and resume, you should use the same font and, for the core text on the resume, the same text size as used in the letter. This way, the two documents look like a pair. Quite often, in contrast, they look like a mismatched couple. When the letter-resume combination shown at the beginning of this appendix (Figures B2 and B3) was originally created, the author had wildly different styles for the letter and resume, which had been prepared months before the letter. The letter is reproduced below (See Figure B18) along with the original resume (Figure B19) to demonstrate the disparate styles. After the resume was revised to match the letter, (Figure B3), the combination represents a suitable couple. The "standby" resume, written long before the author's first cover letter was drafted, contained so many of the aforementioned format flaws that it needed to be retired even if it had not been so stylistically different from the letter.

After matching your fonts and core text sizes, you can attend to text sizes of other elements. Text sizes can change on a resume but not for the same type of information. That is, your name may be larger than your list of language skills. But within your list of skills, all language skills must be in the same text size. Each time you list your city and state of employment or the dates of employment, furthermore, you must be sure to not switch text sizes. Figure B19 (the "mismatched" resume just discussed) has many of these inconsistencies. Overall, resumes containing unusual format choices are difficult to match with letters (recall Figures B13 and B14). For this reason, among others, such resumes are best avoided.

B.8 Guideline #8: Use Strong Verbs for Accomplishments and Responsibilities

Verbs on your resume are the basic building blocks. With verbs, you give the particulars that showcase your abilities and qualifications, based on past or present experiences. So, the advice on verbs is threefold: (1) use verb-based phrases rather than either noun-based phrases or full sentences, (2) avoid weak verbs, and (3) ensure the correct tense.

B.8.1 Use Verb-Based Phrases

When you discuss the experiences at employment, student projects, or volunteer positions, you want to use brief, verb-based phrases. This is a resume convention. You do not use full sentences or noun- or adjective-based phrases:

- NO: I was responsible for overseeing volunteers at this organization who participated in a river-based erosion study
- NO: Responsible for overseeing volunteers for river-based erosion study
- YES: Oversaw volunteers for river-based erosion study

You should present as many of these verb-based phrases as possible to reflect your full range of experience, without becoming redundant or over-packing the sections with accomplishments and responsibilities at the expense of other sections. A balance is needed, but you can tilt toward verbs because of their value in proving your qualifications. The sparse resumes shown previously (Figures B11 and B12) suffer significantly from a lack of verb-based qualifications. Lastly, verb-based qualifications are so fundamental to a resume that they require no sub-heading to precede them, such as "responsibilities" or "duties" (see Figure B16 for a negative example).

B.8.2 Avoid Weak Verbs

I suggest avoiding all weak verbs, such as these: learned, worked, interned, attended, gained (experience), participated (in), and studied. You are advised to concentrate on primary accomplishments rather than tasks where you played a peripheral role (observed, went to meetings, or learned as an understudy). Similarly, it should go without saying that you must not put any item that is anticipated for the future (except for college graduation).

Instead, you want verbs that denote actual work performed by you, not observed or witnessed or planned but not yet executed. The list of strong verbs is endless, such as analyzed, balanced, and coordinated. Run through the alphabet from a to z and look for the best, strong verbs that characterize your accomplishments. This writing task is as simple as knowing your ABCs. Over the years, I have compiled a list of verbs for resumes (Figure B20).

1234 Ash Street
Ann Arbor, MI 48105
(617) 321-9876
September 2, 2017

Russell Brady
Extreme Blue Lab Manager
IBM
10330 Wimbledon Dr.
Charlotte, NC 28262

Dear Mr. Brady:

I am writing this in response to IBM's online post about the Extreme Blue internship program. I am a junior at the University of Michigan studying Computer Science Engineering with a minor in Business, planning to graduate with my BS degree in December 2019. After talking with an IBM representative at my university's career fair, I believe that the Extreme Blue internship would provide me with an unparalleled opportunity to create and design an innovative software solution. I am confident that my background in programming, big data management, and team leadership makes me particularly well suited for the Extreme Blue position.

Through coursework, I have developed the necessary web development and programming skills that would enable me to effectively contribute to an Extreme Blue team. Within my Entrepreneurial Design Class, I collaborated with a group of fellow students to develop an application by using a standardized design methodology, such as interviewing anticipated customers and conducting user research. I not only gained valuable user-interface design techniques, but also specific programming skills in HTML, Angular JS, Flask, and several third-party APIs. This project gave me experience in solving a problem with programming strategies.

Furthermore, through my position as a Laboratory Research Assistant, I have obtained experience with big data analytics. Specifically, I analyzed over 50 papers and articles on various topics in machine learning, particularly natural language processing and computer vision. Building on this literature research, I created a script using Microsoft Computer Vision API to generate feature data from a set of photos. I utilized Python to create a visualization of the data in order to communicate the results to non-technical stakeholders. During the Extreme Blue position, I will be able to incorporate these innovative technologies into any project.

In addition, I have experience as a team leader, with strong interpersonal, planning, and communication skills, as evidenced by my leadership roles in Theta Kappa fraternity (TK) and Palomar Emerging Leadership Experience (PELE). As TK's recruitment chair, I organized events for over 100 members. Furthermore, through PELE, I learned how to work effectively with new colleagues by identifying and leveraging each team member's strengths.

I believe the Extreme Blue team position will benefit from my ability to create software products that leverage big data management. I have enclosed my resume for your consideration, and I will check with you in three weeks' time to confirm receipt of my resume and answer any questions you may have. I hope to be invited to an interview at your earliest convenience.

Sincerely,

Benjamin Farley

Benjamin Farley

enc.

Figure B18 Sample letter written separately from author's "standby" resume (B19)

BENJAMIN FARLEY COMPUTER SCIENCE ENGINEERING

OBJECTIVE

Seeking a software development internship.

CONTACT

bfarley@umich.edu
bfarley.github.io
617.321.9876
1234 Ash Street, Ann Arbor, MI 48105
in
https://www.linkedin.com/in/benfarley943@lld4
bfarley

EDUCATION

University of Michigan, College of Engineering
B.S Computer Science Engineering 2018
Minor In Business, Ross School of Business

SKILLS

PROGRAMMING: C++, Python
PYTHON: Flask, Pelican, Networkx, Matlibplot, Pandas, Seaborn
WEB DEVELOPTMENT: HTML, CSS, Flask, Angular JS, D3.js
TESTING: Selenium, Protractor, TestNG, Jasmine

EMPLOYMENT

University of Michigan, Ross School of Business — Ann Arbor, MI
Assistant in Research — Apr 2016 to Jun 2016, Sep 2016 to Current
- Conducted and presented survey on state of art computer vision and machine learning techniques and methodologies utilized for feature extraction and image processing.
- Researched, analyzed, and implemented various computer vision algorithms using python and Microsoft's computer vision SDK for a collaborative project under statistics professors.

Kronos Incoporated — Chelmsford, MA
Software Performance Engineering Intern — Jun 2016 to Aug 2016
- Developed protractor testing framework using page object model for maintainable E2E testing.
- Implemented timing function to efficiently record and visualize performance of multiple actions using Node.js and Python.
- Created automatic build process using Grunt and Jenkins.
- Gained experience following Agile and sprint methodologies and industry best practices with git.

PROJECTS

Reach (Team)
- Utilized design methodologies to design and develop web app that alerts users when their friends are within a certain distance to increase social interactions.
- Implemented prototype using HTML, Bootstrap, Flask, Angular JS, and Google Maps API.

Related Artist Network (Personal)
- Created interactive network visualization of related Artists using python, Spotify API, NetworkX, HTML and D3.js to aide discovery of new music.

City of Newton Student Commuter (Personal)
- Collaborated with City of Newton to estimate amount of CO_2 produced by students' commute.
- Provided data analysis using Python, Pandas, Google Maps Distance Matrix and Geolocation API.
- Developed and presented visualization with Leaflet.js to support proposal of new bike lanes and sidewalk repair.

Accenture U.S Innovation Challenge (Team)
- Led interdisciplinary team to research, construct and present a business strategy for the prevention of food waste by grocery retailers.
- Selected as runner up in national quarterfinals.

Blog (Personal)
- Implemented and maintain a blog using the Python pelican framework.

ACTIVITIES

Serial Innovator Camp (Procter & Gamble, Intel, Microsoft) — Feb 2015
- Advanced problem definition and solving techniques while working with a interdisciplinary team.

Petrovich Emerging Leader Experience (PELE) — Sep 2015 to Dec 2015
- Selected as 1 of 34 students to participate in sophomore year college leadership program.

Theta Chi — Oct 2014 to Oct 2015
- Elected to oversee 3 recruitment periods, receiving 20 new members after narrowing down over 250 interested participants.
- Selected and managed a committee of nine.
- Developed strong interpersonal and communications skills.

VOLUNTEERING

Accelerate CS (Google) · Instructor — Sep 2015 to Current
- Lead weekly lesson at Ann Arbor middle school to teach student's computer science skills.

Figure B19 Sample "standby" resume that does not match its companion letter (B18)

B.8.3 Ensure the Correct Tense

A simple rule here suffices. With current jobs or experiences, the tenses are simple present (not gerunds), and at past jobs or experiences the tenses are simple past. Authors tend to prefer gerunds when giving present actions or duties, but these verb-derived nouns are unnecessary:

Bad:

> *Internship*, June 2015 to Present:
> Expanding augmented reality mobile app for use in hospitals, designing and implementing new features, and collaborating with software engineers and business developers

Fixed:

> *Internship*, June 2015 to Present:
> Expand augmented reality mobile app for use in hospitals, design and implement new features, and collaborate with software engineers and business developers

In experience that is finished and in the past, the ideal verb tense is past:

Good:

> *Internship*, June 2015 to June 2016:
> Expanded augmented reality mobile app for use in hospitals, designed and implemented new features, and collaborated with software engineers and business developers

Complex verb tenses are unnecessary. Two simple choices are the best options. Inconsistencies arise when trying to use a variety of verb tenses. In the worst-case scenarios, the inconsistencies can evolve into errors in accuracy. Such occurs in the resume discussed already with overlapping columns (Figure B16).

Action-Oriented Verbs for Resumes

Accomplish
achieve
adapt
address
add
adjust
administer
adopt
advise
advocate
aid
allocate
analyze
apply
appraise
apprise
approve
arbitrate
arrange
assemble
assess
assign
assist
attain
audit

Budget

Calculate
calibrate
care (for)
change
check
clarify
classify
coach
code
collate
collect
communicate
compare
compete
compile
complete
compose
compute
conduct
confront
consolidate
construct
control
coordinate
correspond
counsel
create
critique

Decide
define
delegate
deliver
demonstrate
describe
design
detect
determine
develop
devise
diagnose
differentiate
direct
discover
dispatch
dispense
display
dissect
distribute
document
draft
draw

Earn
edit
educate
effect
eliminate
encourage
enforce
enlist
establish
estimate
evaluate
examine
exhibit
expand
expedite
explain
express

Facilitate
finance
forecast
formulate
fulfill

Gather
generate
guide

Handle
hire
hypothesize

Identify
illustrate
imagine
implement
improve
increase
induce
influence
inform
initiate
inspect
inspire
install
institute
instruct
integrate
interpret
interview
introduce
invent
investigate
involve

Judge

Launch
lead
lecture
lobby
locate

Maintain
manage
map
measure
mediate
mentor
model
monitor
motivate

Navigate
negotiate

Observe
obtain
operate
order
organize
originate
oversee

Participate
perform
persuade
photograph
plan
predict
prepare
present
print
prioritize
process
produce
program
project
promote
proofread
propose
provide
publicize
publish
purchase

Question

Raise
recommend
reconcile
record
recruit
redesign
reduce
refer
refine
regulate
rehabilitate
reorganize
repair
replace
replenish
report
represent
research
resolve
respond
restore
retrieve
revamp
review
revise
rewrite

Schedule
secure
select
sell
serve
simplify
sketch
solve
sort
speak
streamline
strengthen
study
succeed
summarize
supervise
survey
synthesize
systematize

Teach
test
theorize
train
transact
transcribe
transfer
translate
treat
troubleshoot
tutor

Update
upgrade
use
utilize

Verify

Write

Figure B20 List of strong action verbs for resumes

B.9 Guideline #9: Either Avoid or Capitalize Upon Personal and Private Items

Personal affiliations are not required resume elements. Because employers cannot ask about your religion, race, marital status, political party membership, and other such personal items, you should not think of those items as primary resume or letter information. But, sometimes a qualification is intertwined with an *affiliation*, as in these examples: vice president of a Greek fraternity, tutor in Bible group, events manager for campus libertarians, and treasurer for the Society of Black Engineers. In such cases, you may opt to share your affiliations and interests, as those activities convey additional aptitudes and work-related experiences you bring to a new employer. These can also worry some future employers (though you may prefer to not work for any employer who would be troubled by your affiliations).

By law, no employer can discriminate against you because of marital status, religion, race, ethnicity, and so forth. Nonetheless, some may unlawfully discriminate (which may be hard to prove and litigate). Keep in mind that these affiliations are not your primary qualifications anyway. Primary qualifications are your professional competencies, education, and experiences at work and on projects.

Although the private, personal items should be considered neutral and never negative, they can also be positive in some situations regardless of how they increase your aptitude for the job. Personal affiliations and private interests might benefit you in helping break the ice in an interview or impressing a particular employer. You cannot predict reactions, either positive or negative. Thus, many advisors recommend avoiding these controversial items altogether.

My advice is more nuanced, as intimated above and repeated here: (1) include such items when a *qualification* is intertwined with an affiliation, (2) include such items if they are very important to you, and (3) include such items when your resume is otherwise sparse of employment, projects, and extra-curricular experiences, and you need to provide a fuller picture of yourself.

B.10 Guideline #10: Avoid Employer-Centric Items in Perpetuity

Do not confuse specific items emphasized by a current recruiter as permanently required information for a letter or resume. Some recruiters will want to see college courses taken or grade point average, for example. These are optional items for a resume, so they certainly can be included under your "Education" heading. Nonetheless, they are not required items. A resume is not grossly incomplete if it omits your GPA from college five years ago. Work experience after college supersedes the importance of GPA very quickly. The accomplishments during employment are much more applicable to your next employer than your distant academic record. The single strongest item on your resume, as a qualification, is employment experience. Other items are good predictors of how you might perform on the job. But work experience is a direct equivalent.

Similarly, you might be asked to provide your immigration status. This can be removed when not specifically requested and when no longer an item of concern.

B.11 Quick Review of Required Information

In addition to the ten guidelines covered above, I want to provide some assistance with selecting the basic information to be included for your primary segments: objective, education, employment, and extra-curricular activities.

B.11.1 Objective Line

Although the cover letter should state your employment objective, this information can be briefly expressed on your resume also. This is absolutely necessary if you are not submitting a cover letter. Indeed, some organizations may not want a cover letter; if so, do NOT provide one. (Follow instructions for all reports: see Chapter 1.1.) In such cases, your resume's Objective line is crucial.

My advice for an Objective line is fourfold:

1) Indicate both the work specialty or position name (if known) and the broad business area, as in all these brief examples:
    - chemist in agriculture science, accounting for non-profit finance, website design for social media, management consulting to retail

industry, assembly language programming for computer hardware, chemical engineering for alternative fuels, computer networking for public library, ground-water modeling at governmental environmental agency, psychology researcher in women's mental health, fundraising and development for college programs, social worker to veterans, art instructor for adult education, and quality assurance engineer in aircraft manufacturing

2) Indicate the position type: permanent, internship, part-time, full-time, etc.

3) Indicate position time range when **not** available immediately and permanently, as in these two examples:

   - Objective: Internship in automobile display design, January to May 2018
   - Objective: Neurology research assistant, fall term 2018

   If available now and on a permanent basis, no time range is needed, as follows:

   - Objective: Permanent position as signal hardware testing engineer

4) Keep it to one line whenever possible: eliminate all unnecessary terms, including "looking" and "seeking" and indefinite articles ("a" and "the").

You actually do not want to keep looking; you want the job. Unlike the game of mating, where the chase may be half the fun for some people, the job hunt is really only satisfying when the conquest is made. Thus, see the change below:

Too Long

- Objective: Seeking a full-time position as a biostatistician in cancer-related research areas—starting in three weeks

Just Right:

- Objective: Full-time biostatistician in cancer research

Importantly, do not use generic, hackneyed, corporate buzz words when describing the work you seek. If you want to work on diesel engine testing, do not call it "challenging work in a dynamic environment." If you want to be a physical therapist, do not call it "helping to grow an organization of like-minded team members." These cutesy business phrases say little about your expertise and make the Objective more than one line. See the following example:

- Objective: Looking for a position with challenging, dynamic prospects in engineering or business where I can use my problem-solving and team-management skills

This is vague, too broad, and too demanding—an unhelpful triumvirate. Let me dissect it, as follows:

Challenging? What if the company of your dreams offers you a mundane, monotonous, paper-pushing position? Would you scoff at it? Would you say, "No thanks"?

Dynamic? Who are you to tell them you want to be constantly stimulated by challenging and dynamic assignments? My advice: Take any job and find a way to make it challenging. Besides, if you are new to the field, and you were working professionally, most jobs will be challenging, so you do not have to add that adjective. Compare these:

Long

- Objective: Challenging full-time biologist's position researching and developing the low-cost Zika virus immunization programs

Just Right

- Objective: Permanent biologist's position focused on Zika virus immunization

Problem solving? You probably will not be asked to solve problems immediately; more likely, you will be given time to learn the lay of the land, as they say. You will do a lot of orientation and training, and you will be asked to do grunt work. You are not likely to hear the following at your new job:

- "Hey, kid, you new here? I thought so. Great. See that machine there, the Super X50 Bim Bam Thwopper; it's been out of alignment for 6 days. Fix it, will ya? Thanks. Oh, yeah, and the Department of Defense called. They want to know the reason we are 8 months late in delivering the Stealth 7000 thingamajig. Call them back and appease them, and check with production and get those things off the assembly line by this Friday!"

Problem solvers are rare: Bob Lutz, Lee Iaccoca, and Wolf from the film *Pulp Fiction*. They have paid their dues and they know their business like it's nobody's business. They solve problems. Most people starting their careers enter data, organize files, and run tests.

Team-management skills? Of course you will work on teams with other people. No one is going to hire you and tell you, "Wander around and find things to do, and be sure to never coordinate with anyone else." You will work with a team and you might get the opportunity to lead a team for some project, but you do not have to state it. Unless the job you are seeking is explicitly associated with team leadership or management, you do not need to include this fundamental work component as a part of the Objective. It is tantamount to saying you will be happy to use phones and computers also. Imagine the following:

Empty Objective

- Looking for a challenging position where I would use phones and computers to execute communication tasks among people in a dynamic organization

So if you wisely remove all these platitudes and generic phrases, you are left with the three items mentioned at the start of this subsection and my advice to edit the Objective to one line. Here are three good examples:

- Objective: Summer internship in civil engineering specializing in traffic management and modeling, available June 1
- Objective: Permanent position as accountant for media and arts companies
- Objective: Co-op in journalism with focus on sports and local news

The Objective line with the letter-resume combination shown at the beginning (Figure B3) is quite good. In contrast, the Objective line on one of the sparse resumes (Figure B11) is weak. The crowded, poorly formatted, and unusual resumes (Figures B13-B16) lack Objective lines, despite having so much information. This poses difficulty when the authors are unlikely to provide complementary job cover letters. The Objective line on the resume with an effective column-based format is well written (Figure B17).

B.11.2 Education Block

The best approach to take with your Education block is to make it short, clear, and simple. Representing your higher education requires very little data, so it should not be obscured by too much supplementary information. Aside from the basic required information, almost all other items I see in this block are often superfluous, redundant, and trivial. Figure B21 is a resume with unnecessary data in the Education block; Figure B22 is just right in this respect. The basic items are these:

- School and its location if not self-evident
- degree and major
- graduation date

That is it.

Here is a basic, good example:

EDUCATION
University of Michigan
Bachelor of Science in Aerospace Engineering, April 2014

All of the following items add clutter:

Attended September 2010 to April 2014

Will graduate in April 2014

Classes as follows: aircraft design, structures, controls and avionics, instrumentation, senior design seminar

Transferred with 54 credits from De Anza Community College (attended 3 terms)

Location: Ann Arbor, Michigan

Andy Wasserstein

(848) 402-4000
andywass@gmail.com

1945 Geddes Ave.
Ypsilanti, MI 49107

OBJECTIVE
Full-time engineering position in hardware design

EDUCATION
University of Michigan — Ann Arbor, MI
Majoring in Computer Engineering — September 2016—April 2018
GPA: 3.5381 / 4.0

Course Highlights:
Microprocessors, Programming and Algorithmic Thinking, Digital Logic, Physics—Electricity and Magnetism, Humanities 489 (seminar)

Delta Tau Delta Fraternity:
Active member of Philanthropy Committee
Organized fundraising of $25,000 during 2015-2016 academic year

EMPLOYMENT
QuickFix, Inc., — San Carlos, CA
Smartphone Technician — Nov. 2014–Dec. 2016

- Performed fast, effective repairs on damaged phones, tablets, and computers
- Assisted clients in comprehending technical aspects of repairs
- Completed sales and refunds using point-of-sales system
- Oversaw daily operations of retail and service location, including opening, closing, and taking inventory

Wells Fargo Credit Corp — San Carlos, CA
Loan Servicing Representative — Jun. 2013–Aug. 2013

- Processed customer payments and managed accounts
- Contacted insurance companies and oversaw verification of customer's policies
- Updated loan information in computerized record system

RESEARCH
University of Arizona Water Quality Research Lab — Tucson, AZ
Student Researcher — Sep. 2013–Dec. 2013
Assembled moisture sensor to measure water content in soil
Conducted field research using sensor, while working with team
Analyzed effects of extreme drought on arid ecosystems

COMPUTER SKILLS
Languages: C++, Matlab, Verilog
Application: Altera Quartus, Mathematica

*Figure B21 Sample resume with **unnecessary** data in the Education block*

Andy Wasserstein

(848) 402-4000
andywass@gmail.com

1945 Geddes Ave.
Ypsilanti, MI 49107

OBJECTIVE
Full-time engineering position in hardware design

EDUCATION
University of Michigan
BSE in Computer Engineering, April 2018
GPA: 3.5/4.0

EMPLOYMENT
QuickFix, Inc., San Carlos, CA
Smartphone Technician Nov. 2014–Dec. 2016

- Performed fast, effective repairs on damaged phones, tablets, and computers
- Assisted clients in comprehending technical aspects of repairs
- Completed sales and refunds using point-of-sales system
- Oversaw daily operations of retail and service location, including opening, closing, and taking inventory

Wells Fargo Credit Corp San Carlos, CA
Loan Servicing Representative Jun. 2013–Aug. 2013

- Processed customer payments and managed accounts
- Contacted insurance companies and oversaw verification of customer's policies
- Updated loan information in computerized record system

RESEARCH
University of Arizona Water Quality Research Lab Tucson, AZ
Student Researcher Sep. 2013–Dec. 2013
Assembled moisture sensor to measure water content in soil
Conducted field research using sensor, while working with team
Analyzed effects of extreme drought on arid ecosystems

COMPUTER SKILLS
Languages: C++, Matlab, Verilog
Application: Altera Quartus, Mathematica

AWARDS
University Honors and Dean's List, Winter 2016

LEADERSHIP
Delta Tau Delta Fraternity:
Chairperson, Philanthropy Committee
Organized fundraising of $25,000 during 2015-2016 academic year

Figure B22 Sample resume with ***necessary*** *data in the Education block*

The city/state is not necessary for well-known universities, especially when the city/state are part of the university's name. You can add city/state for less-known colleges and for ones where the name itself might make people think of a different college or location.

Some students ask whether "Ann Arbor" should be listed because University of Michigan has other campuses, namely Flint and Dearborn. Some leeway is available here, so Ann Arbor can be listed if you prefer. Nonetheless, the Ann Arbor campus is the primary campus, so "University of Michigan" implies Ann Arbor. The other campuses must be explicitly indicated. This is the same with the University of California system. The Berkeley campus is original and primary, so "University of California" implies Berkeley. On the other hand, a graduate can choose to indicate the particular location to be explicit:

- University of Michigan, Ann Arbor
- University of California, Berkeley
- University of Michigan at Ann Arbor

These well-known, historical colleges do not require the city and state to be listed:

- Harvard University, Cambridge, Massachusetts
- Yale University, New Haven, Connecticut
- Stanford University, Palo Alto, California

The following not widely known colleges, and others like them, should be listed with city and state:

- Kennedy College (Orinda, California) NOT Massachusetts
- University of Redlands (Redlands, California) (NOT Colorado or Utah)
- Carlton College (Northfield, Minnesota)
- Hope College (Holland, Michigan) (NOT a southern Baptist, but a Michigan Calvin)

If you really feel compelled to provide more information in your Education block, you certainly can choose to do so. Here are items that can enhance without creating too much clutter:

Course Highlights (list three classes most germane or most impressive to employer)

Honors (Dean's List Fall 2019 or *Cum Laude*)

Special Projects (senior-level design or honors thesis)
Electrical system leader for record-setting solar car (Jan. 2016 to Dec. 2017)
Primary graphics designer for Mars landing mission design project (Sept. to Dec. 2015)

B.11.3 Employment Block

This is the most important item on a resume. All other items are good predictors that you will be a good employee; education, academic awards, community involvement, special interests, clubs, hobbies, and so on suggest you can perform on the job. But work experience, especially episodes of notable duration, are exact testimonials and direct proof that you have been or are currently a good employee. Some job seekers may wish to include work experience above education if it might be more powerful than their degree(s).

I had a student who had worked for Ford for the previous three summers and now was graduating with a Bachelor's in Industrial and Operations Engineering. She put her degree at top, then her work experience, which began with a current job in the University's food services division. Her first application was going to Ford, naturally. I suggested she deviate from the standard order and put her Ford experience first. As a result of this move, her Education block, which appeared near the bottom of the resume, was simply icing on the cake. She asked me how to handle the current job that is less applicable to her desired work at Ford. One solution is to divide experience into two sections: Relevant Experience and Other Employment.

As with the Education block, the Employment block requires a small set of information, and you should not include much beyond these standard items. The critical information for work experience is the following:

- Employer's full name (no abbreviations)
- Employer's city and state
- Job title (only your ending or current title)

- Job duties or achievements (recall: Verbs!)
- Dates of employment (month/year to month/year)

For duties, students ask me, "Can I stretch this?" The answer is, yes, but you cannot lie. If you were the dishwasher at a restaurant, you must include "washed dishes," but if you helped select the new dishwasher after talking with vendors and observing models at other establishments, certainly you may add "selected professional kitchen equipment after interviewing vendors and testing models." Highlighting a set of duties and achievements should be done to show broad skills (taken together with all your jobs on the resume), increasing responsibility, problem solving (if applicable), specialization, and overall intelligence and maturity.

B.11.4 Extra-Curricular Activity Blocks

Such information is not required on a resume, but this data can stimulate conversation, forge connections between you and an interviewer or other people who review your resume, and show a more holistic applicant than education and work alone can show. (Deciding whether or not to include truly personal information is discussed in subsection B.9. Presentation of such data is the focus here).

The biggest problems that arise when presenting such extracurricular data are (1) the "hodge-podge" effect and (2) formatting snafus. Hodge-podge is simply that authors give some fact about one activity and another kind of fact about another activity. Look at the example here:

- Sigma Gamma Tau (2001–present)
- Habitat for Humanity, Volunteer
- Temple and playing guitar

These three items are mismatched. One has a duration, one has a position name, and one combines a religious affiliation with an artistic hobby presented as a gerund, "playing." The solution to the hodge-podge problem is to make the items *parallel* in grammatical style and information provided, even if it means removing some information while adding new information. (Details on parallelism in Appendix C, Section C.6.1.)

Revised:

- Vice President, Sigma Gamma Tau—Aerospace Student Honor Society
- Volunteer, Habitat for Humanity—Charitable Community Organization
- Participant, University of Michigan Hillel—Jewish Campus Synagogue

The choice to leave off dates is done for simplicity; that is, something has to be removed due to space constraints and the need to standardize the three-item list, and the duration was the least important. Who would say to you, "I like your qualifications but, frankly, you have not been helping with 'Habitat' long enough to suit my needs"? Or, "We rarely hire someone unless she has put in a full four years at Hillel. I can tell you are a high-holiday Jew."

I also dropped playing guitar. Why? Well, it's simply hard to fit in with the others though it is an impressive item. It would have to look like this:

- Player, Guitar Group—My Friend's Garage

The dates for guitar playing would probably be incompatible with the others; the hobby requires no formal membership, while the other three do; and the hobby is itself the position and organization. You can re-insert guitar playing if either of the following were true:

- Guitarist, Solar Quartet—Semi-Professional Jazz Band
- Guitarist, University of Michigan Hillel—Jewish Campus Synagogue

This leads to another good maxim when developing resumes, and a good place to end this section: always be flexible enough to change the specifics of an entry while leaving the essence the same.

B.12 Thank-You Notes

A thank-you note to express gratitude for having been invited to an interview is an excellent vehicle for reminding the employer about you and setting yourself apart from other applicants, namely those who do not send a follow-up note after the interview. It can be an opportunity, moreover, for emphasizing one of your strengths and to make a good impression. Not only does it show that you are interested enough in the position to bother to write a note, but it also shows that you are capable of business correspondence and relationship maintenance, both of which are necessary attributes for most jobs.

A thank-you note does not need to be long. You do not even need to fill a full page. A short paragraph or two should be sufficient. It is nice to send such a note through the regular mail, using paper and formatting that matches your cover letter and resume, but sending a thank-you note via email is acceptable and should be sufficient to impress most employers. (If you perceive urgency and a rapid pace of hiring, by all means you should quickly send a thank-you note via email or text, as appropriate for the employer.) Furthermore, you should send a note to each person who interviewed you and who likely has a say in the hiring decision. You do not need to send thanks to persons you were merely introduced to or chatted with briefly on tours or while waiting to be interviewed.

You can follow a basic template for thank-you notes. Other than the standard information (see below), supplemental information can be provided on an as-needed basis. Examples of supplemental information include providing specific information that was asked of you (references or sample written work) and answering any questions that you did not fully answer in the interview. Certainly, you should feel free to personalize a note to reflect the particular situation. Otherwise, the basic elements of a thank-you note are as follows:

Standard Thank-you Note Elements

- Express gratitude for the interview
- Recount something the interviewer told you to both flatter and demonstrate that you were paying attention
- Offer one more qualification that you did not mention or emphasize at the interview
- Adapt one of your previously stated qualifications to something you learned at the interview but had not expressed at the time
- Reassert your interest in the position and your availability
- Provide your phone and email so you can be reached easily

Two sample thank-you notes are provided in Figures B23 and B24.

From: **DeShawn Harris**< desh@emu.edu>
Date: Fri, Oct 31, 2018 at 8:45 AM
Subject: Additional Professional Reference
To: Seungwon Yen <syen@umich.edu >

Dear Ms. Yen:

I wanted to thank you for taking the time to interview me yesterday. I enjoyed talking with you as well as learning more about the research. I am extremely interested in the DEFENS Study, and I wanted to reiterate that I would be able to accommodate the occasional weekend shift that you mentioned. Furthermore, I wanted to use this communication to pass along an additional professional reference:

Kendra Choi
(212) 987-4561
kchoi@mit.edu

Ms. Choi is the manager at the Chemotherapy Lab, where I currently work as an assistant. She can verify that I have handled blood samples and understand the protocol with personal protective equipment (PPEs).

You can reach me at desh@emu.edu or 313.543.4589. Thank you again,

DeShawn Harris

...
****
DeShawn Harris, RN
desh@emu.edu
313.543.4589

Figure B23 Sample thank-you note with reference data

Appx B

From: **Bill Gratz** < billgrat@umich.edu>
Date: Fri, Jun 7, 2018 at 10:25 AM
Subject: Recent Interview at your Office
To: Karen Richardson <ker@fitbit.com>

Dear Ms. Richardson:

Thank you for inviting me to an interview last week. I enjoyed learning about your operation systems and meeting some of the team. I wanted to reiterate my strong interest in the position and mention that I did additional work with haptic feedback beyond the project I highlighted during the interview. Specifically, during an independent study course with one of my design professors at the University of Michigan, I helped write the code for the touch-screen interface for a laptop-like learning device for children with cerebral palsy. My combined experience with advanced interfaces, along with my passion for Fitbit's products, encourage me to pursue the opportunity with your company. Please let me know if you need any further information from me, such as additional personal references, or if I can answer any questions. Thank you again for the friendly welcome and informative interview at your headquarters.

Sincerely,

Bill Gratz

• • •

****

Bill Gratz
billgrat@umich.edu
(313) 242-5566

Figure B24 Sample thank-you note with new information

Appendix C

Formatting Essentials

Each genre has a certain look. A reader can tell immediately that she is holding a letter or a memo, or a procedure manual or a proposal. Each has its tell-tale signs. The very first page should suggest the genre. In addition, the formatting and style of the whole presentation should reinforce that first impression. And that first impression must be a positive one. Ultimately, formatting techniques have a threefold objective. First, the techniques ensure your format complements the genre you are using. Second, they make the document inviting to read and professional looking. Third, they help readers see both the recurring patterns that foster quick reading, random access, rapid memorization, and holistic comprehension. In other words, formatting techniques help readers scan the document, read selectively, and take in the overall message quickly.

Genre-specific formatting advice is presented in Chapter 5.1. In this appendix, I cover some additional formatting essentials applicable across genres that will make your document look its best. Specifically, a document should be eye catching when necessary, neat and clean, and helpful to readers.

Besides assisting readers with good formatting, you are helping yourself: a professional-looking format suggests professionalism of content and of the work behind it. It enhances your reputation. All professionals in ABERST are expected to be competent with computers, word processing, and graphics software. Sharp-looking documents and visual aids are the standard. The shortcuts and quick texting available with smart phones and electronic document delivery have not eliminated high expectations at the professional level in ABERST. Academics, business managers, engineers, researchers, scientists, and technologists must write and deliver neatly formatted documents.

Some professionals mistakenly ignore formatting with the following rationalizations:

- "Only my resume is important."
- "My company or publisher will decide the format."
- "My readers will not care about format."

These are falsehoods. Not everyone will admit this, but we all love an easy-to-read, nice-looking document. With good formatting, everyone reads more quickly. Everyone remembers more fully. The author's reputation and image are enhanced. And, the author brings a bad tradition of sloppy formatting, if and where it has persisted, to an end.

How do you create a nicely formatted document? The essentials fall into six areas:

Formatting Essentials

- ▶ White space: horizontal and vertical
- ▶ Highlights: bold, italics, underline, all capitals, and color
- ▶ Variations: character size, typefont, and line length
- ▶ Headings
- ▶ Numbering systems
- ▶ Bullets

C.1 White Space

The single most effective technique of formatting is the total absence of everything—words, symbols, highlights, and ink—period. When a spot on a page is empty, it is considered "white space." White space gives readers a rest. In addition, it signals divisions, breaks, and ends, and these are crucial to revealing structure and assisting readers in scanning, selective reading, and subsequent referral. White space takes two forms: vertical and horizontal.

Both forms of white space show subordination and hierarchy. By using vertical white space between items, an author indicates separation, which can be further elucidated with other formatting techniques. Vertical white space can be used to separate paragraphs, sections, headings (from text), visual aids (from text), footers/headers, footnotes, and bulleted items. A standardized type of vertical white space is the spacing between lines of text, which is so basic it may seem outside one's control. Nonetheless, it is something that can be adjusted and has its special function: tighter for documents where space is a premium, looser for documents that are drafts (for easy annotation), college papers (for grading notes), and documents meant to be widely distributed to the public or lay audience (for favorable reception). This white space has a special term: leading (coming from the age of the printing press—more on that below). Leading, or, line spacing, does effect the "look" of the document and should be carefully matched to the genre, purpose, and audience of the document. Leading is measured in points, as is character size (see more below). You can pick a leading point size for high-profile documents where you want absolute control of all spacing choices, or alternatively you can use default settings, such as "double spacing" or "single spacing."

Vertical white space is essential for top and bottom margins, which are mandatory on all documents. Primary text should never extend into the margins, as that gives no rest to readers and inhibits the strategic use of headers and footers in the margin space, at top and bottom. Text on a page must always be placed within a framed border of white space on all four sides, to look neat and professional. The frame of margins "protects" the text, so it is never obscured by any type of binding or smudged by oily fingertips. Also, any attempt to print text too close to the edge of the paper will produce print errors and chopped-off text (although this should not be attempted for aesthetic reasons even if it could be done from a technological standpoint.) These important side margins rely on horizontal white space, about which I have more to say below.

In addition to creating side margins, horizontal white space indicates subordination. Text that is indented to the right is subordinate to preceding text that aligns to the left. Furthermore, placing horizontal white space on both sides of text can serve to highlight it as a title, heading, caption, or cell entry in a table's column.

To illustrate the power of white space, I compare two pages from two different documents. To obtain the first example, I took a page from an insurance disclosure form (Figure C1). These disclosure forms are written in small print. Few people read them. They usually arrive with one's policy and premium notice (ugh, another bill). Insurance companies save money by printing these in double-sided manner, on inexpensive paper, in black and white. If you try to read it, your brain will be fried after two pages, and these go on for dozens of pages. I actually was surprised, however, to find an excellent example of using white space effectively. You can fault the insurance company for its fine print and all its exceptions, caveats, and exclusions, but you cannot say they intentionally made the document difficult to read.

The other example (Figure C2) is taken from a large corporation, packed with employees who have advanced degrees from the world's best universities. These people are involved in evolving technology that literally changes the face of our planet, entailing billions of dollars annually. And, the document is hard to read.

On the poor example, you will find very little white space. The extensive use of underline takes away from space. No horizontal white space is evident, just some limited vertical white space. Readers cannot rest.

On the good example, you find vertical white space after the title and the introductory paragraph, as well as before and after the major section headings. Horizontal white space, moreover, is used to subordinate information within three numbered sections. In Section 3, three levels of indentation are used, so horizontal white space is effective in creating a hierarchy on four levels: physical damage coverages, added coverage, leased vehicles, and two further sub-points.

6097DC LEASED MOTOR VEHICLES (L

This endorsement is issued by the State Farm Mutu
Fire and Casualty Company, as shown by the compa
changes this endorsement makes, all other provisio
endorsement.

1. **DEFINITIONS**

   The following is added:

   Lessor means the ***person*** or organization who leases ***your car*** to ***you*** or ***your*** employer for ***your*** regular use, but only if that ***person*** or organization is shown on the Declarations Page immediately following the title of this endorsement.

2. **LIABILITY COVERAGES**

   Additional Definition – Liability Coverage and Limited Property Damage Liability Coverage

   Insured is changed to include the ***lessor*** for the ownership, maintenance, or use of ***your car***.

3. **PHYSICAL DAMAGE COVERAGES**

   a. The following is added:

      Leased Vehicle

      1. Any Comprehensive Coverage, Collision Coverage, Limited Collision Coverage, or Broadened Collision Coverage provided by this policy applies to the ***lessor's*** interest in ***your car***. Coverage for the ***lessor's*** interest is only provided for a ***loss*** that is payable to ***you***.
      2. If the policy is cancelled or nonrenewed, then ***we*** will provide coverage for the ***lessor's*** interest until ***we*** notify the ***lessor*** of the termination of such coverage. This coverage for the ***lessor's*** interest is only provided for a ***loss*** that would have been payable to ***you*** if the policy had not been cancelled or nonrenewed.

         The date such termination is effective will be at least 10 days after the date ***we*** mail or electronically transmit a notice of the termination to the ***lessor***. The mailing or electronic transmittal of the notice will be sufficient proof of notice. However, this 10 day notification does not apply

Page
©, Copyright, State Farm Mutual A

Figure C1 Good use of white space

Appx C

Subject: Annual Project Report - Structural and Solid Mechanics

Introduction

The objective of these three projects is to develop and demonstrate advanced analysis techniques in the areas of structural and solid mechanics. Use of these techniques should reduce product development time while improving product quality and reducing product cost.

Project 14-102 has centered on finite element analysis (FEA) of elastomer components. An elastomer FEA program, developed and tested under this project, is now on CAECAM's PROVE library. Methods for determining elastomer material properties have been developed.

Further development of boundary element analysis (BEA) methods has been completed under project 14-103. A second version of the general purpose stress and heat conduction program is being readied for release on CAECAM's PROVE library. A second version of the BEA-based beam section properties program is also about to be released on the PROVE library. Also developed under this project was the capability to define two-dimensional BEA models in terms of design parameters.

Project 14-107 has dealt with characterizing the loading on analytical models, particularly the use of a random load description with analytical models to predict structural fatigue life.

.55

Figure C2 Poor use of white space

As valuable as white space is, it can be detrimental if used thoughtlessly and too generously. For example, you should not leave an empty chunk at the bottom of a page (unless it is the last page). An empty portion at the bottom of a page may suggest to some readers that the report is finished, and they will not turn the page to read your next section, which may be, even for you, a low-priority one that the reader would not anticipate.

Often writers leave a blank chunk when a visual aid will not fit. What are the possible solutions? One, resize the visual aid to fit, assuming this is possible without rendering it too small to read. Two, bring forward the text that is below the visual aid and use it to fill the bottom of the page. Often this latter solution is doubly useful: The text that was formerly below the visual aid might be helpful if it precedes it.

As much as I am warning against leaving a large, empty space at the bottom of a page, I must also emphasize the overall need to ensure that white space is provided to mark the end of text on each page. Indeed, you must maintain properly sized margins on all sides, including the bottom. Too little white space in the margins is as bad as too much.

One likely cause of shrinking margins is cluttered headers and footers. Sometimes the excessive use of headers and footers takes away from the required empty frame around the text. If the footer runs up against the last line of text, the necessary space is missing and the page looks too crowded and unframed. Another problem arises with overly busy headers and footers: the clutter obscures the truly useful information in the header or footer. Company names, logos, author names, project titles, notes such as "draft," contract numbers, and the date are not necessary to repeat on every page. With all this "fluff," an important item, such as a page number, is often buried or lost in the turmoil. See Figure C3 for a negative example.

This pitfall with headers and footers no doubt explains the reason that standard business practice for organizations with letterhead is to only use the letterhead for the first page of correspondence. As much as an organization likes to use its proprietary letterhead and has invested money in its design and printing, if a letter runs beyond one page, the letterhead is not used for the continuation pages. A plain piece of paper is preferred and normal.

Results Report ***Testing Division*** **Speed Technology Corporation**
January 10, 2019 Draft v. 3.2

of the experiment methodology was to compare the experimental wind tunnel results against the theoretical results the tunnel should be producing. By taking measurements of pressure, area, and geometry, the Testing Division (the three authors) was able to obtain several parameters produced by the wind tunnel that could be compared to theoretical values.

This Division evaluated the tunnel by placing a wedge model of known geometry inside the test section. The wedge model created patterns of Mach waves, which were visualized using a Schlieren optical system. Appendix A gives a detailed explanation of workings of the Schlieren optical system. The static pressure was measured at different points, or stations, within the tunnel using a mercury manometer. The total pressure was also determined using a barometer. The flow area at each pressure station in the tunnel was also measured for future analysis. The calculations and results section (below) outlines the method for determining the values, which were compared experimentally and theoretically. In particular, Mach numbers, static pressures, and wave angles were determined and compared against their theoretical counterparts. Figure 1 (on page 6) describes the experiment setup.

Calculations and Results

Pre-Shock Calculations - Stations 1 through 7

Mach Number
Two techniques were used in order to acquire the data used to validate the tunnel's feasibility and desirability: a geometric technique and a pressure technique.

The geometric technique consisted of calculating the Mach numbers from analyzing the area at different stations and the tunnel throat area, or station four. The areas are tabulated in column two of table 1 in Appendix B and Equation 1 (below) was used to calculate actual Mach numbers in the nozzle prior to the test section. The Mach numbers are tabulated in table two of Appendix B. The areas used in this equation were those experimentally determined, as described above.

Equation 1:

$$\left(\frac{A}{A*}\right)^2 = \frac{1}{M^2}\left[\frac{2}{\gamma+1}\left(1+\frac{\gamma-1}{2}M^2\right)\right]^{\frac{\gamma+1}{\gamma-1}}$$

The pressure technique consisted of using the measured static pressures and total pressure to calculate the Måch number. The pressures are tabulated in column three of table one in Appendix B, and Equation 2 in Appendix B was used in calculating the Mach number. The Mach numbers calculated are tabulated in table two of Appendix B. Thus, the Testing Division now had two Mach numbers from independent relations.
Pressure

As mentioned, experimental pressures were measured during experimentation and are tabulated in Appendix C. Equation 2 was used to determine theoretical pressures: we solved for static pressure, using the measured total

Draft for Internal Circulation p. 5 Contact: djlmst24@stcorp.com
Questions to FAQ: www.stcorpfaq.com

Figure C3 Example page with small margins and crammed header and footer

C.2 Highlights

Highlights added to text are another essential formatting technique. Highlights can be made with any of the following: bold, italics, underline, all capitals, and color. Sometimes a combination of two or more highlights is helpful. One strict warning, however, is necessary before you become too excited about highlighting:

- *Do* NOT overdo IT, or **readers** WILL GET *tired eyes* and **lose patience***!!!!!*

A little highlight goes a long way. One of the best strategies is to combine one highlight with white space, and no other highlighting.

A standard and effective approach to make special items stand out from the rest of the page is to use highlights for these special items:

- Headings
- Titles
- Bullet lists
- Warnings
- Headers and footers
- Signatures
- Notices
- Special mentions

C.3 Variations in Character Size, Typefont, and Line Length

Selecting the character size is an important decision. You want just the right size, and you do not have much wiggle room to play with. If you pick a size too small, readers will be frustrated and may not read the document. They certainly will struggle through it and find it difficult to read. If you pick a size too large, readers may find the document either unserious (casual or childish) or overwhelming (obnoxious or domineering). Size is closely connected to the genre and the specific part of the document. For example, titles and headings on formal documents (proposals and procedural instructions) can be large, while these are usually normal size for emails and business letters. Similarly, document body text is often larger for internal memoranda and formal reports than for journal articles and status reports. Other genre-specific differences exist, so you must look at good samples to develop a feel for appropriate character size.

Specifically, you pick character size using a measurement called a "point." This unit originates with printing presses, and it has remained useful to this day. A point is 1/72 of an inch, so a 72-point character is 1-inch high. Typically, for the main body text, a size between 8 and 14 is suitable, depending on the typefont and printer. Some people might say that the best size is always 12. I am not so sure this can be a universal rule. The variation derives from the different typefonts.

This is probably a good moment to say something about typefonts. Hundreds of them exist. You have many, many options. For most ABERST documents, however, you are advised to stick with a standard, no-frills, business-looking typefont such as Times, Times New Roman, Book Antiqua, Bookman, Century Schoolbook, and Palatino (*serif* fonts), and Geneva and Helvetica (*sans serif* fonts). You can choose from the other options for special highlights in particular documents, but you must be circumspect. *Serifs* are the little hatch marks at angles extending from the top and bottom of letters. These help readers distinguish letters more quickly than when the serifs are absent (*sans serif*), but the absence of *serifs* can offer a clean, simple look that many people admire, especially for titles, headings, tables, and other visuals.

Stylistically, width of any single character varies across both characters and typefonts: some typefonts use thick strokes, others use thin; some have *serifs*, some do not; some are monospace and some are proportional, as detailed below:

- *M* and *W* have the widest set width
- *i* and *l* have the narrowest set width
- **Bookman Old Style** typefont has wider letters, generally, than does Times New Roman and `Courier`: Here is the acronym BMW in three typefonts but the same height in points (12):
  - **BMW** in **Bookman Old Style**
  - BMW in Times New Roman
  - `BMW` in `Courier`

- Courier is monospace, where each character receives the same full-width allotment.
- Times is proportional, where the width of each character is adjusted to the character.

As mentioned, a point is a measure of a character's height (not width), so the height is the measurement that is selected when you pick a type size. A character's width is preset, to be attractively proportional to the height. Thus, due to varying letter widths across typefonts, with equal character size (that is, height), line length (measured in *picas*) varies between typefonts. Depending on the type size and typefont, line length may be too long (too many characters per inch and per line) to be considered easy to read (see good and bad examples below). Therefore, an author must combine decisions of typefont (style), character size (points), and margin width (inches) to produce an easily read line of type.

Bad Sample:

is line has too many characters per inch and runs fully across the page, extending into the space of margins.

Good Sample:

This line is ideal in terms of characters and length. It fits within the margins.

One other issue that impacts line length and readability concerns spacing. Spacing between characters (*kerning*) in different typefonts is set differently, and similarly so is spacing between words (*quads*). In addition, if you set the margin on the right to be justified (all the lines end exactly at the margin edge), the spacing will be adjusted to help create this perfectly "full justified" right margin. For ABERST documents, ragged right is acceptable, and you do not need to use full justified. Unless required by a customer, publisher, or other constraints, you should avoid full justification, which produces unwanted spaces and excessive hyphenation.

Do not always stick to a favorite character size, such as 12 point.

To end this section, I will offer my primary axiom on this topic: Do not always stick to a favorite character size, such as 12 point. Appearance may differ as you change typefonts, so you might want to adjust character size as you change typefonts. Also, overall readability shifts as margin widths are changed, so you would do well to look at line length along with character size, typefont, and margin width. Furthermore, the printed version may be different than you expect because appearance may differ as you change software or hardware. Lastly, do not hastily combine two or more printed documents into one final submission. Typefont, character size, and line length may not match even if nominally the same.

C.4 Headings

Organizing your document into a hierarchy of sections, from the start to the end, is a rhetorical necessity, and it serves as a beneficial formatting technique as well. With sections that are demarcated with headings, the document becomes easy to navigate and less intimidating than a solid block of text. Complex and dense material requires clearly marked divisions, and the headings offer an opportunity to announce your main categories of information. You reinforce your message and its subpoints with headings, and you create a pleasing and attractive document. Thus, headings are doubly advantageous: they are informative, and they create a clean and sharp look.

To review, the basic rules of headings are these:

1. Use multiple levels of headings, from the main divisions, down to subsections.
2. Use specifically informative, or message-based, headings. (Legal writing is marked by full-sentence headings.) The headings below for the same subsection are in order of worst to best:

   Economics
   Cost Considerations
   Re-design Cost Considerations
   Blade Vibration Re-design Costs
   Overcoming Blade Vibration Cheaply

3. Avoid generic, essentially meaningless headings that can be moved to any report, for example, "Discussion" and "Details" (with the exception of the journal article genre).
4. Use a numbering system for headings, either in all-Arabic or in a Roman-Alphanumeric combination, as in the following samples:

Excerpt of Headings in All-Arabic:

3.0 Experimental Method
  3.1 Equipment Setup
  3.2 Calibration
    3.2.1 Non-steady calibration
    3.2.2 Steady calibration

Excerpt of Headings in Roman-Alphanumeric Combination:

VII. Disadvantages of Candidates
  A Disadvantages of Carbon Filtration
    1. Ongoing Maintenance
    2. Disposal Costs
  B. Disadvantages of Reverse-Osmosis
    1. Energy Dependence
    2. Ancillary Equipment

C.5 Numbering Systems

In addition to your heading numbering system, you also will benefit from using numbering systems in other respects. Importantly, you should number the pages of the document. Various styles are used, so you should follow any assigned requirements. If the choice is up to you, either of these options is acceptable:

- page 2 of 9
- -3-

You can also number certain lists of information in your document, so readers can easily see the items that constitute a group and how many items are in the group, such as this example:

Follow these three steps to enjoy next Saturday:
1. Withdraw cash at bank
2. Drive to nearby amusement park
3. Have time of your life

C.6 Bullets

Writers love bullets. Perhaps readers do as well. If done well, a short-ish list of items in bullet form effectively emphasizes those points, by making them set apart from the rest of the text, easy to read, and highlighted. This assumes they are written in parallel grammatical form (required) and are concise (highly recommended). But, bullets are not a one-stop shop. Even assuming they are parallel and concise, they are not always clear. Furthermore, they may not be the best way to present the information. We can look at these four factors separately: parallelism, conciseness, clarity, and presentation.

C.6.1 Parallelism

Bullets should match each other in grammatical form; this is known as parallelism. This is essential for written documents, but keep this guidance in mind for a slide deck for oral presentations as well (more in Appendix E). Indeed, the same rules apply for documents and slides, and they are arguably more important for slides because the bulk of the slide deck is, by intention, bullet lists. Parallelism requires understanding and adherence to basic composition rules of grammar and syntax. Look at the numbered bullet list from the section just above, rewritten in poor, non-parallel format:

> Follow these three steps to enjoy next Saturday:
> 1. Cash withdrawal
> 2. Nearby amusement park (2-hour drive)
> 3. Have the time of your lives

The first point comprises a two-word phrase with a modifier (cash) and a gerund (withdrawal). The second point is a two-word noun (amusement park), preceded by an adjective (nearby), and followed by a parenthetical comment—the only bullet to have such an addendum. The third point is an imperative verb-based phrase with a plural ending (lives), introducing the audience as more than one person.

In the original list, each bullet is an imperative verb-based phrase (withdraw, drive, and have) with a prepositional phrase for modification (at, to, and of).

C.6.2 Conciseness

After ensuring that bullets are parallel, you should edit them to be concise—not just one or two of them, but all of them. Audience members and readers do not want to see a bullet point that is longer than one or two sentences, which defeats the nature of the "bullet," which implies brevity. In documents, bullet points that are a few sentences or more are acceptable, but they are no longer true bullets; they must be considered indented paragraphs, and the normal ideal of conciseness applies as well. Therefore, whether a true short bullet or an indented paragraph, tight and succinct prose is always the gold standard.

In the special case of presentation slides, moreover, you can design your bullet points with awareness that a speaker will be elaborating upon each bullet. This enables you to use concise bullet points, and, again, parallelism is the best format. You do not want your audience reading the equivalent of a full document while you are speaking. You also do not want to make them stop and ponder the change in form when they should be absorbing as much content as possible from each concise bullet, effortlessly. So bullets on slides will not be fully meaningful alone, but they will effectively reinforce the spoken words. Nonetheless, without making them too long, some key words can be added to ensure they are parallel, concise, and as meaningful (clear) as possible. Clarity is the third expectation of bullets.

C.6.3 Clarity

Here is a sample projection slide placed near the beginning of an oral presentation slide deck. The items are neither parallel nor meaningful to all audience members, once you consider that this is presented at the start of a talk (not the end), and the audience will not consist exclusively of members of the team working on the project. It will contain people not familiar with the day-to-day operations of the project. The slide below is an overview of the talk about to be presented, on a proposed new factory for a chemical engineering company, using certain feed materials to react to produce a desired output of commercial product(s):

<u>Our Talk's Agenda</u>
Motivation of our project
Scope
ISBL/OSBL difficulties
ASPEN
Literature
Tasks/Timeline
Costs

People listening to this presentation are not likely to know the acronyms ISBL, OSBL, or ASPEN. They may not have any idea how literature will fit into the talk or whether "Costs" refers to the project team's development costs or the capital cost required to build the proposed factory.

Therefore, this bullet list could be rewritten to be both parallel and meaningful, while still being concise, which is a quality it did have from the start. If specialized terms are to be used at all, they should be added in parentheses. (Remember the axiom to put familiar and general terms before unfamiliar and particular terms, Sections 2.2.1, 4.3.2, and 4.5.1.)

<u>Our Talk's Agenda</u>
Project Motivation
Project Scope
Factory Design
Factory Simulation (ASPEN)
Building Costs (Factory and Site Costs)
Key Reactions (Theoretical and Experimental Literature)
Future Tasks and Timeline

The same lesson on clarity applies to written bullet lists in documents. Here is a sample bullet list, placed under an introductory heading:

Restaurants
- Jay's Burrito Palace
- The Vineyard
- Blue Rock Top

You may generously say in the context of reading the bullet list above, the point the author wants to make is perfectly clear. But, that demands readers are immersed in the document. Bullets, however, might very well "pull them away" from the document, as savvy readers think they can focus on the bullets, read them alone, and save time from reading everything else. We can ask about the bullet list of restaurants, the following questions:

- Did they fail a recent health department inspection?
- Are they places that have private, banquet rooms?
- Are they the author's personal favorites in Chicago?

So, the author has not made this crystal clear in the bullets alone. If he further relies on a reader's prior knowledge or assumptions that his point is obvious, he may not even provide the explanation that makes this list convey meaningful, purpose-related information. If he has provided the meaning in surrounding text, that text is essential, and the bullets could just as easily have been integrated into the normal paragraph text style, as shown in the alternative below.

Indeed, to fix the problem of unclear bullets, an author need only add a bit more information to the bulleted segment, as follows:

Three restaurants failed the latest city health inspection:
- Jay's Burrito Palace
- The Vineyard
- Blue Rock Top

If using the paragraph form it would look like this, with less emphasis on the restaurant names and more emphasis on the overall message:

> The city recently conducted its annual restaurant inspections of all dining establishments with seating for 10 or more persons.

> Most restaurants, although having small health code violations, passed the inspections. Three establishments, however, failed the inspection: Jay's Burrito Palace, The Vineyard, and Blue Rock Top. These were given 14 days to come into compliance or they would be forcibly shuttered.

C.6.4 Presentation

If you cannot find a way to make your bullets parallel, concise, and clear, perhaps you should use a different presentation approach altogether. Maybe the information requires sentences combined into a paragraph, or perhaps a table or chart would be best. Bullets are not essential to a report. You are not required to use them. They can be beneficial, but sometimes a better presentation method can be found.

The following bullet list is an example taken again from a chemical engineering plant design. This is a substantial project with millions of dollars at risk. The main findings of the economic analysis were presented with these bullets, with an introductory paragraph that emphasized a lower rate of return on investment than desired (which is not even mentioned in the bullet list):

- Revenue from sale of chlorine, caustic, and hydrogen: $4,200,000
- Cost of raw materials such as NaOH, HCl, and NaCO3: $2,375.000
- Breakeven Point: not reached within 20 years, possibly due to high capital costs
- Net profit: $1,844,000/year
- Capital cost: $37,954,000

These bullets are neither parallel nor concise. Moreover, they present too much information, hardly any of which is clearly explained in the preceding paragraph. A better presentation would be a table:

Item	Cost/Expense ($)	Year(s)	Notes
Capital cost	37,954,000.	Zero	Only once
Revenue	4,200,000.	1-20	Chlorine, caustic, and hydrogen
Raw material cost	2,375,000.	1-20	NaOH, HCl, and NaCO3
Net profit	1,844,000.	1-20	Annually
Breakeven point	N/A	N/A	Never reached

C.7 Final Words on Formatting

As the five parts of this book have demonstrated, technical writing in ABERST involves a handful of crucial strategies, skills, and techniques. It cannot be reduced to knowing whether to place a figure title above or below the image, or which font is my favorite. Such are instances of missing the forest for the trees. If you are trying to please someone by putting the page numbers in the right place, you are either trying too hard or neglecting the important issues.

Only the obsessive-compulsive are worried about the tiniest formatting issues. Readers and other professionals in ABERST care about the main message, argument and support, theory and limitations, mathematical accuracy, intellectual rigor, fundamental science, best business practice, due diligence, human health and safety, and sound engineering. They will not obsess over formatting choices if the purpose of the document is stated clearly at the start and fulfilled through the body.

Therefore, do not vacillate over formatting choices. If you cannot make a decision about the placement of the page numbers or figure titles, how are you ever going to pick a proper pipe size, furnace type, piezoelectric actuator, satellite skin, and so on? My final advice is this regarding formatting: Make some decisions, look the document over, and if it looks OK, leave it alone. The cousin corollary to this is consistency. Here is a rule to follow: Use punctuation and highlights consistently.

Changes in format that have no rationale or purpose are merely embellishments, accoutrements, or oversights that detract from the writer's reputation, distract the reader, and create document analysis that is unnecessary. Inconsistencies—in page number style, titling, highlighting, indentation, spacing, and spelling—are to be avoided as sure as computational or analytical errors are to be avoided. More severely, however, are inconsistencies in terminology, which is a special case that deserves some additional emphasis, as is covered in Appendix A, Section A.3. In the end, select some formatting techniques and use them consistently.

Appendix D
Conference Posters

A poster depicting research, design development, project planning and implementation, or any other endeavor in ABERST, whether in progress or complete, is a special genre. Professionals in academia make posters as often as once or twice a year, as they attend conferences within their special areas. For professionals in business, they may design and present posters at academic conferences alongside traditional academics, or they may create posters for conventions, trade shows, or meetings with customers and clients. In addition to those people who identify as members of academia or the business community, all other professionals whose work touches on engineering, research, science, and technology are likely to create a poster at some point in their careers.

Posters often are developed by professionals who have responsibility for preparing other technical documents critical to implementation of a project, namely proposals, final reports, and journal articles. Fortunately, these various genres can support one another and share both text and visual aids. If a poster has been created, and a final project report remains to be done, very likely much of the poster material can be integrated into the final report, expediting the effort on the latter task. And the opposite is true: if a proposal, final report, or journal article has been written, and a need arises to create a poster, these previous documents should be very helpful as sources of text and visual aids for the poster.

I cannot provide a perfect poster template, nor can I say, "Here is a guaranteed award-winning poster style, so simply fill it in with your information and graphics, and you will have a complete poster in 30 minutes." Posters should be unique. Each one should be somewhat customized to the particular amount of text and graphics that must be conveyed. Posters are an opportunity to work within a genre that allows some thinking outside the box and letting inspiration overtake you. That said, some normalcy and predictability might be helpful too. People viewing posters are familiar with a certain style; they have some common expectations. They want to spend time learning about your work, not deciphering your symbols and puzzling over your formatting. Within the confines of this conventional paradigm, you must convey your information clearly and attractively. More precisely, you must achieve three objectives with a poster:

- Attract
- Interest
- Communicate—inform or persuade

Most other ABERST genres start with the last objective: communicate. With most genres, aside from formal reports and proposals that require book-style covers, an author's primary concern is simply to communicate right from the start and through to the end. Posters, however, have two additional objectives that come into play prior to the commencement of communication *per se*: attract and interest. A poster must attract a viewer to it or the author may not have even a chance at communicating to that viewer.

This appendix provides guidelines for designing a conference poster so that it achieves all three objectives. With posters, the emphasis is on how to attract and keep interest, with communication taking a tertiary role. The guidelines below emphasize design techniques for fulfilling the first two objectives, which are unique to this genre. I also offer some poster-specific tips for fulfilling the third objective, communicating. For additional help on fulfilling the third objective, you can consult the preceding five main parts of this book. Before presenting the design guidelines, I start with a rationale for the guidelines. Lastly, at the end of this appendix, I offer some thoughts on hosting your poster while it is displayed at a conference.

D.1 Rationale for Design Recommendations

The social and environmental context for a typical poster helps to explain the design expectations of this genre. In brief, posters are meant to be displayed and presented at conferences or similar large-scale gatherings where other posters are also displayed, usually all under some "umbrella" theme or topic, though some posters may veer somewhat far afield if the organizers are so inclined. Even assuming a common "umbrella" theme, the sub-topics within that theme may be vast and plentiful. Unless the conference is small and super specialized, the posters presentations are usually immensely diverse in terms of specialty, focus, and objectives.

To put it another way, posters are not a typical genre used for one-to-one communication between one writer and one recipient. They are rarely used for meetings and events where a single poster is the only one presented. On the contrary, posters represent a genre that, by definition, has the inherent context of being intended to provide a great deal of information to as many people as possible in the shortest possible time. We must safely assume the audience consists of people mostly unfamiliar with the content of the poster, although they have some connection to the overarching "umbrella" theme of the conference. This social and environmental context creates the two primary reasons that a poster must be designed to attract its audience and catch that audience's attention, rather than simply jumping right in with trying to communicate a message. The two reasons are the (1) non-optimal viewing conditions and (2) non-optimal audience characteristics.

Non-optimal viewing conditions are the following:

- Crowds, distractions, and noise surround the poster. Take a look at a typical conference in Figure D1.
- Numerous posters are vying for a viewer's attention at the same time.
- The time made available for viewing posters is limited, as determined by the event organizers.

If you are not convinced that poster conferences are crowded and noisy, with a general feeling of rush, rush, rush, take a look at Figure D2.

Figure D1 Posters and people crowded together at a conference

Figure D2 Poster Conference for the American Association of Cancer Research, with at least 23 sections, and each section contains dozens of posters.

Non-optimal audience characteristics are even more numerous and influential on the design requirements:

- The audience is not captive as is a recipient of a written document, who knows the document is for her and must be read.
- At conferences, the viewers are free to move past, as no single poster is addressed to them.
- The audience comprises quick viewers—people who might spend 1 to 7 minutes per poster.

The typical viewer will ask questions for further details, which is a welcome change from looking for details in long written reports. For many, a quick conversation with a poster's author is an efficient way to learn more or clarify an unknown. The conversation, if it is to occur, requires that one or more poster authors be on hand at the poster, but this is usually feasible and expected at poster conferences.

Lastly, the audience has high expectations. They are looking for the "Wow Factor."

D.2 Design Objective #1: Attract

Attracting viewers to your poster is done in these four ways:

- ▶ Using color
- ▶ Adding visual elements
- ▶ Creating a pleasant layout
- ▶ Spotlighting a clear and informative title

We can look at good and bad examples to represent each of these design techniques.

D.2.1 Using Color

In terms of color, my suggestion is to use a singular color theme in the poster's framework, which comprises the borders, headings, highlights, and background.

Black and white will seem too under-designed (see Figure D3). People expect color with the prevalence of color printers and colorful PowerPoint slide decks at nearly every meeting. Sometimes a color theme is just one color, pure and simple (see Figure D4). A color is also known as a "hue." A modest embellishment from the single-hue theme is to add a variation of that pure hue, by changing the chroma, which is to make it more or less intense, by either subtracting or adding grey, black, or white. The color with a low chroma will look somewhat dull or muted, and with a high chroma, more intense. This is similar to thinking of both light and dark versions of the color. See Figure D5 for an example of a well-designed poster with chroma variations. In the example, the dark- and light-blue color theme is perpetuated throughout the border box, section heading banner, and visual aids.

Another option is to use two different hues, in which case they should be either nicely complementary or opposite, which will create a striking contrast. If one uses such a multi-hue theme, it should be used consistently and throughout the poster, with no additional color variation tossed in to create chaos. Color, whether one hue or several, is fine if it looks intentional and deliberate. If the color makes your poster look like a bunch of flyers tacked up on a community bulletin board, you have not created a singular, color theme. Excessive color with no plan is not much better than the absence of color.

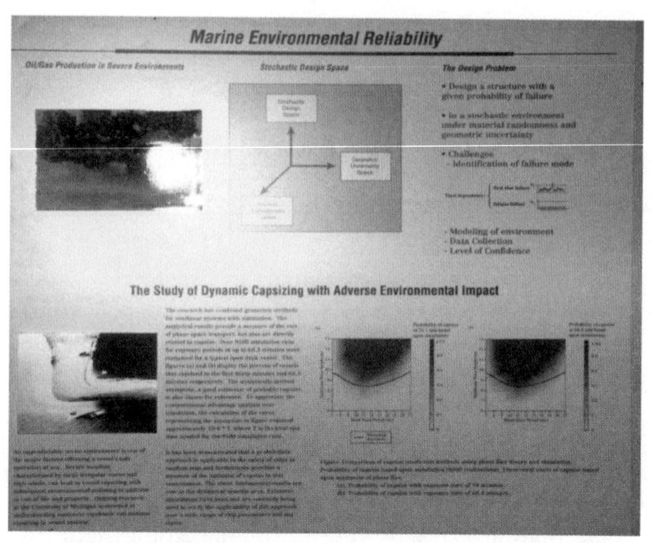

Figure D3 Professional poster that suffers for lack of color

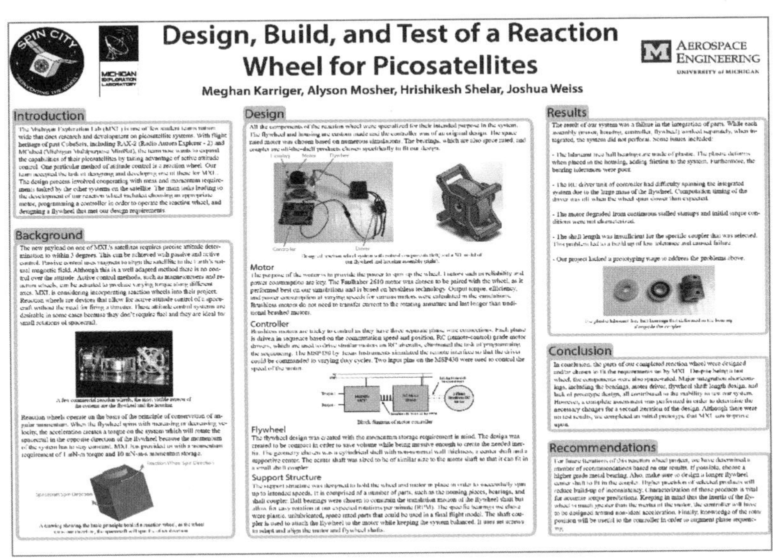

Figure D4 Poster with a single thematic color

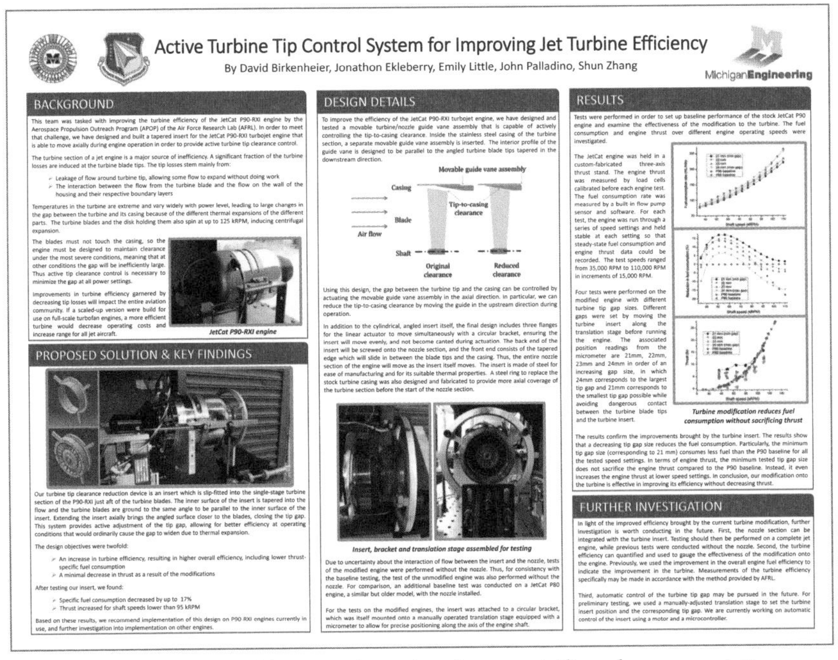

Figure D5 Poster with attractive color theme entailing chroma variation

One common technique that requires caution is to place an enlarged image in the background of the entire poster, with text and visual aids on top of this image. Sometimes an organizational logo is the chosen image, sometimes a natural scene (see Figure D6 for poster with corn field in background), or sometimes a product or component. All of these choices can be very distracting or so nearly invisible as to be meaningless but still slightly distracting.

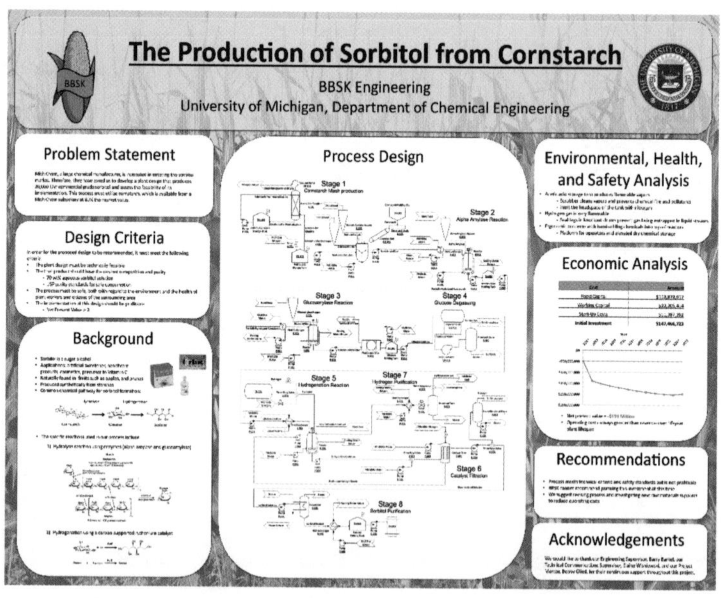

Figure D6 Poster using enlarged image in background

Another area for caution involves excessive and unnecessary design flourishes, such as ribbons, vines, stripes, repeating shapes, symbols, and so on. See Figure D7 for an example of misplaced use of flourishes from certificates, invitations, and other formal declarations.

D.2.2 Adding Visual Elements

Communicating your message with visual elements of any type–graphs, diagrams, schemata, photographs, or tables–has the dual purpose of displaying details of your project and attracting viewers. Thus, you are wise to plan your poster to accommodate a healthy number of your best visual aids to draw in busy people with just a glance. Adding colorful visual elements, therefore, is a second method of attraction.

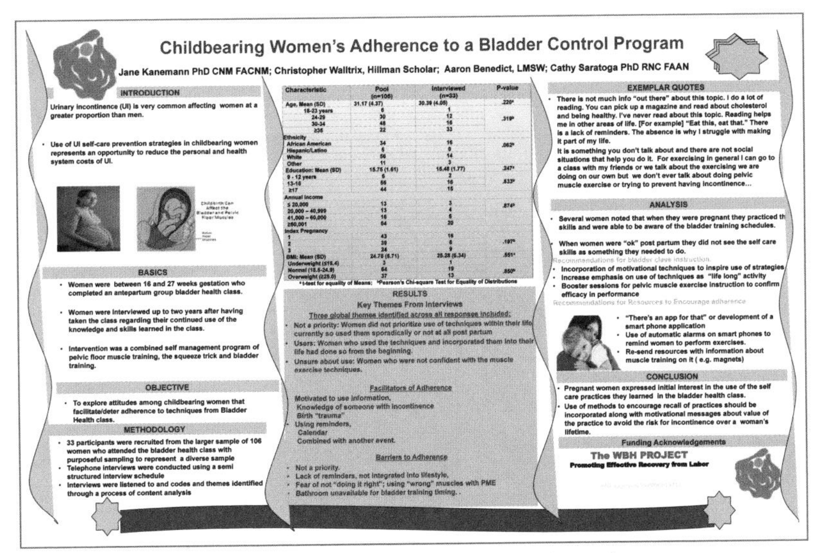

Figure D7 Poster with unnecessary flourishes

With visual aids, I have two reminders: (1) follow all the principles and guidelines presented for this specialty genre in Chapter 5.2, and (2) add color to attract busy and demanding viewers. Indeed, if you have made graphs and tables previously for a different document, and they are in black and white, you will want to add color to them for a poster. Color can be placed into the labeling (text), axes, data (curves or values), shading/fill, and symbols. The color added to these visual aids will add sparkle to the poster, and you can play off the color used for the overall theme. Both borders and internal elements can be adjusted to appear harmonious with the background color(s).

When you do the above and complement the visual aids with text that gives them context and explains the message to be reinforced or displayed in the visual aid, you will create an attractive, interesting, and informative poster. The two-part example below (Figure D8) compares strong and weak versions of the same poster segment. For this one segment, Testing Procedure, a hierarchy of information is ideal, as shown on the left. Also, on the left the visual aids are designed well and help attract viewers. On the right, in contrast, the use of visual aids is incomplete and mediocre. In the good example (left), three headings are provided, each one associated with some paragraphs of text, accompanied by two helpful visual aids, one a schematic and the other a photograph. The gold color theme permeates the background and the visual aids. This example demonstrates the way to make text and visual aids (in color) work together for successful communication. On the right, only two headings are used, along with one simple visual aid, providing fewer attractive elements.

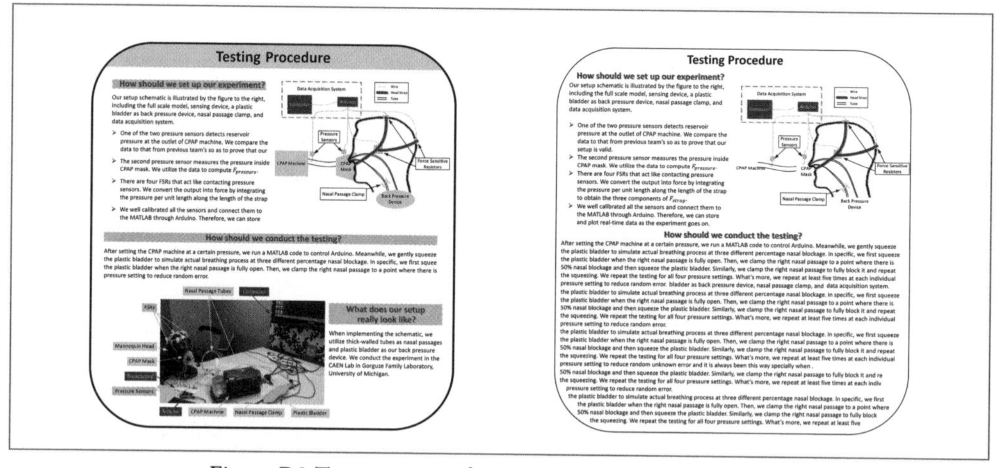

Figure D8 Two versions of a poster segment—one visually enhanced (left), the other lacking visuals and color (right)

In terms of visual aids, a poster is no place to hold back. You want to show all your cards, so to speak. If you have something like a finite element analysis of a jet aircraft (See Figure D9), you want to use it, likewise a Schlieren photograph like the one taken by Harold Edgerton (Figure D10).

One caution is necessary at this point: As much as you are advised to include alluring drawings and charts, you must remember that sufficient and well-structured text will be needed also. Relying too heavily on visual aids will create, most likely, a sparse and uninformative poster. A poster with nearly all visual aids may look intriguing, but it may not be clear to viewers (See Figure D11). As discussed further below in section A.4, clear and concise text explanations must complement the visual aids, and the poster must convey a hierarchy of information, presented pleasingly, which is the guideline discussed next.

Figure D9 Stunning visual image to enhance a poster

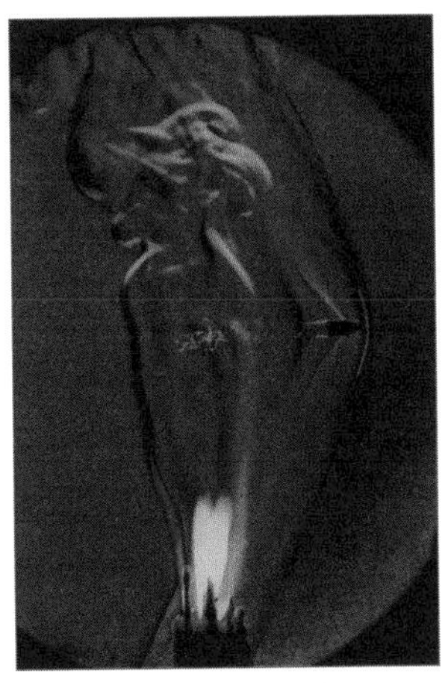

Figure D10 Powerful photograph to enhance a poster

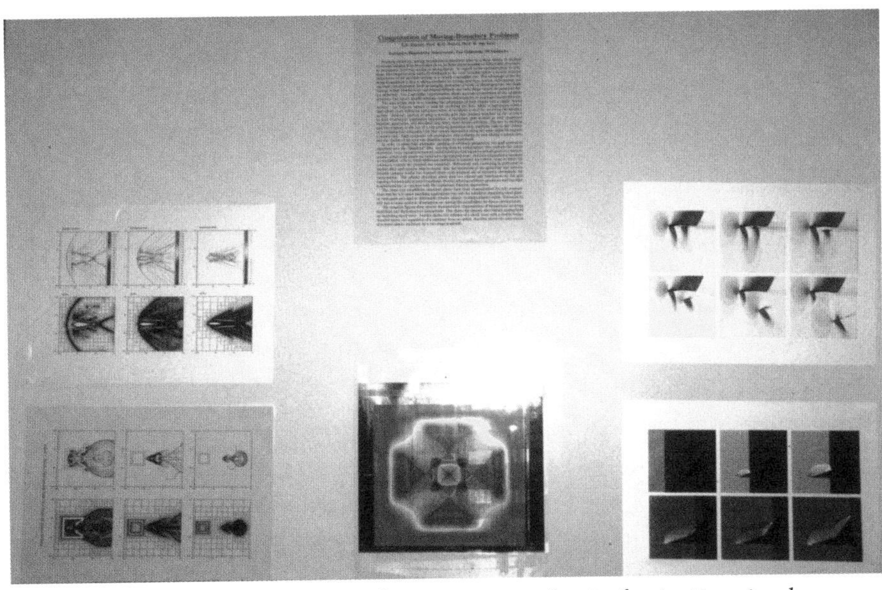

Figure D11 Potentially uninformative poster despite fascinating visuals

D.2.3 Creating a Pleasant Layout

Layout refers to the arrangement of the material on the poster. Material is either text or visual aids. Arrangement refers to the overall placement and groupings of the material. Placement and groupings are elements of a theme explored throughout this book, namely, organization. Similar to any well-organized document, a poster should be hierarchical and structured. Organizing portions of text together into natural topical and sub-topical divisions has benefits, rather than running text into one undifferentiated monolithic whole (see figure D11 again). Furthermore, interspersing visual aids with the text is usually better than bunching all the text into one location and all the visual aids in another. This means a poster is best divided into sections with integrated visual aids, and the sections' order and proximity to one another constitutes the layout. (Placement of the poster's title is covered in the next section, A.2.4.) The goal of a pleasant layout is fourfold:

1. Organize and segregate the poster into a hierarchy of sections, suitable to the content, not too many, with integrated visual aids
2. Make the sequence for reading and progressing through the content obvious to viewers, even before they read a single word on the poster
3. Leave adequate white space to minimize stress and pressure on viewers
4. Achieve balance across the poster

The benefit of fulfilling these four layout-related goals is that the poster, which may be fully packed with hundreds of pieces of information, will seem less overwhelming than it would with a poor layout. A pleasant layout invites viewers to peruse in a predictable sequence a handful of easy-to-understand and easy-to-anticipate sections.

To reiterate the goals and the benefits of a pleasant layout, keep the following five points in mind, and details follow in the four, goal-based subsections.

- The poster must have an easy-to-follow flow.
- Viewers must not be given whiplash caused by moving their eyes (neck and head also) all over the place, or back and forth, or randomly from here and there.

- Viewers must easily see the section breaks capped with headings and subheadings, as those elements will be set apart from competing elements by white space.
- The poster must appear spacious and easy to scan, neither packed, dense, nor crowded.
- The poster must be balanced in all directions, with attractive elements and something of importance in each quadrant, top to bottom, left to right, and center.

Goal #1: Organize and segregate the poster into a hierarchy of sections, suitable to the content, not too many, with integrated visual aids

In terms of layout, the specific content of the sections and subsections is not of importance. Content is addressed in the sections below on interest (A.3) and communicate (A.4). The issue of importance for ensuring a pleasant layout is the existence itself of a hierarchy of sections and subsections. As for any technical report, the communication message must be announced and fulfilled, and the main points and sub-points of that message must be organized into a hierarchy. Any presentation of content that is not organized as such will very likely confuse, overwhelm, disappoint, or bore the poster's viewers. Thus, the first goal to achieve in creating a pleasant layout is to plan a hierarchy of sections and subsections, with integrated visual aids. The final wording does not need to be decided. On the contrary, a general outline and bullet points for fleshing out the main ideas and sub-points is sufficient for planning the layout of sections.

Goal #2: Make the sequence for reading and progressing through the content obvious to viewers, even before they read a single word on the poster

Once a general plan has been developed for the section hierarchy and integrated visuals, an overall layout of the material can be selected. This involves the viewing sequence, which determines the direction of the reader's eyes from top to bottom and left to right, and the order the material is absorbed. Unlike a document, which has a sequence of pages, with only one page (or, two pages) visible at any one time, all the material is visible to a reader at each and every moment she stands near the poster. Needless to say, a viewer might read the poster in an order unintended by the author. This can happen out of convenience (parts of the poster are blocked by other viewers), choice (the viewer self-selects a particular section to look at), or confusion (the order is not self-evident). You may not be able to control convenience (blockage) and choice (self-selection), but you must definitely control confusion. You do this by making sure your pre-determined sequence for viewing is self-evident.

Indeed, the sequence for viewing is of paramount importance for layout. In fact, it stands above and distinct from the details of selecting specific sections and sub-sections, so it must be determined alongside developing the hierarchy of sections. In terms of viewing sequence, you have a few options, and I encourage you to stick with one of them and not try anything too radical unless you feel a particular poster and communication context allows for something innovative and experimental. Over the years, I have seen some unusual layouts for the sections of a poster. Some have been intriguing in their own right but not that effective. Some have been more of a puzzle than a poster. Look at Figures D12 and Figure D13.

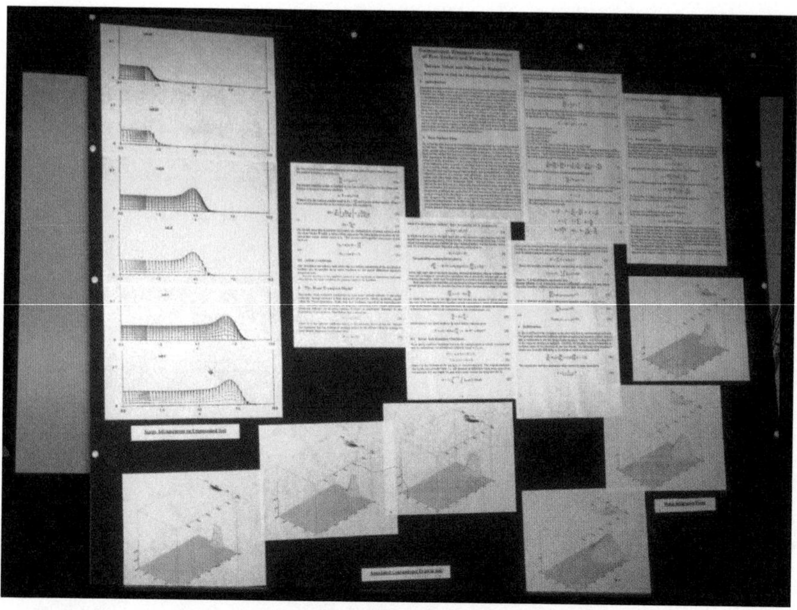

Figure D12 Poster with complex viewing sequence

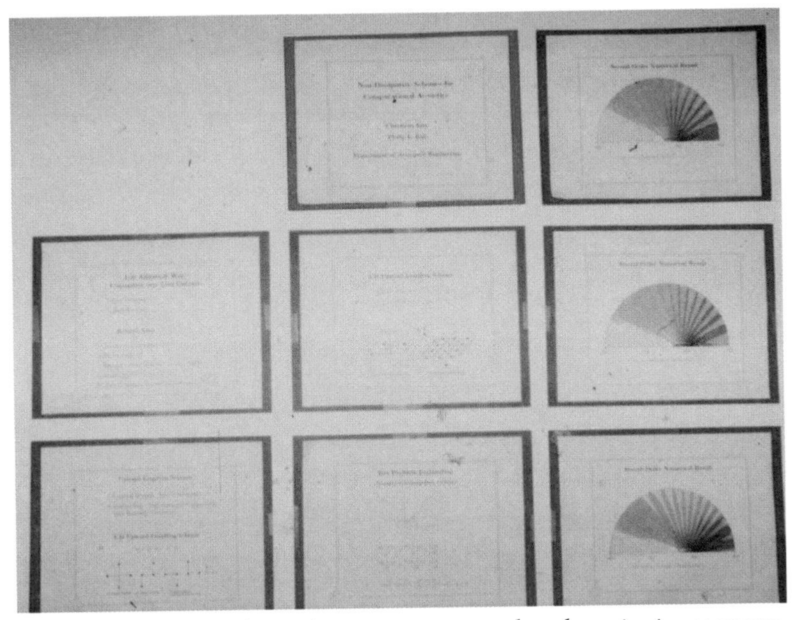

Figure D13 Poster with puzzling empty space and unclear viewing sequence

The ideal approach with a viewing sequence is to avoid drawing any attention to it. It should be intuitive and obvious, rather than something viewers must figure out. This intuitive obviousness involves using one of the common ordering sequences, displayed in Figure D14:

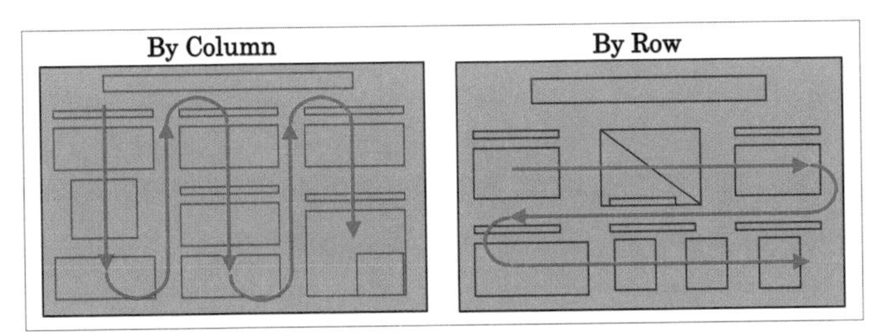

Figure D14 Two effective viewing sequences

The key to this ordering is a concept known as "long lines." In an ordering by column, the long lines are vertical, dividing the columns. In contrast, in an ordering by row, the long lines are horizontal, separating the rows. Long lines are bands of consistent-width white space that extend continually from either top to bottom or side to side, without any obstruction or interference. If you look at the column ordering at left in Figure D14, the possible horizontal long line at the top (under the first row of boxes) changes width as it extends to the second column, and the possible bottom horizontal long line gets blocked fully by a text box in the third column. Both the change in width and the obstruction indicate that the order is not by row, which need to be separated by consistent, horizontal long lines. The obstruction of a horizontal long line is also shown in Figure D15.

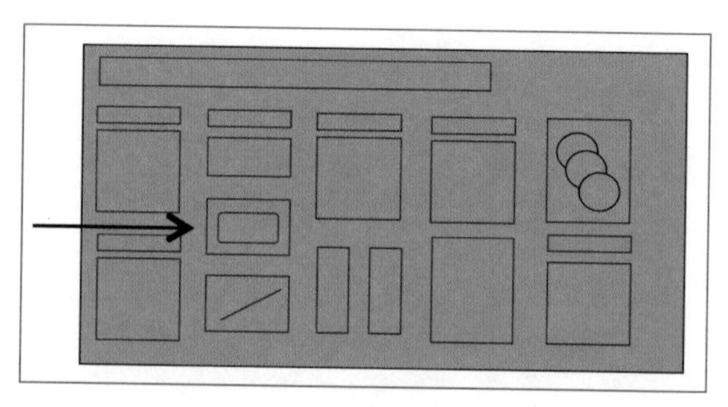

Figure D15 Viewing sequence by column because viewing by row is obstructed in the second column

As the figures above reveal, a key yet simple take-away message about ordering is that the text and visual aids should not play any role in helping viewers see the order. The layout of the sections and white space between them should be sufficient to make the ordering immediately apparent. This type of ordering is "transparent" and it allows the viewer to concentrate on the content, not requiring an effort to determine the correct order for viewing and reading. In Figure D16, the sequence is ambiguous, with viewers wondering if the order is in rows, across the top then the bottom, or in columns, starting down the left side. This poster has an obvious absence of long lines for either rows or columns. The poster has many good aspects but the layout introduces confusion, and the viewing sequence is not transparent.

Finally, another option for viewing sequence that appears in the literature is the "golden rectangle." I am not a big fan of this layout, for reasons explained below. But, I will say that some people feel it can be effective, and possibly it is ideal for certain posters. The golden rectangle layout is shown in Figure D17, with an example in Figure D18.

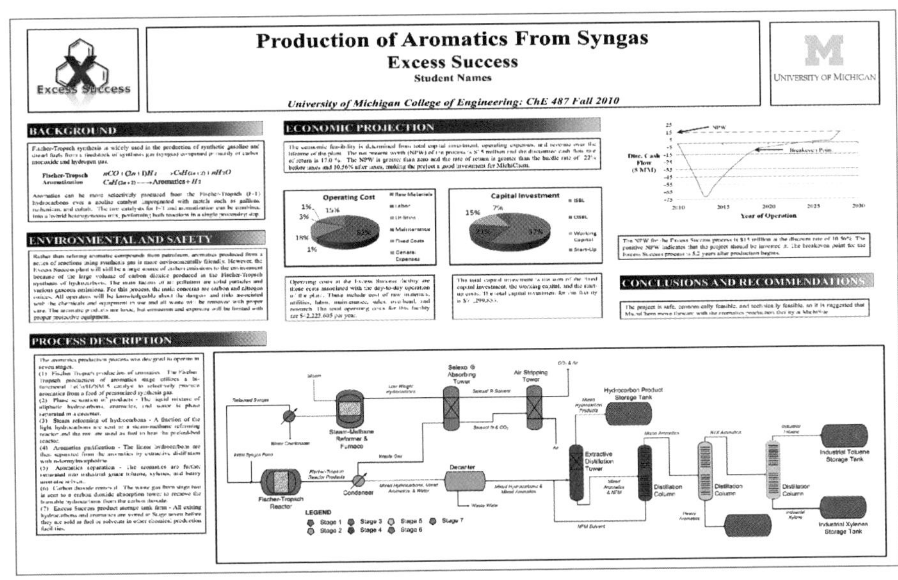

Figure D16 Viewing sequence requires effort from viewers

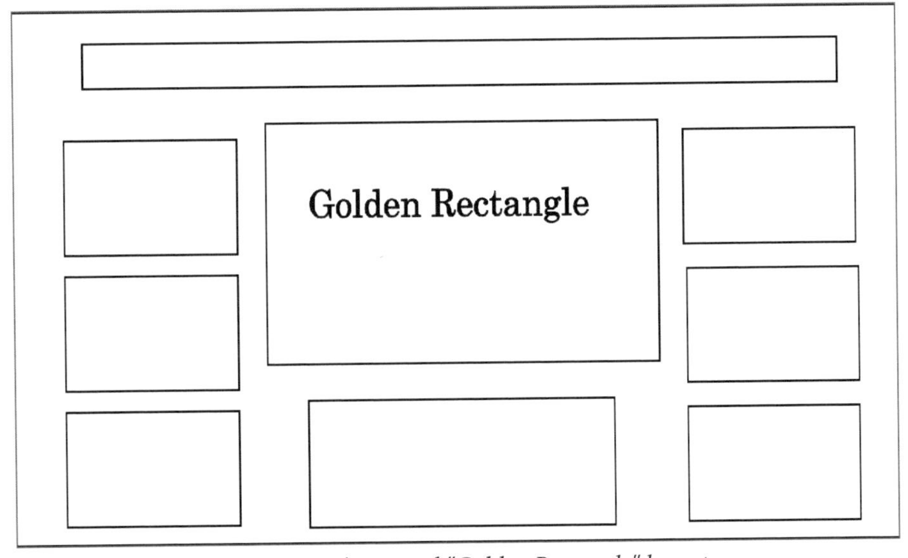

Figure D17 Theoretical "Golden Rectangle" layout

Figure D18 Poster with golden rectangle layout

My concerns with the golden rectangle are fourfold:

First, assuming the golden rectangle is used for text, this layout can contribute to the mistake of providing a monolithic chunk of text, which will likely intimidate readers and chase them away instead of drawing them in. Second, the layout enables an author to separate the text and visual aids instead of integrating these two elements.

Third, the sequence for viewing is not transparent. Even if a viewer starts in the golden rectangle, which I will call a "hub," one can only wonder if the parts of the outer "wheel" are meant to be read in a counter-clockwise rotation starting from top left, or in a clockwise rotation, starting at top right.

Fourth and lastly, as shown in Figure D18, what does a viewer do with items seemingly floating on the poster, neither part of the hub or the wheel? In Figure D18, two chunks of text sit in the bottom corners. They leave me with uncertainty: What are they connected to? When should I read them?

Figure D19 is a final example of a poster using the golden rectangle. It illustrates most of the problems for which I suggest caution before using this approach.

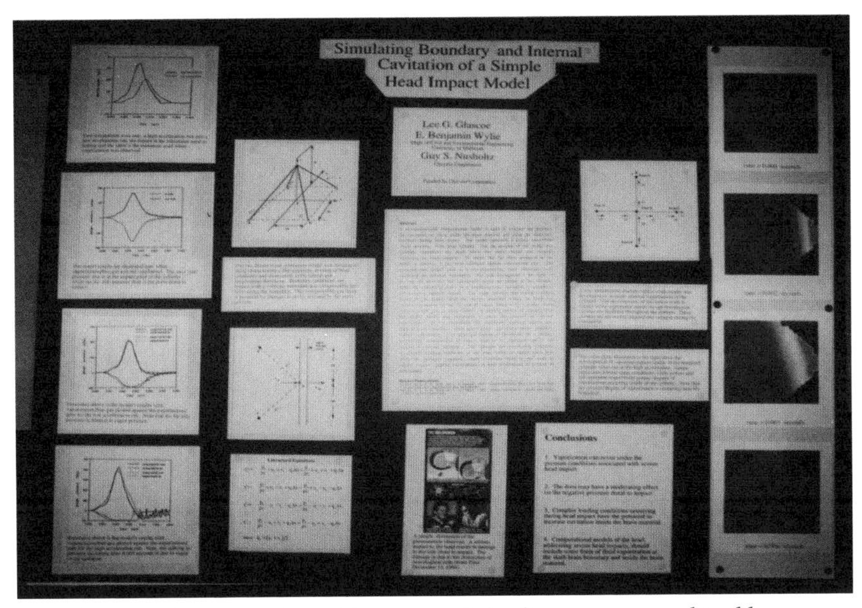

Figure D19 Poster with golden rectangle leading to compound problems

Goal #3: Leave adequate white space to minimize stress and pressure on viewers

You have learned that white space can be utilized in the form of long lines to reveal the sequence of flow. In the next section, you will learn how white space is needed to isolate and spotlight the title from the rest of the poster. In this subsection, focused on Goal #3, I cover ensuring that appropriate vertical and horizontal white space indicate the proper level of each section (first level, second level, etc.) and the relationship of subordination among sections. Importantly, two or more sections matched together as subsections of a higher-level section (nested sections) should be connected through similar white space to indicate their mutual subordination to the higher-level section. Similarly, separation of two high-level sections must use similar white space, somewhat different from that used between nested sections within a subordinate block.

Aside from these three special uses of white space, white space must be dispersed throughout the poster to minimize stress and pressure on viewers. A poster that is covered fully in text and other elements will overwhelm and chase away potential viewers. See Figure D20 for an example.

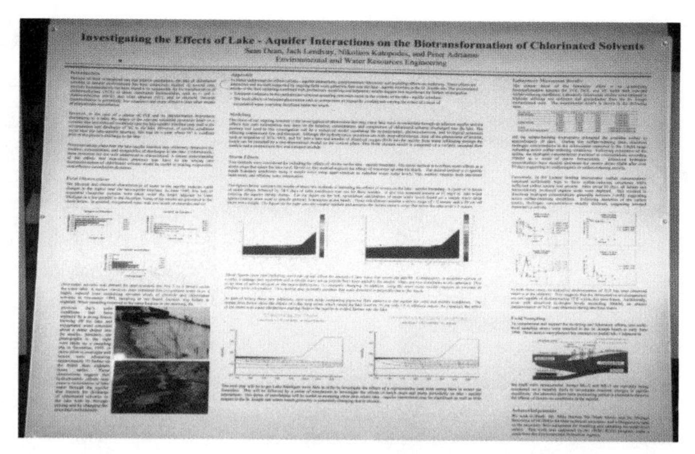

Figure D20 Poster appearing largely white but lacking adequate white space due to overwhelming use of too much small text

I've also discussed that too much (random, purposeless) white space will create a sloppy and unfinished look (recall Figures D11 and D13). Thus, you need just the right amount and no less or more. Think of this in terms of a chef and seasonings: too little, the food is bland; too much, the food is saturated and possibly too salty or too spicy.

In the scholarly literature on poster design, I read that a poster's white space should constitute approximately 50% of the total poster area. Think about that. Half of the poster is empty. That is hard to fathom. It makes sense, though, when one adds up all the space between every element and along the borders. If the white space is bunched up in any one spot, it will look bad. But sprinkled throughout, it makes the poster seem pleasant, concise, and accommodating to busy viewers. Figure D21 is an example with effective white space.

Goal #4: Achieve balance across the poster

Complementing the need to evenly distribute white space, ensuring an overall balance is important and entails a few style preferences, as follows:

- One part of the poster should not overpower any other; all items should be evenly distributed, including the eye-catching items that are critical for attracting viewers.
- Color elements and visual aids should not be bunched together in one portion.

- The poster's bottom portion should not be relegated (constrained) to displaying merely secondary information in small text.

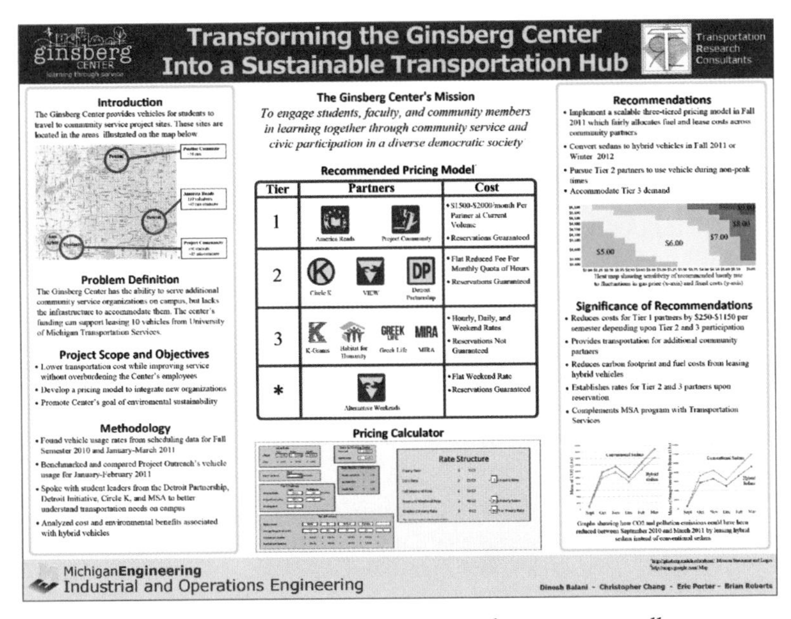

Figure D21 Poster with adequate white space overall

D.2.4 Spotlighting a Clear and Informative Title

Aside from one or two clear statements of your main message, the poster's title is the most important portion of text. Like naming a book or a film, creating a title for the poster is an important task that can significantly contribute to communication success or just as significantly hinder it. With a clear and informative title that is easy to notice and read from a distance, you will give yourself another "hook" by which to catch a busy, non-captive viewer. The title is the last tool that contributes to attracting an audience, but with a clear and informative title you will not only attract but also catch a viewer's interest, so it's a fitting place to end this section and move into the next section on maintaining the audience's interest. Before moving on, I offer some tips for creating and presenting a clear and informative title.

A good title involves decisions of both content and format. In terms of content, I offer two guidelines: (1) indicate the work in context of its contribution to the community and field(s); and (2) avoid all acronyms or jargon. If only the work is named, it may be unfamiliar and esoteric to most people at a poster conference. Work is specialized and specific, involving chemicals, machine parts, theories, equipment, diseases, genes, regulations, methods, and so on. By titling your poster with nothing more than a description of the work, as you think of it, you are building up a situation of ineffective communication except to a very tiny group of experts. A poster needs to communicate to non-experts; you can start with your title.

Consider the following poor titles:

- Stator Blade Cooling Designs
- Chemical Reaction Modeling of FAEE using CFD
- Development of Phantom Nodules for Subtle Lesion Detection

In these examples, the work is described as the authors understand it, but few people would know what this work pertains to, why it matters, and even the field of study within which the work is contained. The items missing are the contribution to the community and the field, both of which provide badly needed context for the work.

Below are the improved titles for all three:

- Cooling of Turbine Engine Stator Blades for Better Fuel Efficiency of Small Airplanes and Projectiles
- Reaction Modeling of Fatty Acid Ethyl Esters to Improve Development of Synthetic Biofuels with Lower Greenhouse Gas Emissions
- Teflon Nodules Lead to Improved Radiography for Early Cancer Detection

In terms of format, the title must be both prominent and easy to read from a distance. For most purposes, the title is best placed at the top, either in the center or in the left corner. A large type size, such as 72 point, must be used so the words are large enough to be read by someone standing at a distance from the poster. Experts say that 72-point type can be seen from 20 feet. You can experiment with variations smaller and larger than 72 point to find an ideal size for your poster.

D.3 Design Objective #2: Interest

You probably wonder how I can spend 22 pages talking about posters and say nothing about required content, aside from a title. Of course, the poster must convey some information. But, it will be useless and go unread if the poster is not attractive. Hence, all the above guidelines are primary. On a secondary level, you will indeed need content.

For that, I am not worried about your ability to discuss your project. You can always do that. The poster's goal is to get other people to be interested in your project. That's a huge difference. You know your project so well. You find it fascinating. You value its theoretical, scientific, and creative aspects. No doubt the problem-solving element is innovative and elegant. But, that's you. How can you get someone else as interested in the project as you are? After all, these people are hurried and possibly intrigued by other posters nearby.

What is interest? Is it memorizing all the facts? Is it seeing an equation? We don't know. For one viewer, an equation may spark interest. For another, a diagram of equipment may be the secret sauce. All you can do is play your best cards. I suggest you use some tried-and-true techniques for helping non-experts and laypersons, along with peers, colleagues, and professionals in your chosen field or specialty, to find your work interesting. In this sense, "interesting" is a broad term that captures some other characteristics: valuable, useful, important, unusual, innovative, and necessary. You convey these characteristics in the detailed portions of text and visual aids, in an order that facilitates comprehension by people unfamiliar with the project. Thus, once the sequence of viewing has been selected and laid out, you can work on goal #2: maintain interest.

This involves two components:

1. Ensure that the hierarchy of sections is logical and simple.
2. Answer as many of the following questions as possible:
    - Did you improve a test? What is the test for and why is it needed?
    - Did you improve a product or device? For what use?
    - Did you develop or confirm theory? For what reason?
    - Did you acquire new knowledge? For what purpose?
    - Who does this help?

- How does it help them?
- Does it make something...
  - Smaller
  - Faster
  - More durable
  - Brighter
  - More accurate
  - More reliable
  - Safer
  - Cleaner, or
  - More efficient?

These questions address the general and big-picture aspects of your work. You must cover this. People not involved in a project personally cannot find an entry to the new and unfamiliar through esoteric details and arcane points. But, if a person enters a new topic through broad concepts such as those covered in the aforementioned questions, they might be able to see how the subordinate and specialized details fit into the big picture. Again, this is the concept of putting general information before particulars. Basically, because of people's innate curiosity and concern for social and economic issues, a topic is understandable if its far-reaching benefits, purpose, impact, and value are covered clearly and succinctly in strategic spots throughout the poster, including the first section.

The first section can be titled anything such as "Introduction," "Background," "Purpose," "Opening Statement," "Project Motivation," "Overview," "Problem Statement," etc. The word chosen does not matter. The important part is to cover the usual items that should be placed at the beginning of any ABERST report (review Chapter 1.2). Once you have started well, you can shape the poster for all readers as covered in Chapter 1.4, with the exception that posters do not have a Documentation segment. A poster is limited to only two parts of the structural pyramid: Overview and Discussion. Tertiary information is unnecessary. More importantly, the key aspect of shaping that applies to posters is to create a hierarchy of sections just as you would for any other technical genre. This is the technique of subordination as covered in Chapters 1.4 and 3.4. The specifics of your hierarchy will be prompted by your message and your need to communicate some particular set of information drawn from your work. This is the focus of the next section in this appendix.

Five further tips are applicable to catching and maintaining a viewer's interest.

1) Be sure to use numerous visual aids and explain them. This helps attract viewers to your poster (subsection A.2.2), but it also keeps their interest because images offer so many advantages (review Chapter 5.2). These advantages include displaying the details of how natural and man-made things look, the trends and inflection points of important data sets, the organizational and spatial relationships of things and people, and the most important aspects of your message.
2) Be sure to use popularized vocabulary and not specialty terms; or, if you must use jargon, define or explain these unusual terms. Look at the following pairs and notice the difference between jargon and common language:
   - Negative example: "We are developing an ornithopter."
   - Positive example: "We are developing a flapping-wing aircraft, modeled after a bird."
   - Negative example: "We want this new robot to ambulate, not roll on wheels."
   - Positive example: "We want this new robot to walk, on two legs, like a human being."
   - Negative example: "I suggest implementing on-line inspection."
   - Positive example: "I suggest implementing a check on manufacturing quality during production, not after."
   - Negative example: "The customer wants us to study an inverted, cambered airfoil."
   - Positive example: "The customer wants us to study an automobile spoiler."

   I heard a researcher describing elephant seals recently, and she said this particular species of seal is known for its large, dangling proboscis in the front of its face. She quickly added, "It's a nose, and it's similar to an elephant's trunk, hence the name 'elephant seal.'" This was clear communication that maintained my interest because I understood it.
3) Another way to keep interest is to add some element that is three-dimensional. The vast majority of posters are printed with a plotter in color, and they are 2-dimensional, and that is fine. All the guidelines presented so far are intended to help you make your 2-D poster as attractive and interesting as possible.

But, if you wish to add a little flair and creativity, and you have the time and resources, you can attach a 3-D item or two. One simple method of making your poster "pop" is to attach all your visual aids to a piece of foam core and lift them off the surface (see Figure D22). Another option is to attach small samples of your materials, such as polymers, alloys, or composites; 3-D printed prototypes; test tubes with fluids; molecular models; and so on.

4) Unless absolutely integral to the poster's purpose, you can leave out a References list. A poster ideally presents your work and findings. You are not only the poster's author, but also the primary *authority* behind the poster. As a medium, posters are suited to showing one person's or team's current achievement. It is not the place to present a literature review or an evolutionary account of your specialty. As such, a References section is unnecessary and uninteresting. You do not need take up valuable space with bibliographical entries, including publishing data and dense abbreviations. Aside from rare exceptions, few viewers need this data. Perhaps, as an exception, your poster is a review of other people's work or you are presenting an overview of recent, interrelated developments in your research area. In such cases, you certainly need to mention the various authorities by name and the year of their contributions. Integrating that information into the poster's main text, however, is more interesting than slapping on a detailed References section. Because you should include the other authorities in the poster's narrative, a References list is avoidable. On posters that highlight the author's own work, a References list is additionally superfluous. The credit for the information on the poster goes to the author, not other published sources. If a viewer wants to learn about the related literature, this can be handled in conversation. In short, on most posters, a References list of dense data can be replaced with interesting text or visuals.

5) I offer this final suggestion with caution, and I have seen it used on posters only rarely. As with oral presentations, the inclusion of a cartoon, humor, or reference to popular culture may be an effective way to maintain the audience's interest. That said, this technique may be inappropriate for most poster conferences and may be a blemish on an author's credibility. If used by a senior professional or by a person whose integrity and ability are well established, it could be an added flourish to a poster. See one example below (Figure D23), where a hand-drawn cartoon is used to illustrate the science in a light-hearted manner.

Figure D22 Poster with extra texture

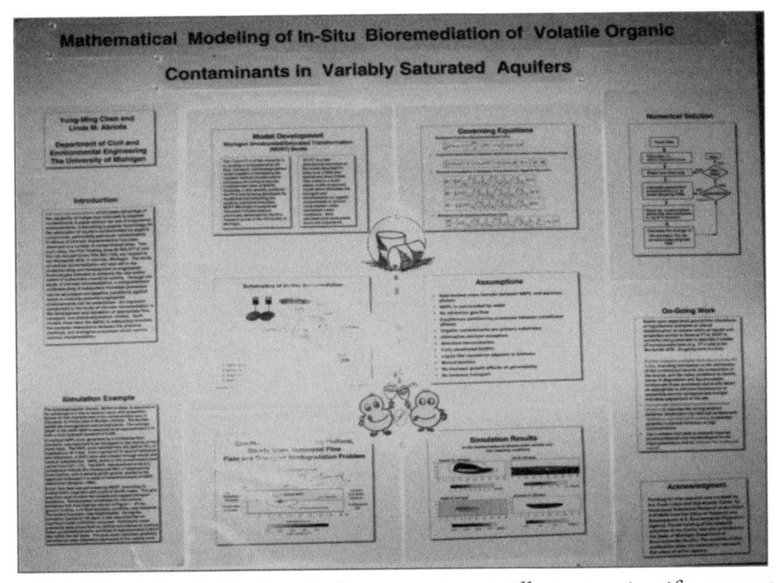

Figure D23 Poster with cleverly drawn cartoon to illustrate scientific concept

Appx D

D.4 Design Objective #3: Communicate—Inform or Persuade

One important foundational concept to remember when assembling the content for your poster is that you cannot expect viewers to have looked at any of your previous documents. The poster must be clear and complete on its own, and that includes being independent of any items you have placed around it, such as videos, equipment, and models, if the conference allows such additions (and most do not). This rule of "independence" is true for all ABERST technical documents, and it easily makes sense with regard to posters, which are intended to be a first, and often only, exposure for most of the wide and diverse audience members attending the conference.

Starting with the premise that a poster must be independent adds pressure to include only the essential information, especially when considering that you must be concise as well. In order to be both complete and concise, you must start with your main message. At the very least, the communication message must be announced and fulfilled (review Chapter 3.1). The content to be presented and the specific sections are determined by this message, providing text and visual aids that fulfill the author's communication objectives. Although each poster is unique, a common approach is applicable to the vast majority.

First, follow all the guidelines from Part 3 (Fulfill Your Communication Purpose) and Part 4 (Persuade Your Reader), if persuasion is applicable. As you recall from those parts of the book, the information used in all technical reports is hierarchical, some of which is primary, some secondary, and so forth. That principle is extremely crucial for posters because you are limited in space to one page, albeit a large page. So, you must be selective. Thus, as much as you need to assemble the right information, you need to omit a good deal of information too. As mentioned previously, a poster has no Documentation segment, so tertiary information can be ignored. Similarly, much secondary information can be left out also. Overall, omitting non-essential details is a crucial step in ensuring the poster is neither too crowded nor packed, as covered in Section A.2.3.

Second, while selecting the key information to include on the poster, you are also completing simultaneously another critical task: organizing that information into the ideal hierarchy of sections and subsections. Selecting implies organizing, so these two aspects of communicating your message, with both main and support points, should be done together. As you start to place the various sections within the dimensions of your full poster, you will also plan the viewing sequence, per Section A.2.3.

Third and finally, one last suggestion will add some difficulty to completing your poster, but it should not be ignored. As much as you have limited space and should avoid redundancy, you do need to repeat the few key ideas twice. This is a tall task, but it must be done if you are to ensure that your audience retains your main message and main points. Following the instructions given above (Chapter 3.4 and Section 4.5.2), you should present your main ideas in the poster's introductory sections and again revisit them as you present support points for those main ideas. Similarly, you need to ensure that ideas and details presented with visual aids are also reinforced with text, and vice versa, so that viewers are not left scratching their heads in wonder at what something looks like but is not shown, or what something shown means.

I will address one technique often offered by others to address the need to omit information and keep the poster succinct. Some guidance for posters recommends limiting the text to bullet points only. I understand the spirit of this advice, namely, to keep the poster concise and easy to read. Nonetheless, if the bullet points do not communicate the message fully, including main and support points, the poster will be vacuous and unclear. Thus, bullets can be used in places as appropriate, but I do not believe a hard-and-fast rule of "bullets only" needs to be followed.

Whether you need to inform or persuade will depend on the status of your project and your personal desires for your poster. If you simply wish to inform, you will borrow from heuristics presented in Part 5 for whichever genre(s) is closely aligned to your communication purpose and message. If you want to present an argument, including a thesis and support points, you will draw on the guidance provided in Part 4.

One formatting trick that you can use strategically to help increase the communicative effect of your poster is to utilize different type sizes to your advantage. When you occasionally use a small type size (but not so small to be illegible to a viewer standing 3 feet away), you will free up some space to include more information at a higher level, presented in a larger type size. A scheme that involves a variety of type sizes is best: use largest sizes (72 to 50 point) for titles and headings, use medium sizes (28 to 18 point) for body text, and use smallest sizes (16 to 12 point) for detailed items that are not immediately important to viewers but can be read if someone wants to look closely, such as axes labels on graphs. Assuming 20-20 vision, various text sizes are visible at the distances listed below.

1-inch type (72 point) is viewable from 19 feet
.8-inch type (58 point) is viewable from 16.7 feet
.5-inch type (36 point) is viewable from 9.5 feet

28 point is readable from 8 feet
20 point is readable from 6 feet
18 point is readable from 5 feet (basic text)
14-point type is readable from 4 feet (1.23 meters)
12 point is viewable from 3-1/2 feet (1.06 meters)

One informal test that can be used to determine if your choices of text size will be effective and if your smallest text size will be legible is to print a draft of the poster in 8-1/2- by 11-inch format and see if it is readable at arm's length. If it is readable, the text sizes are sufficient.

D.5 Poster Presenter Behavior

At a typical poster conference, talks and oral presentations are not required to accompany a poster. Instead, the author is expected to host the poster, standing alongside it to greet viewers and be available for questions. If a team has created the poster, at least one team member should be near the poster at all times. The whole team does not always need to be present, unless the conference organizers have requested it. In fact, a possible benefit arises when team members take turns hosting the poster: space is left for viewers rather than the team crowding them out. This crowding is one of the common problems posed by authors: by standing in front of the poster talking amongst themselves or speaking to one viewer, they exclude other viewers from getting a glimpse. Those potential viewers may just walk on past if they cannot see the poster through the throng. My advice is stand near, but not in front of, the poster, so you do not obstruct a clear view of the poster.

Furthermore, you should be friendly and courteous, and you should offer to answer questions. When a viewer comes over to look, they do not need a speech or a narration from you. Instead, greet them with a smile and "hello," and let them know you will answer any questions. The poster speaks for itself (and you have put the effort into making it so). Give viewers the chance to look over and read your poster before pushing a dialogue upon them, or—even worse—a monologue. Resist the urge to treat a casual viewer like a captive audience. If you pounce on them and force a conversation upon them, they might walk away. (They are free and not captive.) Dominating or suffocating a viewer is neither courteous nor effective.

As you stand by, you can wait until a viewer initiates the conversation. If they give you a compliment, say "thanks," and leave it at that. Let the viewer continue the conversation. If the viewer asks a question, you have the privilege of answering and taking the opportunity to explain, clarify, and defend your work. This is an awesome moment, so play it cool and go lightly on the viewer. Here are some tips for answering questions:

- Be brief with answers. Break your answer into "chunks," by giving a good, general answer only, without launching into a monologue.
- After giving your concise, initial answer, you can ask if the viewer would like to hear more.
- If you know something about the viewer, you can emphasize the relevance of your work to that particular person.
- If you want to tailor your answer to the viewer but need more information about her, you can politely ask for such information and explain that it could help you with your answer (such as the person's field, specialty, office location, prior knowledge, etc.).
- Resist the urge to show off technical jargon and acronyms. Assume the viewer does not know these. Few people do.

You want to thank all viewers who come by, and you can offer your business card if you are inclined to use the opportunity to network. You can also ask for their cards. (Electronically, you can exchange contact information with your smart phones.) I have seen some poster presenters use a guest book to collect name, address, phone, and email from each guest. Such personal information may come in handy one day, as more and more people collaborate across universities and organizations.

I hope you find my advice helpful and enjoy making a poster. Good luck at conferences.

Appendix E

Oral Presentations and Projection Slides

In ABERST, giving an oral presentation, whether to a small group or a large audience, is a normal part of the job. In some situations, the speaker is helped with some sort of speaking aid, such as a projection slide deck, poster, tangible prototype or product, or flip chart. In other situations, the speaker must rely on her words alone, just as one speaks conversationally, naturally, and spontaneously. Situations in between these two extremes involve speaking with the help of notes, giving a prepared speech from script or teleprompter, or presenting a talk from memory. No matter the situation, speaking venue, or type of assistance, the job of giving an oral presentation is always difficult. Professionals in ABERST, usually not trained in acting, oratory, or entertainment arts, are not likely to relish this task. In fact, "oral presentation" is a euphemism, intended to water down the truth. The cold hard truth is that I am actually talking about public speaking, and public speaking makes most people uncomfortable.

Aside from conversations with our peers and co-workers, all other oral presentations on the job are a type of public speaking. Even if the audience is just one person, the briefing could be a type of public speaking. If you are giving an update to someone who needs the information and might judge you and can impact your career from a position of authority, you have to make a good impression. Perhaps a presentation to a small handful of people may have more impact on your career at one point than a speech to a room full of strangers. In another instance, it might be the big speech that could mark a turning point in one's career. Either may be a cause for anxiety, and that means preparation is required.

You can always ensure successful public speaking by getting prepared for each and every oral presentation, both your intended spoken words and your accompanying material, such as projection slides. Usually you will be given time to do so, and this appendix offers help for preparing your overall presentation generally and your projection slides specifically.

One situation will still fall outside the scope of this appendix: In the worst-case scenario, you might be surprised, without advance warning, that you must brief a supervisor. If a manager or senior member of your organization approaches you unexpectedly and wants some information about your project, you will not be able to prepare. In this scenario, you must rely on the fact that you are knowledgeable and capable of explaining the situation because you are immersed in the day-to-day project. You may not have any prepared remarks, but you should be able to converse fluently about your project. You might even pull up something on a computer, spontaneously, if you feel you can quickly find and use a visual aid effectively. If you cannot do that, you are on your own to speak clearly and eloquently. Do not worry. Just be confident that you will speak well when such a situation arises and do your best. The knowledge and skill you will obtain from studying this appendix will help you in such situations also.

Fortunately, most of the oral presentations you will be required to give will be scheduled, and this allows you to prepare. This appendix offers some guidelines for (1) organizing an oral report, (2) preparing supporting projection slides, and (3) speaking to a group.

E.1 Organizing an Oral Report

I use an overall fivefold strategy for public speaking that was taught to me many years ago:

- Make it short
- Emphasize the main point(s) quickly
- Make the organization obvious
- Make the ideas simple and vivid
- Summarize and invite questions

In this strategy, you will see elements that overlap with the fundamental guidelines for a written document: (1) Be as short as possible, leaving out information that is not absolutely essential; most items that would go into an appendix, for example, would not be needed in an oral presentation. (2) Have a main point or points and make these easy to find. (3) Make the organization obvious, just as a written report's overview forecasts its body/discussion.

Moreover, many ideas in ABERST are complex, so the more you can help your audience see and grasp in their minds your ideas, in a way that is meaningful to them, the higher the rate of comprehension. If you have projection slides, you can be sure to show diagrams, graphs, tables, maps, and so on (as discussed in Chapter 5.2). Remember, as with a written document, you must assume that your audience members during an oral presentation are non-experts and less knowledgeable than you about your project. Therefore, (4) being simple and vivid is necessary to fulfill your communication purpose. Lastly, (5) the only part that is different from a written document is to summarize and invite questions, which is self-explanatory.

The aforementioned fivefold strategy guides the organization of an oral presentation, especially the first three components. First, a short presentation helps you and the audience. If you have been given a defined amount of time, you can aim to leave 10% of the time unused, as buffer. No one in the audience will complain if you do not consume all of your allotted time. Attempting to make a short presentation demands that you distinguish primary information from secondary and tertiary information. Second, a short presentation demands that you have a quick and clear introduction, followed by getting to your main points right away. Third, as you promptly reveal your main points, you will inherently be forecasting your organization. The body of your presentation is simply composed of a hierarchical presentation of all your main points and necessary sub-points to inform and persuade your audience.

Structurally, an oral presentation can be divided into four segments:

- Introduction
- Body
- Closing
- Questions and Answers (Q and A)

This structure allows a speaker to implement the final two components of the fivefold strategy. Fourth, simple and vivid ideas are used from beginning to end, in all four segments. You might start out with a clear illustration of the problem that is driving the project, for example, cracking and pitting highway concrete, and you might end with a review of a visual aid as you answer a question from the audience, for example, comparison table of three highway repair plans, with varying concrete types. Fifth and finally, in the Closing segment, you succinctly summarize your presentation and invite questions, initiating the last segment.

The organization of the Body, which needs to be revealed early on, will be adapted to each report purpose, type, and content. The Closing will follow from the information presented in the Body, and the content of the Q and A session is anyone's guess. Thus, only the Introduction can be mapped out generally for most oral presentations. An Introduction usually involves four components, and ideally one or two slides can be devoted to each of them. As shown below, moreover, each component has multiple elements:

Introduction
- Title: Speech Title (Topic and Focus), Presentation Purpose, Speaker Name(s) and Affiliation, and Date
- Background: Motivating Problem and Justification of Work
- Objective: Primary Work Objectives, Intermediate Goals, Questions to Answer, and Key Phases of Work
- Summary: Substantive Main Points

A sample of both a title slide (Figure E1) and an objective slide (Figure E2) are provided below. With these two samples as initial illustrations of slide design, I turn to some detailed instructions for slide preparation next.

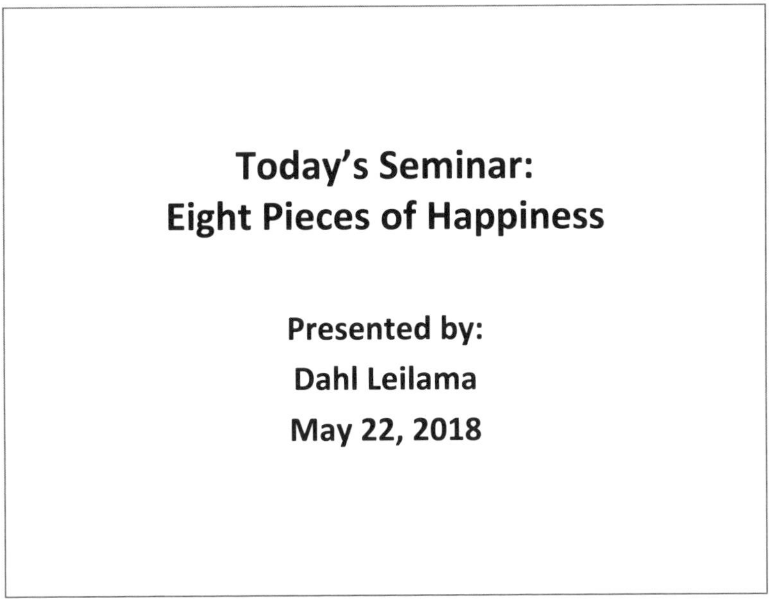

Today's Seminar:
Eight Pieces of Happiness

Presented by:
Dahl Leilama
May 22, 2018

Figure E1 Title slide

Project Objectives and Goals

- Design Optimized Reaction Wheel System
  - Select Wheels
  - Select Motors
  - Build Customized Housing
  - Assemble Whole System
- Characterize Fundamental System Properties
  - Test Motor
    - Nominal Torque
    - Maximum Momentum
  - Test System
    - Vacuum, Thermal, and Harmonics

Figure E2 Objective slide

E.2 Preparing Supporting Projection Slides

For all four segments of an oral presentation, projection slides prepared in advance serve two purposes: (1) they record and display the outline of all main and subordinate points, as well as all visual aids you wish to show your audience; (2) they serve as speaking cues, to foster an extemporaneous style for the speaker without fully reading the points on the slides to the audience; you glance at the point on project slides and elaborate upon it; each projection slide represents a little portion of the spoken talk, in a concise, abbreviated format. Thus, projection slides help a speaker both prepare the speech and deliver it.

You will not always use projection slides. Here is a sampling of types of oral presentations with a parenthetical note whether slides are used or not:

- 1-minute elevator pitch (no slides)
- Impromptu "boss at your desk" briefing (no slides)
- Quick update to group at small meeting (usually no slides, maybe a handout)
- Invited contribution to a meeting (possibly slides depending on meeting purpose, time allotted to you, and status of your work/knowledge)
- Design unveiling with easel or physical/tangible display (no slides)
- Progress/status report for management/client (slides or handouts, or both)
- Proposal for funding or contract (slides)
- Final Report with design, solution, or recommendations (slides)
- Conference paper (slides)

Other books and authors cover slide design in greater detail than I do here, and some experts offer new ways of adapting this genre, such as Michael Alley's assertion-evidence approach. I offer a basic foundation that is widely applicable to various presentation situations. To begin, slides can be divided into two categories:

- *Structural Slides* that outline talking points and maintain flow
- *Content Slides* that provide details, typically visually or quantitatively

Structural slides are fairly simple, primarily consisting of bullet points, arranged in a hierarchy. Eight basic slide design rules for structural slides are these:

1) Place a slide title at top, centered or left aligned.
2) Use large type size and different sizes but always larger than the size used for printed documents. A safe range is 20- to 40-point text, but aiming for a range between 30 to 50 points is not prohibited. See samples in Figure E3a, with 28, 20, and 16 points, and E3b, with 44, 32, and 26 points.
3) Use terse phrases, not sentences. See a comparison below, Figures E4 and E5.
4) Align first-level phrases at left margin.
5) Indent to tab stops for subordinate phrases.
6) Place phrases into parallel grammatical form, as explained shortly below. (See also Appendix C, section C.6.1.)
7) Limit redundant information in recurring headers/footers, as this is not substantive content and the audience does not need to see it the whole time. However, information that can be helpful if repeated on a succession of selected slides indicates either (1) a particular phase of work or (2) a segment of the presentation being covered with those slides.
8) Leave adequate white space on each slide.

Let me say something additionally about slide titles. Certainly, a slide title belongs at the top, containing the main point or most general phrase that characterizes the information on the slide, followed by a small group of bullet points or phrases. (A common variation is to present only one point on a slide, under the title, for emphasis.) If you find that you are putting the same title on numerous slides, you have two options: (1) keep that main title and add a subtitle or some other note (cont, cont'd, more, second, 2, etc.) to differentiate the different slides with similar titles; or (2) change the title from its current broadly repetitive version to something unique and specific for each slide.

For example, perhaps your deck has four slides that are titled, "Economics Outlook." To distinguish between these four similarly titled slides, I suggest two options. Option 1 is to add numbers: Economics Outlook 1, Economics Outlook 2, etc. Option 2 is to change the titles to focused sub-parts of your economics outlook: Projected 5-Year Revenue, Anticipated Labor Costs, Taxes, and Earnings and Profits.

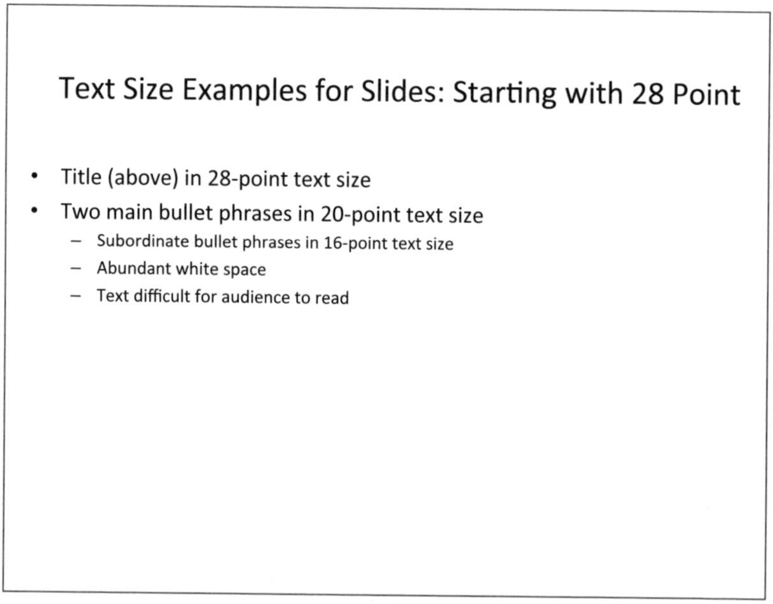

Text Size Examples for Slides: Starting with 28 Point

- Title (above) in 28-point text size
- Two main bullet phrases in 20-point text size
  - Subordinate bullet phrases in 16-point text size
  - Abundant white space
  - Text difficult for audience to read

Figure E3a Small text size

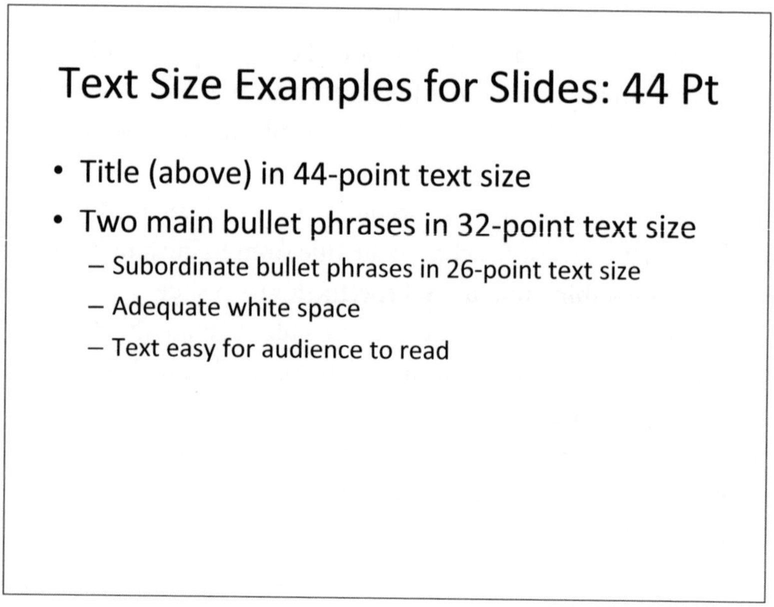

Text Size Examples for Slides: 44 Pt

- Title (above) in 44-point text size
- Two main bullet phrases in 32-point text size
  - Subordinate bullet phrases in 26-point text size
  - Adequate white space
  - Text easy for audience to read

Figure E3b Large text size

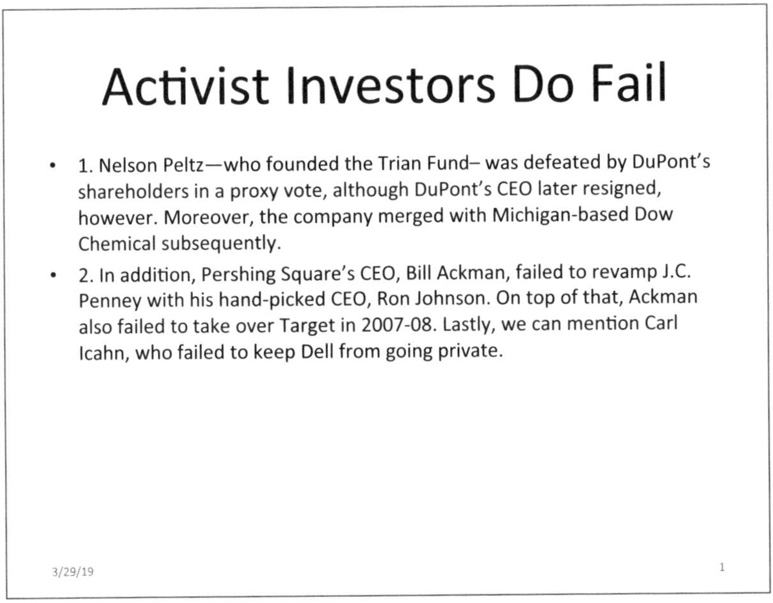

Activist Investors Do Fail

- 1. Nelson Peltz—who founded the Trian Fund– was defeated by DuPont's shareholders in a proxy vote, although DuPont's CEO later resigned, however. Moreover, the company merged with Michigan-based Dow Chemical subsequently.
- 2. In addition, Pershing Square's CEO, Bill Ackman, failed to revamp J.C. Penney with his hand-picked CEO, Ron Johnson. On top of that, Ackman also failed to take over Target in 2007-08. Lastly, we can mention Carl Icahn, who failed to keep Dell from going private.

3/29/19 1

Figure E4 Sample with too much text

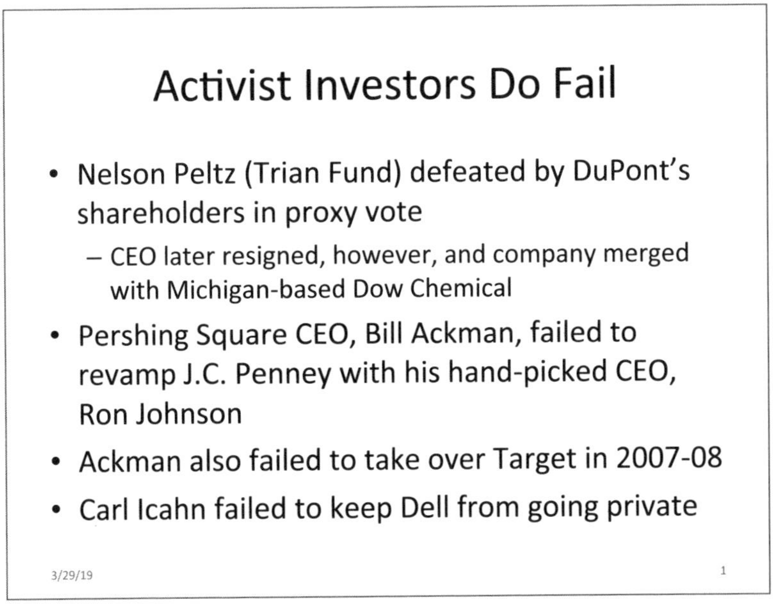

Activist Investors Do Fail

- Nelson Peltz (Trian Fund) defeated by DuPont's shareholders in proxy vote
  – CEO later resigned, however, and company merged with Michigan-based Dow Chemical
- Pershing Square CEO, Bill Ackman, failed to revamp J.C. Penney with his hand-picked CEO, Ron Johnson
- Ackman also failed to take over Target in 2007-08
- Carl Icahn failed to keep Dell from going private

3/29/19 1

Figure E5 Sample with good text phrases

Parallelism is the characteristic of matching phrases in grammatical form and punctuation. The best way to understand this is to see two versions of the same phrases:

Not Parallel

We imposed these constraints on our economics analysis:

- Raw materials are assumed purchased at bulk prices
- Used equipment had a salvage value of zero
- Saleable products will be sold at competitive market prices for bulk quantities
- Profit to be estimated using discounted cash flow rate of return analysis

Parallel

We imposed these constraints on our economics analysis:

- Raw materials purchased at bulk prices
- Used equipment depreciated to zero salvage value
- Saleable products sold at competitive market prices for bulk quantities
- Profit estimated by discounted cash flow rate of return analysis

A few positive examples of structural slides are shown in Figures E6 and E7.

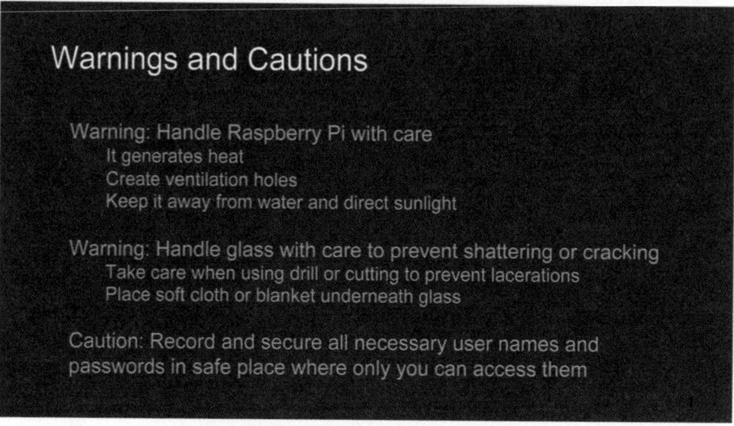

Figure E6 Structural slide sample

Eight Pieces of Happiness:
Seminar Summary

- Purpose
- Strengths
- Optimism
- Gratitude
- Self compassion
- Mindfulness
- Body and soul
- Community

Figure E7 Structural slide sample

Content slides are devoted entirely to a visual aid of one type or another. Just as you have a variety of visual aids at your disposal for written documents (Chapter 5.2), you have those same options for projection slides. If you have a table, graph, map, or diagram to present to your audience, a projection slide will enable this. A couple reminders apply to these content visuals: Use the slide title or subtitle to convey the message you want to provide with the visual aid; ensure that all parts of the visual aid are visible and legible to an audience looking at it on the projection screen from a distance (namely, the back of the room); design the visual aid in compliance with the principles of efficacy, honesty, autonomy, and simplicity; and use a pointer to highlight particular parts of the visual aid as your talk about those parts, to guide the audience's eyes to the right places. Some positive examples are shown in Figures E8 and E9.

Different slide themes or colors can be used for distinct sections of the report if that is helpful, but the default is to use a common theme and color (theme design, background color, text color and font, footers, and headers) for the whole slide deck. Choices for theme designs and colors are endless, so I cannot provide a single recommendation. Some organizations have an established slide deck style, and you may be required to use it. This may include logos and other headers/footers that maintain standardization throughout the organization.

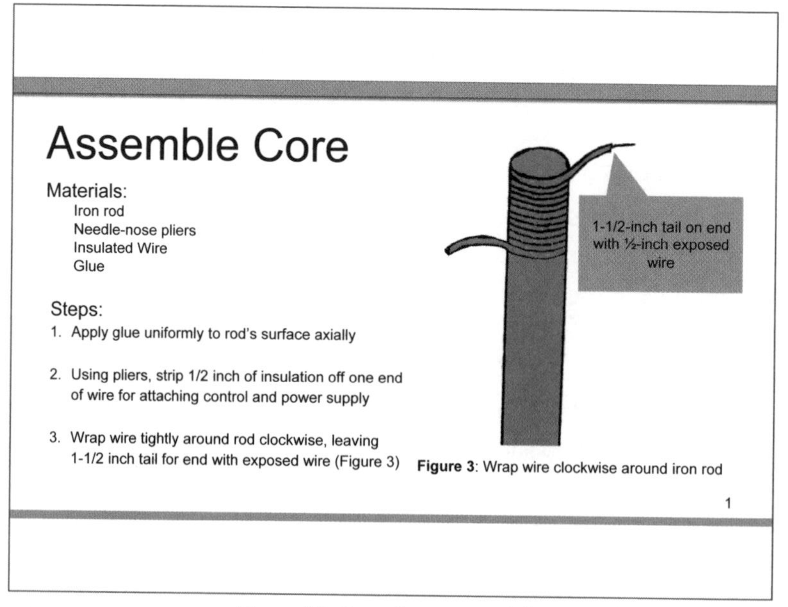

Figure E8 Sample content slide

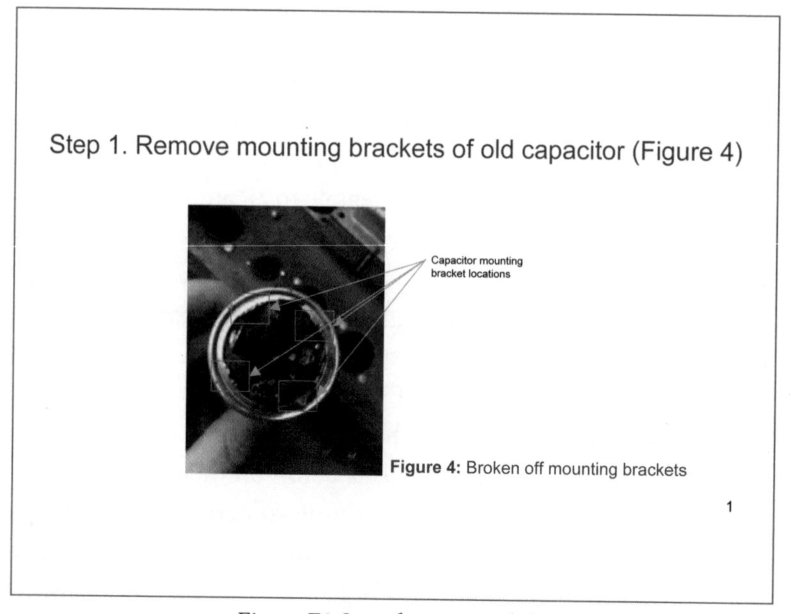

Figure E9 Sample content slide

Some organizations encourage creativity, so you can choose a ready-made template design or develop an original one, depending on your inclination and schedule. Two slide themes with organizational logos are shown to indicate the endless possibilities here (Figures E10 and E11).

A couple of general rules for design are the following:

Ensure that the text stands out from the background so it is legible. A contrast must be apparent. Ideally, the text should be much darker than the background, hence the regularity of black on white; the inverse works also: white or yellow text on black or blue. See comparisons of good and bad contrast in Figures E12 and E13.

Be careful with bullet animations and slide transitions, as they can sometimes be slow and tiring to audiences; These were popular when first introduced as software evolved, but they tend to be unnecessary and mostly a "party trick" for novices. Simplicity with your slide show may be best. That said, you can use them strategically if they complement your presentation. Bringing in bullets and other items on your slides deliberately, in a careful and delayed sequence, as you speak (and advance the presentation with a mouse click) can be effective.

One final tip for slides is to number them, so a specific slide can be referenced by its place in the sequence, and if handouts are made from your slides, they will be easy to put into order if the pages should become unsorted.

The final slides of a presentation will support your Closing segment. You will need these:

- Substantive conclusion (one slide)
- Presentation wrap-up (one slide)

Opinion varies as to the need for a wrap-up slide. Some authorities say you should not write "thanks" or "questions?" on a final slide. Others say you can certainly indicate what comes next:

- Questions and Answers (15 minutes)
- Reception (30 minutes)

And still other authorities like to see contact information on a final slide. As usual, opinions differ. My final word on this final slide: If you have made your oral presentation short, clear, and well organized, as I advised at the outset, your last slide will not be all that critical. You have succeeded with your audience by that time. Do what you feel is best for you.

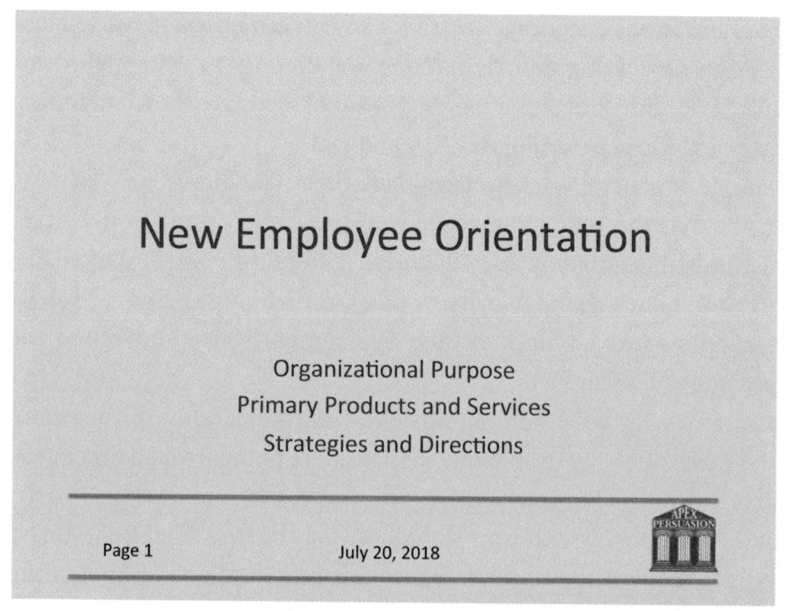

Figure E10 Sample slide theme with organization's logo

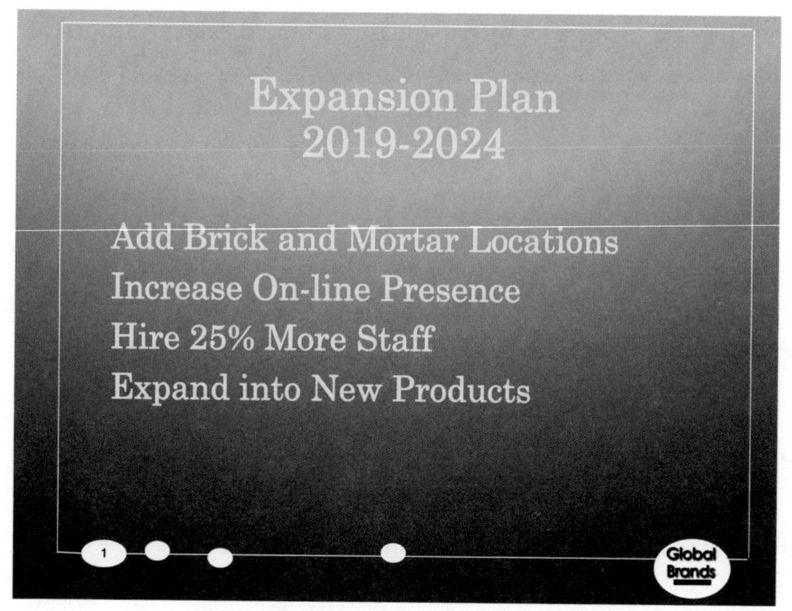

Figure E11 Sample slide theme with logo

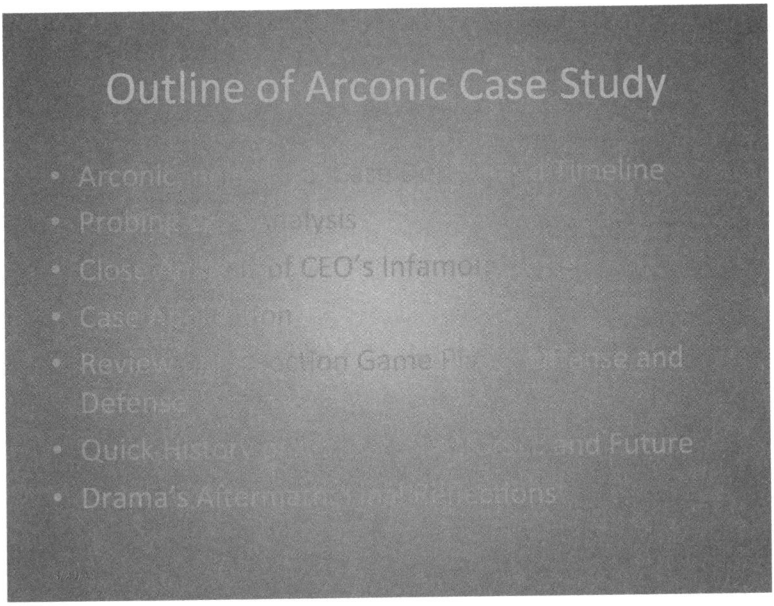

Outline of Arconic Case Study

- Arconic ... Timeline
- Probing ... alysis
- Close ... of CEO's Infamo...
- Case ...
- Review ... ction Game Pl... ffense and Defense
- Quick History ... and Future
- Drama's Afterma...

Figure E12 Poor Text Contrast

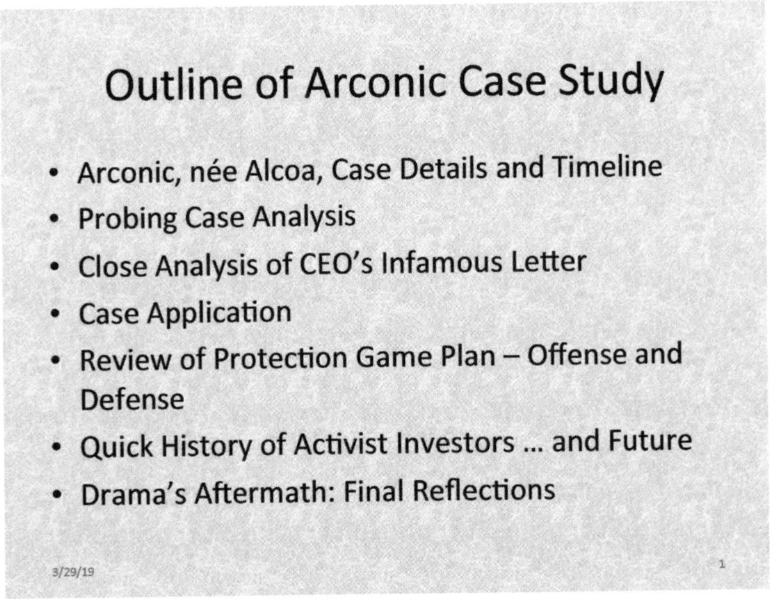

Outline of Arconic Case Study

- Arconic, née Alcoa, Case Details and Timeline
- Probing Case Analysis
- Close Analysis of CEO's Infamous Letter
- Case Application
- Review of Protection Game Plan – Offense and Defense
- Quick History of Activist Investors ... and Future
- Drama's Aftermath: Final Reflections

3/29/19 1

Figure E13 Good text contrast

E.3 Speaking to a Group

The optimal style of speaking for an oral presentation is extemporaneous. This means to speak in a spontaneous manner, as if you are speaking off the top of your head, but in reality you are following a prepared outline. If you were to give the same presentation two or more times, each time would be a little different due to your extemporaneous approach. This is different from *ad lib*, which means truly made up at the moment, off the lip. *Ad lib* is also called "off the cuff" or "impromptu." If you do well at this spontaneous form of oral presentation, you are gifted. If you do not do well, you might be described as "shooting from the hip," and you are not alone. Many people find it hard to speak impromptu. Fortunately, you only do this type of speaking when you are caught unexpectedly by a request to brief, update, or answer questions posed by a manager or other senior person.

Most of the oral presentations you will be asked to deliver, however, can be prepared in advance. But remember to deliver them extemporaneously. The alternatives are undesirable. Specifically, two highly discouraged styles of speaking in ABERST are to read from a prepared script or to memorize and deliver the speech word for word from memory. (These are used in other professions, however.) Both of those are laden with as many hazards as an *ad lib* speech. If you read from a script, you fail to gain rapport with your audience. Most, if not all, of the audience will doubt your competency and sincerity, and they will wonder why they bothered to show up when the speech could be read on their own time, if they were given a copy of the script. With memorization, you are susceptible to three dangers: One, you are likely to forget where you are at some point, then you will need to repeat yourself until you find your flow again. Or, two, you might skip portions that you forget and thereby not deliver your whole speech. Also, three, you may sound like a robot when delivering a memorized speech. Audiences do not want to listen to an automaton, who acts mechanically and unemotionally.

Your overall speaking tone is important. In addition to being extemporaneous, you must also be professional, pleasant, and competent. You must convey confidence, knowledge, and honesty. Do not make apologies. Do not admit to being nervous. You have no reason to tell the audience anything outside of your main message and the support points. You have no need, and it will not be on topic or useful, to apologize, such as the following remarks:

- We didn't understand the assignment at first
- We started late
- We didn't get all the measurements we wanted

In addition, you should not blow your audience away with esoteric ideas and jargon. If you talk over their heads, you will not communicate to them. So, you must remember that you are the expert and they are hoping to learn from you. (You absolutely do know more about your project than your audience; this is true even if supervisors and managers are present.) Be clear and helpful.

A few additional remarks can be offered regarding team speaking. I have listened to many team-based presentations, so I have seen both positives and negatives in this format, enabling me to make the following suggestions. If your organization or team chooses to do something different, I would hope that you have good reasons. Otherwise, you might benefit from the handful of tips below.

Speakers should not trade-off, back and forth, in "tag-team" manner during the presentation. Exchanges of speakers take up valuable time and are distracting. Each time a speaker changes, the audience must readjust to that person's voice (tone, rate, volume, accent, elocution, and so on). Once the audience adjusts, they would just as soon have that same person continue and finish. But, a new speaker inevitably arrives. The readjustment must occur again. This happens sequentially, and it challenges the audience. Thus, the changes of speaker should be kept to a minimum. Each speaker should speak only once. Each person should speak for a suitable (usually, evenly apportioned) amount of time, only once. If speakers are not specialized and the team is egalitarian, the time should be divided approximately evenly. Another way of division might be according to each person's specialty. Speaking times may be uneven in such cases. Another exception might be that the same speakers starts and finishes for the team.

Each speaker should stand in the same place as the previous speaker, with the other "dormant" speakers off to the side. Audiences will be confused if speaker location changes. Ideal locations are the following: at a podium, if there is one; at center stage; or at the side of the projection screen, if there is one.

Speaking skills may not be even among team members. Experts often provide advice for managing this disparity, such as "start with your second-best speaker and end with your best speaker; squeeze weaker speakers into the middle." My advice is different. I suggest that you practice together. Rehearsals can improve the skills of all speakers. If the talk is prepared well and projection slides are designed sufficiently, each speaker can become skillful with some rehearsal, so you do not have to accommodate "weak" speakers.

Starting and ending well are important components of the oral presentation. For starting, always do these tasks at a minimum: greet the audience, introduce your team and each speaker (full names), and introduce your presentation with a title that encompasses both topic and communication purpose. Example: "Good morning. We are The Chapter 11 Team, charged with identifying a plan for bringing this organization out of bankruptcy. We are Nadine Roth, Salman Oz, and Joyce Munro. Today we are here to present the four-stage plan we have developed over the past 2 months."

For ending, a speaker needs to be prepared for one important contingency: to leave out information and skip to the closing if the time limit is quickly approaching. This can be handled smoothly, without anger or anxiety. Here is a suggestion: "I see that our time is nearly up, so I will skip a few points on bonds and interest rates, which you can read about in the handouts. In closing, I want to summarize our three main points . . . . .thank you . . . . We will answer your questions now."

If you have time to present all of your prepared talk, do so and bring the presentation to a close. Indeed, the last task is to indicate clearly and overtly that the presentation is finished and invite the audience to ask questions. Here are some suggested phrases to indicate the end has arrived:

- In conclusion . . .
- To sum up . . .
- We would like to leave you with this point, namely, . . .
- This concludes our presentation . . . .

If you are at a loss for words while nearing the end, try to recall this simple and short closing remark: "Thank you. Do you have any questions?"

The Q and A portion is another opportunity to emphasize your message and main points and make a good impression. Sometimes, however, you may be unable to answer a particular question. It might be outside the scope of your work or simply something you do not know. In such instances, you can admit you cannot answer the question. If you feel inclined to research an answer at a later date, you can tell the audience you will provide an answer via written communication at a later time. If in a team presentation, the appropriate team member (expert) should answer the question. If a visual aid in your slide deck would help you with answering, you can take a moment to find it and project it.

Before I bring this appendix to a close, I have a few more suggestions for speaking to a group, starting with speaker transitions.

A transition provides a link between speakers, to ease the audience's adjustment to listening to a new voice. The transition is an opportunity to achieve one multi-faceted rhetorical goal in addition to helping the audience become familiar with the new speaker before new and complex information is provided. Accordingly, the incoming speaker is responsible for this. (The outgoing speaker can simply finish providing her information to the audience.)

The transition's rhetorical goal that can be achieved is threefold: remind listeners about the main message of the talk, remind listeners about the immediately previous point, and introduce a new point, which the new speaker intends to embellish upon. The new speaker must walk a fine line here: she must reiterate main and prior point(s) without being overtly redundant. Main points may be any of these:

- Overall finding
- Overall purpose for doing work
- Overall objective to fulfill or question to answer
- Overall information unknown to make known

A sample transition statement is this: "In pursuing an optimal reduction to our engine particulate emissions, we identified the most efficient catalytic conversion method, and kept in mind the goal of adding as little additional weight as possible. The system's weight analysis is as follows."

Transitions are one place to demonstrate skillful speaking. Other positive aspects of public speaking involve the following expert techniques:

- Make eye contact with audience
- Display enthusiasm and expertise
- Keep your body erect and stable
- Keep your hands at sides or in "ready" position, except when using them for a specific task, such as pointing at a spot on a slide or illustrating something visual
- Modify your rate, volume, and articulation for emphasis

Eight likely problem areas with arms and hands are depicted in Figure E14, for the reasons given. To avoid these eight problem areas, you must make a deliberate effort to control your arms and hands in an acceptable manner. Some positions for non-distracting limbs are the following:

At side, alongside outer thigh

- Arms and hands will not attract attention
- You will not appear stiff because audience will listen to your words and watch your eyes
- Arms and hands will indeed move at specific moments, intentionally, to point to screen and advance slides

Lightly clasped, over belly button

- Like a ball and socket or teepee with fingertips touching
- Not clenched (will cut off circulation)

If the previous suggestions seem uncomfortable to you or if you have a strong tendency to move your arms and hands randomly, known as a "fidgety speaker," an accepted alternative is to place your hands in your pockets. With this approach, however, you must be careful that you do not jangle keys and coins.

With all of the above guidelines now at your disposal, you are ready to deliver an impressive oral presentation in any situation. Good luck and do your best; I am sure all will go well.

The Hand Cuff
Suggests you're about to offer a last statement before sentencing

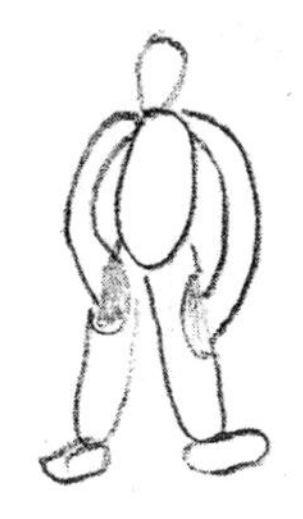

The Hands in Pocket
Suggests you're waiting for something, perhaps someone to give you a thesis

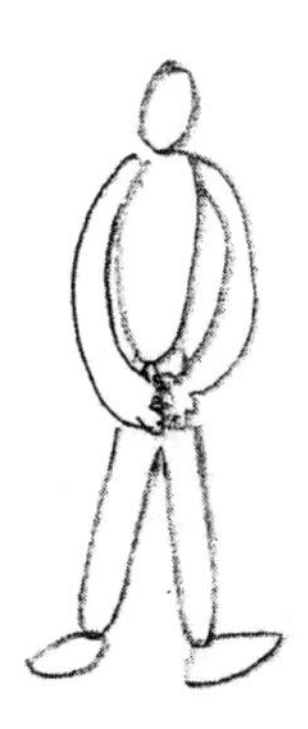

The Fig Leaf
Suggests that you're Adam or Eve, embarrassed and guilty

The Treading Water
Suggests you're doing all you can to survive the experience

Figure E14 Arm and hand positions to avoid, with reasons

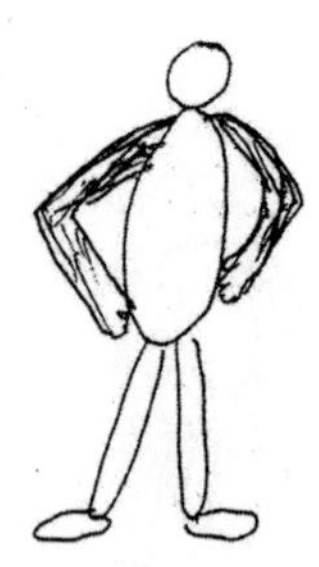

The Bouncer
Suggests you're not going to let anyone else get even close to the podium. (Confrontational, also known as "arms akimbo.")

The Hair Dresser
Suggests a greater concern for one's appearance than for the issues under discussion

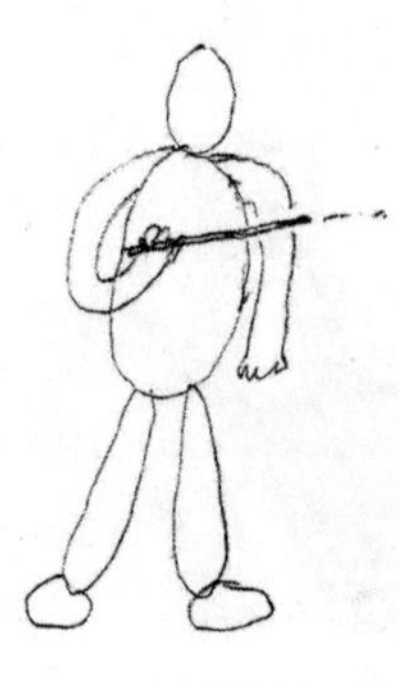

The Swordsman
Suggests a secret desire to duel with the audience, or force them to agree with you

The Paper Clutch
Suggests a serious dependency on crutches and notes, or failure to prepare the final version

Figure E14 Arm and hand positions to avoid, with reasons

Acknowledgments

Many books have been useful to me over the years as I tried to improve my writing and my teaching of the subject. Those resources are listed in the Bibliography. All of them are helpful, and you may choose to consult them if you want further elucidation on a topic I have omitted or covered without sufficient depth. In addition to the books listed, I have had special guidance from my colleagues at the University of Michigan. In particular, Rod Johnson and Leslie Olsen hired me many years ago and mentored me as I learned to teach and approach technical writing from a pedagogical perspective. Professor Olsen's textbook, written with Thomas Huckin, informed my view, and I mention it in spots in this book. Similarly, my mentors J.C. Mathes and Dwight Stevenson provided much assistance as I was developing my own courses, and their textbook is a classic in the field. The other books listed are superior resources but I cannot say I know the authors personally.

Many of my colleagues have shared their insights, suggestions, and personal writing peccadilloes, and I have enjoyed conversing with my fellow instructors. In particular, a few of them have shared ideas and materials that I adapted for this book, and I want to express my gratitude to them: Peter Nagourney, Elaine Wisniewski, Erik Hildinger, Christian Casper, Mary Northrop, Rob Sulewski, and Angela Violi. I need to give a big shout out to my friend Greta Krapohl. She helped me write this book without even lifting a finger. Just by doing the work she does, she inspired some of my best examples. Other contributors who have helped add various examples include Gregory Ledva, Benjamin Donitz, Emanuela Della Bosca, and Derrick Dominic. I am indebted to them for their generous support.

I am grateful for the contributions made by all my former students. Throughout the years, they asked many questions and presented writing conundrums for me to solve. They also produced report after report as part of their assignments, mostly those from the College of Engineering. From all this interaction with students at the University of Michigan, I pieced together the guidelines and explanations in this book. The students pushed me to find clear illustrations of composition principles and document design choices. In the constant back and forth of the document development and review process, I learned from them possibly more than they learned from me. Indeed, many of the examples in this book are adaptations of text passages and visual aids produced by my students, and they are incorporated here for the purpose of helping the next cohort of students and working professionals.

Several people helped read drafts of this book, and they need to hear that I am very appreciative. Danielle Henderson was an excellent reviewer and improved the book with her insights. A truly exceptional editor, Elise Ann Wormuth at NorthLight Wordcraft, came to me through good fortune. She was selected to prepare the index for this book, which she accomplished with utmost skill. Beyond that, however, she edited the entire book and offered me help with the trickier points of English grammar and modern usage. Other editors include my closest family members: My daughter, Sarah, offered many amazing suggestions. Jacob, my son, was a dedicated proofreader who found mistakes and suggested additions. Finally, my wife, Astrid, was always the trusty editor, and she supported me through every twist and turn. My family provided encouragement throughout the process, and I am indebted to them in perpetuity. To this close-knit clan, I dedicate this book.

I want to express my heartfelt appreciation to Tamra Tuller and Kandy Tobias, at Thomson-Shore Printing, my publishers. They have been committed to this book from beginning to end. Even during the upheavals in the publishing world, when their own futures and careers were uncertain, they never wavered in their commitment to me. The two of them did the work normally done by a team of many more people; I was ever amazed at their resourcefulness and abilities to read my mind. Without them, this book would still be sitting on my computer's hard drive.

Lastly, let me thank you, the reader, for trusting me and selecting this book as your guide and reference. For you, I have attempted an original approach to a classic subject, to make the material interesting and as fun as possible. At times, I may be somewhat cheeky, but above all else I hope my reverence for the subject matter comes through.

Bibliography

Argumentation:

Aristotle, *The Art of Rhetoric*, trans. by Hugh Lawson-Tancred, Penguin Books, London, UK, 1991.

Fisher, Roger, William Ury, and Bruce Patton (ed.), *Getting to Yes: Negotiating Agreement without Giving In*, Houghton Mifflin Company, Boston, MA, 2nd edition, 1991.

Spinoza, Baruch, *Ethics: Demonstrated in Geometric Order*, Matthew J. Kisner (ed.), Cambridge University Press, New York, NY, 2018.

Toulmin, Steven E. *The Uses of Argument*, Cambridge University Press, New York, NY, Updated edition, 2003.

Career Search and Materials:

Bolles, Richard N., *What Color is Your Parachute? A practical manual of job-hunters and career changers*, Ten Speed Press, Berkeley, CA, 2017.

Design and Visual Aids:

Tufte, Edward, *The Quantitative Display of Visual Information*, Graphics Press, Cheshire, CT, 1983 (or 2nd edition, 2001).

Williams, Robin, *The NonDesigners Design Book*, Peachpit Press, Berkeley, CA, 1995 (or 3rd edition 2008, or 4th edition 2014).

Composition, Grammar, and Writing:

Bailey, Edward P., *Plain English at Work: A Guide to Business Writing and Speaking*, Oxford University Press, New York, NY, 1996.

Brohaugh, William, *Write Tight: How to Keep your Prose Sharp, Focused, and Concise*, Writer's Digest Books, Cincinnati, OH, 1993 (or Intercollegiate Studies Institute edition, 2002).

Crews, Frederick, *The Random House Handbook*, Random House, New York, NY, 1992 (or any subsequent edition).

Flesch, Rudolph, *The Art of Readable Writing*, Macmillan Simon & Schuster, New York, NY, 1949.

Fowler, H. Ramsey, *The Little, Brown Handbook*, Little, Brown and Company, Boston, MA, 1980 (or 13th edition, 2015).

Gordon, Karen E., *The Transitive Vampire: A Handbook of Grammar for the Innocent, the Eager, and the Doomed*, Times Books, New York, NY, 1984 (or Pantheon deluxe edition, 1993).

Gunning, Robert, *The Technique of Clear Writing*, McGraw-Hill Book Company, New York, NY, 1952.

Sabin, William A., *The Gregg Reference Manual*, Gregg Division/McGraw-Hill, New York, NY, 1991 (or 10th edition, 2004).

Strunk, William Jr., and E.B. White, *The Elements of Style*, Macmillan Publishing Co., Inc., New York, NY, 1972 (or any subsequent edition).

Williams, Joseph, *Style: (Ten) Lessons in Clarity and Grace*, 3rd edition, Scott Foresman, Glenview, IL, 1989 (or Longman 11th edition, 2013).

Zinsser, William, *On Writing Well*, Harper & Row, New York, NY, 1988 (or 30th anniversary edition, 2006).

Learning:

Brown, Peter C., Henry L. Roediger III, and Mark A McDaniel, *Make it stick: The Science of Successful Learning*, Harvard University Press, Cambridge, MA, 2014.

Ratey, John J., and Eric Hagerman, *Spark: the Revolutionary New Science of Exercise and the Brain*, Little, Brown and Company, New York, NY, 1st edition, 2008.

Oral Presentations and Public Speaking:

Adler, Ronald B., and George Rodman, *Understanding Human Communication*, Holt, Rinehart and Winston, Inc., Chicago, IL, 4th edition, 1991.

Hoff, Ron, *Say It In Six: How to say exactly what you mean in 6 minutes or less*, Barnes & Noble Books, New York, NY, 2003.

Technical Writing:

Alley, Michael, *The Craft of Scientific Writing*, Springer Science+Business Media, New York, NY, 4th edition, 2018.

Mathes, J.C., and Dwight Stevenson, *Designing Technical Reports: Writing for Audiences in Organizations*, Macmillan Publishing Co., New York, NY, 1991.

Olsen, Leslie A., and Thomas Huckin, *Technical Writing and Professional Communication*, McGraw-Hill, Inc., New York, NY, 1991.

Riordan, Daniel G., and Steven E. Pauley, *Technical Report Writing Today*, Houghton Mifflin Co., Boston, MA, 1999 (or Wadsworth 10th edition, 2013).

Index

Page numbers in italics refer to figures and boxed examples.

E

F

P

Q

R

S

Y

About the Author

Jack Fishstrom has over 30 years' experience writing, editing, and presenting technical reports on a wide range of topics within academia, business, engineering, research, science, and technology. His work also includes legal analysis and litigation support. He has been teaching technical writing to engineering students at the University of Michigan since 1993. He also teaches public speaking and poster design. Jack has a background in video and film production as well, and he wrote and directed a feature film, *Voices*. In addition to four fictional screenplays and several educational video scripts, his writing portfolio includes hundreds of non-fiction items, such as proposals, environmental impact statements, public outreach materials, remedial designs, style guides, legal testimony, newsletters, magazines, software documentation, and several published journal articles. In addition to his university teaching, he is a licensed attorney, small-business owner, and consultant to engineering, legal, and medical organizations. He earned a BA in Philosophy from the University of California, Berkeley, *magna cum laude*. He holds an MA in Telecommunication Arts from the University of Michigan, and a JD from the University of Toledo, *cum laude*. In his free time, he meditates, gardens, exercises, reads, cooks, travels, and does absolutely nothing, as often as possible. His website is *www.apexpersuasion.com*.